Post-processing Techniques in Antenna Measurement

Related titles on electromagnetic waves:

Dielectric Resonators, 2nd Edition Kajfez and Guillon
Electronic Applications of the Smith Chart Smith
Fiber Optic Technology Jha
Filtering in the Time and Frequency Domains Blinchikoff and Zverev
HF Filter Design and Computer Simulation Rhea
HF Radio Systems and Circuits Sabin
Microwave Field-Effect Transistors: Theory, design and application, 3rd Edition Pengelly
Microwave Semiconductor Engineering White
Microwave Transmission Line Impedance Data Gunston
Optical Fibers and RF: A natural combination Romeiser
Oscillator Design and Computer Simulation Rhea
Radio-Electronic Transmission Fundamentals, 2nd Edition Griffith, Jr
RF and Microwave Modeling and Measurement Techniques for Field Effect Transistors Jianjun Gao
RF Power Amplifiers Albulet
Small Signal Microwave Amplifier Design Grosch
Small Signal Microwave Amplifier Design: Solutions Grosch
2008+ Solved Problems in Electromagnetics Nasar
Antennas: Fundamentals, design, measurement, 3rd Edition Blake and Long
Designing Electronic Systems for EMC Duff
Electromagnetic Measurements in the Near Field, 2nd Edition Bienkowski and Trzaska
Fundamentals of Electromagnetics with MATLAB®, 2nd Edition Lonngren *et al.*
Fundamentals of Wave Phenomena, 2nd Edition Hirose and Lonngren
Integral Equation Methods for Electromagnetics Volakis and Sertel
Introduction to Adaptive Arrays, 2nd Edition Monzingo *et al.*
Microstrip and Printed Antenna Design, 2nd Edition Bancroft
Numerical Methods for Engineering: An introduction using MATLAB® and computational electromagnetics Warnick
Return of the Ether Deutsch
The Finite Difference Time Domain Method for Electromagnetics: With MATLAB® simulations Elsherbeni and Demir
Theory of Edge Diffraction in Electromagnetics Ufimtsev
Scattering of Wedges and Cones with Impedance Boundary Conditions Lyalinov and Zhu
Circuit Modeling for Electromagnetic Compatibility Darney
The Wiener-Hopf Method in Electromagnetics Daniele and Zich
Microwave and RF Design: A systems approach, 2nd Edition Steer
Spectrum and Network Measurements, 2nd Edition Witte
EMI Troubleshooting Cookbook for Product Designers Andre and Wyatt
Transmission Line Transformers Raymond Mack and Jerry Sevick
Electromagnetic Field Standards and Exposure Systems Grudzinski and Trzaska
Practical Communication Theory, 2nd Edition Adamy
Complex Space Source Theory of Spatially Localized Electromagnetic Waves Seshadri
Electromagnetic Compatibility Pocket Guide: Key EMC facts, equations and data Wyatt and Jost
Antenna Analysis and Design Using FEKO Electromagnetic Simulation Software Elsherbeni, Nayeri and Reddy
Scattering of Electromagnetic Waves by Obstacles Kristensson
Adjoint Sensitivity Analysis of High Frequency Structures with MATLAB® Bakr, Elsherbeni and Demir
Developments in Antenna Analysis and Synthesis vol. 1 and vol. 2 Mittra
Advances in Planar Filters Design Hong

Post-processing Techniques in Antenna Measurement

Edited by
Manuel Sierra Castañer and Lars J. Foged

The Institution of Engineering and Technology

Published by SciTech Publishing, an imprint of The Institution of Engineering and Technology, London, United Kingdom

The Institution of Engineering and Technology is registered as a Charity in England & Wales (no. 211014) and Scotland (no. SC038698).

First published 2019

The Institution of Engineering and Technology
Michael Faraday House
Six Hills Way, Stevenage
Herts, SG1 2AY, United Kingdom

www.theiet.org

British Library Cataloguing in Publication Data
A catalogue record for this product is available from the British Library

ISBN 978-1-78561-537-5 (hardback)
ISBN 978-1-78561-538-2 (PDF)

Typeset in India by MPS Limited
Printed in the UK by CPI Group (UK) Ltd, Croydon

Contents

About the editors ix

1 Introduction to antenna measurement and post-processing techniques **1**
Lars J. Foged and Manuel Sierra Castañer

References 6

2 Source reconstruction using equivalent currents **9**
Javier Leonardo Araque Quijano, Marco Righero, Giorgio Giordanengo, and Giuseppe Vecchi

2.1 Introduction 9
2.2 Formulation of the source reconstruction problem 9
2.2.1 Application of the equivalence theorem 10
2.2.2 Uniqueness of the solution to the reconstruction problem 13
2.3 Numerical solution of the source reconstruction problem 18
2.3.1 Discretization 18
2.3.2 Setup of the linear system 18
2.3.3 Application of existing Method of Moments codes in source reconstruction 19
2.4 Probe correction 19
2.4.1 Reciprocity and the reception formula 20
2.4.2 Formulation of the source reconstruction problem with probe correction 22
2.5 Application examples 23
2.5.1 Application as an investigation and diagnostic tool 23
2.5.2 Application in field transformations 27
2.5.3 Array antenna synthesis 30
2.6 Numerical tests on the accuracy of source reconstruction 35
2.7 Conclusions and outlook 39
References 39

3 Antenna and measurement improvements based on equivalent currents reconstruction **45**
Maria Alberica Saporetti, Andrea Giacomini, Francesco Saccardi, and Lars J. Foged

3.1 Introduction 45
3.2 Application of EQC in the improvement of antenna system design 46

3.2.1 Advanced diagnostic of a wideband dual-ridge horn 47
3.2.2 Design improvement of an X-band 32 × 8 array 50
3.2.3 Verification of complex excitation coefficient of a space array antenna 56
3.2.4 Analysis of coupling phenomena between spacecraft antennas 60
3.2.5 Investigation of spherical higher order modes in the design of NF probes 64
3.3 Application of EQC in the improvement of antenna measurements accuracy 69
3.3.1 Diagnostic and spatial filtering of supporting structure 69
3.3.2 Echo suppression 73
3.3.3 Reduction of truncation errors 78
3.4 Further applications of the EQC technique 81
3.4.1 Power density determination of measured devices 81
3.4.2 Application of the EQC technique in EMC testing 84
3.5 Conclusions and discussion on advantages and limitations of the EQC technique 87
References 88

4 Holographic techniques to improve antenna measurements 91
Francisco José Cano Fácila and Manuel Sierra Castañer

4.1 Use of holographic techniques for reducing errors in antenna measurements 93
4.1.1 Potential of source reconstruction techniques 93
4.1.2 Noise 94
4.1.3 Reflections 94
4.1.4 Measurement scan truncation 96
4.1.5 Leakage 98
4.2 Theory of holographic techniques in electromagnetics 99
4.3 Noise reduction using holographic techniques 101
4.4 Echo reduction using holographic techniques 113
4.5 Extrapolation of the radiation pattern and reduction of truncation errors 121
4.6 Detection and suppression of leakage 126
4.7 Conclusions 133
References 133

5 Antenna measurement improvement using spherical or cylindrical mode expansion 139
Francesco Saccardi, Manuel Sierra Castañer, and Lars J. Foged

5.1 Improvements of antenna measurement accuracy using spherical mode expansion 141
5.1.1 Spherical wave expansion theory 142

5.1.2 SWE as spatial filter 155
5.2 Exploiting cylindrical mode expansion for fast single-cut measurements with improved accuracy 183
5.2.1 Theory of cylindrical waves 186
5.2.2 Single-cut measurements 188
5.2.3 Applications to echo reduction in single-cut measurements 190
5.3 Conclusions 193
References 194

6 Phase recovery in antenna measurements 197
Ana Arboleya, Yuri Álvarez, Jaime Laviada, and Fernando Las-Heras

6.1 Introduction 198
6.1.1 Main phase recovery techniques 199
6.1.2 Interferometric techniques 199
6.1.3 Iterative techniques 201
6.2 Indirect off-axis holography 202
6.2.1 Method 203
6.2.2 Improved schemes 205
6.2.3 Broadband indirect off-axis holography 212
6.2.4 Error analysis 218
6.3 Iterative phase retrieval techniques 219
6.3.1 Fundamentals 219
6.3.2 Phaseless sources reconstruction method 221
6.4 Conclusions 232
References 234

7 Link between simulation and measurements 239
Lucia Scialaqua, Francesca Mioc, and Lars J. Foged

7.1 Introduction 239
7.2 Theory 240
7.2.1 Domain decomposition technique 240
7.2.2 Black box representation of devices radiating into 3D space 240
7.2.3 Definition of The Link 241
7.2.4 Validation of The Link 246
7.3 Antenna in free space 247
7.4 Mounted antennas 250
7.4.1 Reflector fed by a horn antenna 250
7.4.2 Flush-mounted antenna on a flat structure 255
7.5 Practical application of The Link 266
7.5.1 Test object and complex scenario 266
7.5.2 Validation workflow 266
7.5.3 Results 268

7.6 Conclusions 269
References 272

8 Future trends in antenna measurements **275**
Manuel Sierra Castañer and Lars J. Foged

8.1 Summary of the analysed post-processing techniques 275
8.2 From passive to digital antenna technologies 278
8.3 Performance parameters for digital antennas 279
8.4 Antenna measurement systems for digital antennas 284
8.5 Outlook for the new developments in antenna measurement systems 290
8.6 Conclusions 292
References 292

Index **295**

About the editors

Manuel Sierra Castañer is a professor at Universidad Politécnica de Madrid. He is a senior member of the IEEE and the Antenna Measurement Techniques Association (AMTA). He is currently vice-chair of the European Association of Antennas and Propagation (EurAAP). In 2007, he was awarded the IEEE-APS Schelkunoff Prize Paper Award for "Dual-Polarization Dual Coverage Reflectarray for Space Applications".

Lars Jacob Foged is the scientific director of Microwave Vision Group and associate director of Microwave Vision Italy. He is a senior member of IEEE and has been secretary of the IEEE Antenna Standards Committee since 2004. He is a fellow of the Antenna Measurements Techniques Association (AMTA) and recipient of its Distinguished Achievement Award.

The editors have published extensively on scientific topics related to antennas, antenna measurements, measurement uncertainty and post-processing techniques. They have been successive leaders of the "Measurements" WG of the European Association on Antennas and Propagation (EurAAP) and members of the EurAAP Delegate Assembly. They are both board members of the European School of Antennas (ESoA) and organizers of the bi-annual Antenna Measurement Course in Europe. Together, they have organized workshops, courses and short courses on antenna measurement techniques in Europe and Asia.

Chapter 1

Introduction to antenna measurement and post-processing techniques

Lars J. Foged[1] *and Manuel Sierra Castañer*[2]

Antenna and electromagnetic characterization have been challenging topics for more than 50 years. In the very beginning, far-field techniques for antenna characterizations of electrical parameters were introduced. In this sense, radiation pattern measurements, gain characterization, cross-polar and axial ratio measurements have been extensively studied. References [1,2] explained in a detailed way the measurement techniques of the main antenna types in those years. Also, the IEEE standard [3] defined the main antenna electrical parameters to be characterized, and it is until now the reference document for antenna designers. Evans, in [4], summarized the most important techniques used in far-field characterization. These far-field techniques are still used for the characterization of most important medium-sized and large-sized antennas. The concept of far-field distance, depending on antenna dimensions and wavelength, is used for accepting a sufficient accuracy of antenna measurements. $2D^2/\lambda$, where D is the antenna diameter, has become the sufficient distance for characterizing the antenna electrical properties. This distance bounds the phase errors to $\pi/8$ radians, and it is sufficient for the characterization of most of the antennas.

However, when the frequency is increased, or antenna dimensions become larger, other techniques appeared. In 1968, Georgia Tech designed the first compact antenna test range (CATR) using large reflectors. A parabolic reflector transforms the spherical wave in a plane wave, allowing the measurement of far-field antenna parameters in a reduced distance. During the last years, the CATR systems have been improved, introducing serrated of rolled-edge reflectors to reduce the effect of the diffraction caused by the edge of the parabolic reflectors. Also, dual reflectors, using Gregorian or Cassegrain configurations, have been designed to reduce the effect of cross polarization in the antenna-measurement system design.

In 1988, Hansen summarized the spherical near-to-far-field transformation algorithms in a detailed book [5], a theory that appeared in 1973. Yaghjian, in [6],

[1]Microwave Vision Italy (MVI), Pomezia, Italy
[2]Information Processing and Telecommunications Center, ETSI Telecomunicación, Universidad Politécnica de Madrid, Madrid, Spain

summarizes these near-to-far-field algorithms for the three classical configurations: spherical, cylindrical and planar, where the basis of near-field antenna measurements is perfectly explained. Later, in 1988, *IEEE Transactions of Antennas and Propagation* published a special issue for near-field antenna measurement systems [7], where the main techniques are detailed. In 2012, IEEE published a new standard for near-field antenna measurements [8].

During all these years, Antenna Measurement Techniques Association symposium has been the focal point for the main advances in the electrical characterization of antenna or radar cross section. During these years, some of the main challenges have been the reduction of uncertainties, the reduction of measurement time and the increase of the frequency for new applications.

The development of mobile communications since 1990 makes also necessary new measurement systems for user terminals and base stations. Traditional antenna measurement ranges are not the most appropriates for the measurement of these devices since the effect of the positioners, the cable interaction, the difficulty of extracting a reference antenna, among others make difficult classical measurements. Classical roll over azimuth or elevation over azimuth positioners was often replaced by gantry arc systems or multiprobe systems.

During the last years, uncertainty and errors have been extensively studied. Depending on the antenna specifications, antenna size, frequency band or type of measurement setup, different effects are more important than others. For instance, the most important effects in outdoor ranges are reflections; at mm-wave frequencies, dynamic range or phase errors due to positioning errors have to be considered; for small antennas the interaction with positioning system (including cables). Therefore, in general, near-field measurement setups are affected by other errors that do not appear in far-field ranges, and in all the cases, the effect of the measurement error on the final result cannot be immediately deduced. Classical metrology theory establishes that the quantity to be determined differs from the measured quantity, and an uncertainty analysis is required to assess the effect of measurement errors on the accuracy of the results. Most studies have been focused on this problem, trying to give a reliable estimate of the uncertainty in the far-field pattern of an antenna for a given acquisition error (in both near field or far field) or for a combination of different errors. One of the earliest studies on this topic can be found in [9], where Newell and Yaghjian derived error expressions from the equations relating planar near-field (PNF) and far-field quantities in order to deduce how some PNF errors affect the accuracies of the far-field parameters. The errors investigated were truncation due to the finite size of the acquisition plane, position errors of the probe, amplitude and phase uncertainties of the measured data and multiple reflections between the probe and the antenna under test (AUT). The results of this study not only fixed upper bound errors in far field but also specified the form of the error (linear, quadratic, periodic, etc.). Several years later, Newell gave a more detailed study about errors in PNF measurements in [10], based mainly on the work presented previously in [11] by Yaghjian. In this publication, first, the more significant sources of errors in a planar range were identified, creating the "18-term error budget," that has been widely used in later analysis and has been fairly well accepted among experts. Then, the impact of each near-field error on the far-field pattern was estimated. Methods for estimating uncertainties were classified into three categories: mathematical analysis, computer simulation and measurement test. The first alternative is based on the use of the Fourier transform pair relations between the near-field and the corresponding plane wave spectrum (PWS) to deduce error expressions, as was done in [9]. In computer simulation, errors are mathematically modelled and added to every near-field data, computing later the corrupted far-field pattern, which is compared to that one obtained in an error-free simulation. The last option is the measurement test, which is based on changing something on the measurement setup and comparing the results between several results. Regardless the chosen alternative, the magnitude of each near-field error is measured or estimated, then, its effect on the far-field pattern is determined and finally the total uncertainty in the measurement is estimated by properly combining the individual estimates of each error. Therefore, the results obtained in these studies can be used to estimate the impact of a particular error or a combination of errors on the far-field pattern. In addition, the results can be used to deduce the maximum admissible near-field error for a given level of accuracy in the far field or to assess the accuracy of a near-field range, as is done in [12]. The work exposed in [10] has been the basis of multiple studies, and some of them published few years ago, e.g. in [13], where Francis and Wittmann use a similar concept to the "18-term error budget" to define the sources of uncertainty for gain measurement and relative pattern in near-field measurements. These uncertainty analyses were applied to the measurement of an ultra-low side-lobe array antenna. Moreover, due to the success of the analysis presented in [10] to estimate the uncertainties and to evaluate planar measurement facilities, it was extended to the other two most common near-field setups (cylindrical and spherical).

The extension to the cylindrical case was presented in [14], given special emphasis to position errors, probe correction effects and room scattering. After describing the ways to estimate uncertainties in the results obtained from cylindrical near-field (CNF), a comparison of the far-field pattern of the VAST-12 antenna [15] obtained on planar and cylindrical ranges was presented, reaching the conclusion that the differences between both patterns were consistent with the estimates of uncertainty for the two ranges.

Regarding uncertainties in spherical near-field (SNF) measurements, one of the initial studies was presented in [16] by Hansen. Later, in [17], Hess identified a great number of additional alignment-related errors, increasing the number of error terms given by Newell from 18 up to 50. Although Hess proposed this expanded approach to SNF uncertainty analysis, in many situations, most of the sources of errors are negligible and do not contribute to the overall error budget, increasing the testing time. In [18], Hindman and Newell proposed a process to assess the side-lobe accuracy in an SNF range, focusing only on the largest contribution in the "18-term error budget." The effect of all errors is estimated by using self-comparison tests, and the objective of taking only into account the largest errors is to be able to quantify the level of accuracy in a good enough manner with an important reduction of the testing time. The study published by Miller in [19] is also based on the list of 18 errors given by Newell, adapting it for other types of measurements, such as gain measurements in SNF ranges.

All these studies were completed in the work developed in the report written in the frame of the European network "Antenna Centre of Excellence" [20]. In this study, the effects on the antenna-measured parameters of the different elements in antenna-measurement systems are studied and analysed. That document summarizes the previous existing documents, harmonizing all the contributions.

The algorithms discussed in this book do reduce uncertainty terms by post-processing antenna-measurement results, cancelling some errors that appear during measurement. The focus of the book is on how to reduce antenna errors using new algorithms and information of AUT. Therefore, the use of all these algorithms not only has the effect of uncertainty reduction, but also the reduction of systematic errors that are usually included in uncertainty budget. Most of the techniques included in this book are related with the reconstruction of the electromagnetic sources of the AUT, using different algorithms (holographic techniques or integral equation methods). Also, the electromagnetic fields can be expressed using spherical wave, cylindrical wave or plane wave expansions. In all these domains, antenna geometry can be exploded for the improvement of antenna-measurement results.

Therefore, the increase of precision in antenna-measurement systems, the development of new algorithms and the rapid evolution of the computers allow the improvement of the measurement results, reducing the uncertainty and measurement errors, using electromagnetic information, information of the AUT or combining the measurements with simulations. The use of the algorithms presented in this book allows the improvement of measurement results, reducing measurement time and mechanical specifications of the antenna-measurement system.

During the last years, some new techniques based on modal or spatial filtering have been proposed to increase the quality of the antenna-measurement results (and, in general, to improve the quality of electromagnetic measurements). In this sense, large improvements have been done in algorithms for the post-processing of the acquired data. Near-to far-field algorithms have been improved in order to be able to measure in non-uniform grids. Also, there has been a large improvement in different techniques to reduce acquisition time. Finally, there has been a rapid evolution in post-processing tools to improve the results of the measurement, reducing the effect of echoes, noise, leakage among others or extrapolating the radiation pattern to angular ranges without acquired data. Finally, last improvements in the combination of simulation and measurements allow a more accurate and faster antenna design.

All the contents included in this book have had important theoretical and practical developments during the last years. All these methods are based on the translation of the near or far-field data to other domains (as modes or currents on the antenna surfaces). In those domains, the antenna engineers can recognize and correct errors in an easiest and fastest way, having some *a priori* information of the antenna. Obviously, the progress in this field has been supported by a deep study in universities and research centres. Also, commercial antenna diagnostic tools, as

Insight (MVG) or Diatool (TICRA), have been proposed for doing very accurate source reconstructions using integral equation methods. Other commercial tools, as Mars (NSI), Isofilter (MI-Tech) or MVEcho (MVG), use the decomposition of the acquired field in cylindrical or spherical modes to filter out spurious signals.

After this short introduction, Chapter 2 introduces the source reconstruction techniques using equivalent currents. Researchers from Politecnico di Torino explain the basis of the Love's Equivalence Principle and the application to the reconstruction of the electrical and magnetic currents on a volume enclosing the AUT. This can be done if the two tangential components of the electric field, for instance in spherical near-field measurements, on a sphere enclosing the AUT are acquired. This is the theory employed in the software Insight (MVG), and it is based on integral equation and Huygens's equivalent principle.

Once, the currents are reconstructed, Chapter 3, written by Microwave Vision Group, shows the main applications of these techniques. These applications can be summarized in cancellation of spurious signals, being able to separate the information of the AUT from the information coming from noise, defaults of the antenna-measurement systems, spurious signals, etc. In that chapter, not only the applications but also the limitations of this technique for each application will be justified.

Chapter 4, written by researchers from Universidad Politécnica de Madrid, explains the use of holographic techniques to improve antenna measurement. The authors introduce the concept of PWS and the relation to electrical sources of the antenna in a plane. This chapter also details the application of the knowledge of the relation between those electrical sources and geometry of AUT for the reduction of echoes, detection and cancellation of leakage, increasing of signal to noise or extrapolation of antenna measurement pattern to a complete hemisphere for PNF measurements, cylindrical near-field measurements or truncated spherical measurements. In this sense, the main advantage of this method is the simplicity and processing time, since the use of fast Fourier transform reduces the complexity. However, the main limitations come from the reduction of the resolution to half wavelength, the limitation to a hemisphere, or the main application to planar antennas.

Chapter 5, by Microwave Vision Group and Universidad Politécnica de Madrid, extends the concepts of the previous chapter to spherical or cylindrical mode expansion. This chapter shows the application of the decomposition of the near or far field in cylindrical or spherical modes. In these domains (cylindrical or spherical modes), the information of the AUT can be separated from the information of other spurious signals, using information of the AUT as the size and position in the antenna-measurement system. The techniques will be demonstrated with different application examples.

In Chapter 6, researchers from Universidad de Oviedo explain the basis of phase recovery in amplitude antenna measurement. These techniques became important for millimetre and submillimetre wave measurements since positioning errors force large phase errors in acquisition. But, they are also necessary for special measurements where the reference signal is difficult to obtain (as active measurements, cellular phones, MIMO systems, etc.). In these cases, the phase cannot be measured, and only amplitude is available. These methods are more and more important since it is difficult to separate the antenna from the radiofrequency (RF) subsystem or even the signal-processing hardware.

Chapter 7 establishes the basis and includes some examples of the link between simulation and measurements. Microwave Vision Group works on these techniques to extract the parameters for very large antennas, or antennas mounted on complex structures, through the electromagnetic characterization of components of the AUT and the simulation of the antenna integrated in a more complex structure, as airplanes or satellites. The improvement in the capability of the computers allows accurate calculations of the electromagnetic fields generated by these complex structures.

Chapter 8 concludes the book and summarizes the requirements for the measurement of the future antennas in 5G, active antennas, smart antenna, MIMO systems. In those new systems, antenna is difficult to isolate from other subsystems such as RF or even digital processors. Post-processing techniques, phase-recovery techniques, combination of simulations and measurements explained in the previous chapters will be necessary to accomplish the measurement of these more and more complex antennas and RF systems. In fact, for these new systems, mainly related to 5G standard and also to new digital Radars, digital beamformers, massive MIMO systems, etc., it will be impossible to separate antenna part to RF subsystems. However, for the correct performance of the whole system, electromagnetic parameters have to be characterized in order to be able to optimize the performance of the whole system. In this sense, new antenna parameters have to be defined and new measurement techniques have to be studied. Also, new communications systems are not only for long-distance communications but also for short-range communications. Far-field characterization has to evolve to the characterization of the electrical sources of the antenna. In this sense, the algorithms presented in this book, basically designed for helping in the classical antenna characterization, are also useful for the requirements of the future antenna systems. In this last chapter, the editors of the book will present the challenges of the close future in antenna-domain measurement.

References

[1] J. S. Hollis, T. J. Lyon and L. Clayton, Microwaves Antenna Measurements. Atlanta, GA: Scientific Atlanta. 1969, 1970, 1985 editions.

[2] IEEE Std. 149-1979, IEEE Standards Test Procedures for Antennas. John Wiley. 1979.

[3] IEEE Standard Definitions of Terms for Antennas. 1993.

[4] G. E. Evans, Antenna Measurement Techniques. Norwood, MA: Artech House Inc. 1990.

[5] J. E. Hansen, Spherical Near Field Antenna Measurements. IEE Electromagnetic Series. London: Peter Peregrinus Ltd. 1988.

[6] A. D. Yaghjian, "An overview of near-field antenna measurements," *IEEE Trans. Antennas Propag.*, vol. 34, no. 1, pp. 30–45, 1986.

[7] "Special issue on near-field scanning techniques," *IEEE Trans. Antennas Propag.*, 1988.

[8] IEEE Std 1720-2012, IEEE Recommended Practice for Near-Field Antenna Measurements. 2012.

[9] A. C. Newell and A. D. Yaghjian, "Study of errors in planar near-field measurements," in *Antennas and Propagation Society International Symposium*, 1975, pp. 470–473.

[10] A. C. Newell, "Error analysis techniques for planar near-field measurements," *IEEE Trans. Antennas Propag.*, vol. 36, no. 6, pp. 754–768, 1988.

[11] A. D. Yaghjian, "Upper-bound errors in far-field antenna parameters determined from planar near-field measurements, Part 1: Analysis," Nat. Bur. Stand., Tech. Note 667, 1975.

[12] E. B. Joy, "Near-field range qualification methodology," *IEEE Trans. Antennas Propag.*, vol. 36, no. 6, pp. 836–844, 1988.

[13] M. H. Francis and R. C. Wittmann, "Sources of uncertainty for near-field antenna measurements," in *2nd European Conference on Antennas Propagat., EuCAP 2007*, Edinburgh, 2007, pp. 1–3.

[14] A. C. Newell and D. Lee, "Application of the NIST 18 term error model to cylindrical near-field antenna measurements," in *Antenna Measurement Techniques Association, AMTA 2000*, Philadelphia, Pennsylvania, 2000.

[15] J. E. Hansen, "Definition, design, manufacture, test and use of a 12 GHz validation standard antenna," Technical University of Denmark, Lyngby, Tech. Rep. R672, 1997.

[16] J. E. Hansen, "Error analysis of spherical near-field measurements," in *Spherical Near-Field Antenna Measurements*. London: Peter Peregrinus. 1988, ch. 6.

[17] D. W. Hess, "An expanded approach to spherical near-field uncertainty analysis," in *AMTA 24th Annual Meeting and Symposium*, Cleveland, OH, Oct., 2002.

[18] G. Hindman and A. C. Newell, "Simplified spherical near-field accuracy assessment," *IEEE Antennas Propag. Mag.*, vol. 49, no. 1, pp. 233–240, 2007.

[19] P. Miller, "Uncertainty considerations in spherical near-field antenna measurements," in *2nd European Conference on Antenna and Propagation, EuCAP 2007*, Edinburgh, UK, Nov. 2007.

[20] Antenna Centre of Excellence, 6th European Union FrameWork. Contract FP6-IST 508009. "Deliverable A1.2D4 Recommended practices for Near Field measurements". 2005.

Chapter 2

Source reconstruction using equivalent currents

Javier Leonardo Araque Quijano[1], Marco Righero[2], Giorgio Giordanengo[2], and Giuseppe Vecchi[3]

2.1 Introduction

Equivalent currents connect antenna measurements with processing and analysis, shifting the information acquired from the measured near-field (NF) or far-field (FF) samples to a domain where the characteristics of the antenna can be identified more easily, as already explained in Chapter 1.

The technique described in this chapter (in particular, in Section 2.2 for its theoretical aspects, in Section 2.3 for its numerical implementation, and in Section 2.4 for one of its extensions) leverages on very little additional information about the antenna, namely, a lower bound on its spatial occupancy. Measured data and this information are used to determine equivalent currents, on a surface enclosing the antenna, which can be used as a diagnostic tool (as described in Section 2.5.1) to perform field transformation (as described in Section 2.5.2) or to synthesize antenna array from a goal pattern (as described in Section 2.5.3). Section 2.6 presents a study on the accuracy of the equivalent currents using synthetic data.

Chapter 3 will give more details about the applications for antenna analysis, and in Chapter 7, equivalent currents will be used as primary sources in numerical codes, establishing a link between measurements and simulations.

2.2 Formulation of the source reconstruction problem

The source-reconstruction technique discussed allows computing a current distribution with known support that generates a given field. In this chapter, we focus on the problem in the frequency domain. The input data required are as follows:

1. A "measurement surface" Σ_M where the field is known; this is commonly a canonical surface, such as a sphere, a cylinder, or a plane, but the technique poses no constraints on the shape of this surface.

[1]Electrical Engineering Department, Universidad Nacional de Colombia, Bogotá, Colombia
[2]LINKS Foundation, Torino, Italy
[3]Department of Electronics and Telecommunications, Politecnico di Torino, Torino, Italy

2. Complex vector amplitude of the electric or magnetic field at spatial samples of Σ_M, or the result of a known linear operator on it when the probe effect or other interaction with the measurement system is considered.
3. A "reconstruction surface" denoted by Σ_R, most often internal to Σ_M, where the current distribution is sought.

Concerning the location of Σ_R, it could be indeed external to Σ_M, in which case the equivalence would hold inside Σ_M. This could be of interest to attain an equivalent representation of the fields due to sources located *outside* Σ_M; however, this is not pursued in the present discussion.

We initially note that Σ_R should be a closed surface for the theory presented below to hold rigorously; however, with suitable approximation, both Σ_R and Σ_M could be open (e.g., truncated sphere, finite planar domain). Σ_R needs to include all the primary sources (antennas) and secondary sources (obstacles, supports, non-background material) for the solution to be possible, and it could be a single surface or a set of disjoint surfaces.

The result is a surface current distribution on Σ_R that reproduces the field samples given at Σ_M, and theoretically at any point outside Σ_R; however, the numerical nature of the procedure leads to a widely varying field accuracy near Σ_R depending on the equivalence formulation employed, as seen below.

Given the fact that the field is known in complex vector form, and that the reconstruction surface is provided, the source reconstruction problem is linear, in contrast to phase-retrieval or inverse-scattering problems, which lack one or the other. Considering this, the problem for a 3D reconstruction surface Σ_R is formulated in terms of integral equations based on the equivalence theorem, which provides a general framework for all the applications of interest. The resulting integral equations are subsequently discretized and solved via the Method of MoMents (MoM) as shown in Section 2.3. A numeric analysis of the uniqueness of the solution is presented at the end of Section 2.2.2.

2.2.1 Application of the equivalence theorem

The original radiation problem is illustrated in Figure 2.1 (left), where all the original sources and material bodies differing from free space (antenna conductors, supports, obstacles, and the like) are enclosed by the surface Σ_R, while its exterior is free space. Figure 2.1 (right) depicts the equivalent problem, where the original contents of Σ_R are substituted by equivalent currents $\mathbf{J}_{eq}$, $\mathbf{M}_{eq}$ on Σ_R that preserve the original fields external to $\Sigma_R(\mathbf{E}_+, \mathbf{H}_+)$, while the original internal fields $(\mathbf{E}_-, \mathbf{H}_-)$ are in general replaced by other fields, denoted by $\left(\mathbf{E}_-^{'}, \mathbf{H}_-^{'}\right)$. Field equivalence outside Σ_R requires that these equivalent currents obey [1,2]:

$$\left.\begin{aligned} \hat{\mathbf{n}}' \times \left[\mathbf{H}_+(\mathbf{r}) - \mathbf{H}_-^{'}(\mathbf{r})\right] &= \mathbf{J}_{eq} \\ -\hat{\mathbf{n}}' \times \left[\mathbf{E}_+(\mathbf{r}) - \mathbf{E}_-^{'}(\mathbf{r})\right] &= \mathbf{M}_{eq} \end{aligned}\right\} \mathbf{r} \in \Sigma_R \tag{2.1}$$

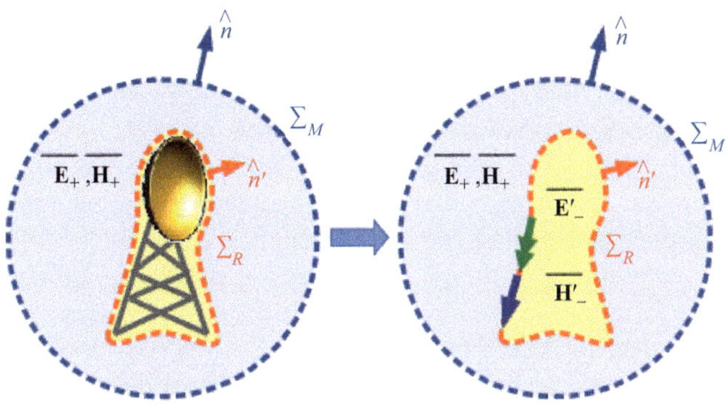

Figure 2.1 Illustration of the equivalence theorem: original (left) and general equivalent problem (right)

From (2.1), it is clear that it is not possible to reproduce both the internal and the external fields, $(\mathbf{E}_{-},\mathbf{H}_{-})$ and $(\mathbf{E}_{+},\mathbf{H}_{+})$, since this would imply zero-field jump across Σ_R and therefore null equivalent sources.

While there is some freedom in the choice of the internal fields $\left(\mathbf{E}'_{-},\mathbf{H}'_{-}\right)$ (as will become soon apparent), due to the replacement of all internal sources and material bodies by the equivalent currents $\left(\mathbf{J}_{eq},\mathbf{M}_{eq}\right)$, in all cases, the latter radiate in free space; hence, the resulting fields can be computed everywhere by using the free-space radiation operator:

$$\mathbf{E}(\mathbf{r}) = -\eta_0 \mathscr{L}\left(\mathbf{J}_{eq};\mathbf{r}\right) + \mathscr{K}\left(\mathbf{M}_{eq};\mathbf{r}\right) \tag{2.2}$$

where

$$\begin{aligned}
\mathscr{L}\left(\mathbf{J}_{eq};\mathbf{r}\right) &= jk_0 \int_{\Sigma_R} \left[\mathbf{J}_{eq}(\mathbf{r}') + \frac{1}{k_0^2}\nabla\nabla'_s \cdot \mathbf{J}_{eq}(\mathbf{r}')\right] g(\mathbf{r},\mathbf{r}')\, ds' \\
\mathscr{K}\left(\mathbf{M}_{eq};\mathbf{r}\right) &= \int_{\Sigma_R} \mathbf{M}_{eq}(\mathbf{r}') \times \nabla g(\mathbf{r},\mathbf{r}')\, ds' \\
g(\mathbf{r},\mathbf{r}') &= \frac{e^{-jk_0|\mathbf{r}-\mathbf{r}'|}}{4\pi|\mathbf{r}-\mathbf{r}'|}
\end{aligned} \tag{2.3}$$

where $\eta_0 = \sqrt{\mu_0/\varepsilon_0}$ is the medium intrinsic impedance, $k_0 = \omega\sqrt{\mu_0\varepsilon_0}$ is the medium wavenumber, $\mathbf{r}$ and $\mathbf{r}'$ are the observation and source coordinates, respectively, and ∇'_s stands for the surface divergence operator $(\mathbf{r} \notin \Sigma_R)$.

According to the equivalence theorem [2, p. 654], the equivalent currents in (2.1) are equivalent in the sense that they produce the original fields $(\mathbf{E}_{+},\mathbf{H}_{+})$ only outside Σ_R, while fields $\left(\mathbf{E}'_{-},\mathbf{H}'_{-}\right)$ different from the originals result within Σ_R. When using equivalence as stated above, the internal fields $\left(\mathbf{E}'_{-},\mathbf{H}'_{-}\right)$ constitute a degree of freedom with the proviso that it must be generated by the equivalent currents as given by (2.2) so that the jump conditions in (2.1) are satisfied. Therefore, the internal field must be due to external sources; in other words, it must satisfy the homogeneous Maxwell's equations inside Σ_R (no sources therein).

In principle, there is an infinity of valid settings for the internal fields and equivalent currents [1, p. 107]; however, due to the abovementioned proviso, some of the choices are better suited for specific applications. For instance, in direct problems, Σ_R represents a material boundary where boundary conditions are to be enforced. On the other hand, the problem at hand simply requires that equivalent currents at Σ_R satisfy the boundary condition on Σ_M given in terms of the known radiated field; no boundary conditions need to be enforced on Σ_R since, in principle, the fields inside it are of no interest [3]. Nonetheless, it should be noted that the physical interpretation of the equivalent currents is determined by the choice of the modified internal fields $\left(\mathbf{E}'_-, \mathbf{H}'_-\right)$. For this reason, we proceed now to discuss choices of internal fields resulting in meaningful equivalent currents (i.e., currents with specific relations with the actual fields); these choices are also viable for the setup of integral equations in direct problems.

2.2.1.1 Zero interior fields—Love's equivalence

By setting the interior fields to zero, we obtain Love's form of the equivalence theorem:

$$\left.\begin{aligned}\hat{\mathbf{n}} \times \mathbf{H}_+(\mathbf{r}) &= \mathbf{J}_{eq}^{Love}\\ -\hat{\mathbf{n}} \times \mathbf{E}_+(\mathbf{r}) &= \mathbf{M}_{eq}^{Love}\end{aligned}\right\}\mathbf{r} \in \Sigma_n \tag{2.4}$$

With this form of equivalence, the fields on Σ_R bear a simple, direct relation to equivalent currents, which is very important in antenna-diagnostic applications. Given the zero fields inside Σ_R, its interior can be filled with any material, which is useful in some situations. For example, filling the interior with a perfect electric conductor (PEC) perfect magnetic conductor (PMC) suppresses one of the two equivalent currents and yet let them remain consistent with the actual fields on Σ_R via (2.4); however, to compute the radiated field, one must use the Green's function appropriate for the surface and boundary condition considered, which is straightforward only in the case when Σ_R is an infinite plane (e.g., as in [4–9]).

2.2.1.2 $\mathbf{J}_{eq}$-only/$\mathbf{M}_{eq}$-only formulation

As an alternative to employing a single equivalent current requiring the Green's function of a body occupying Σ_R (as in Love's equivalence with PEC/PMC filling), an equivalent problem involving only electric (magnetic) current in a homogeneous medium is obtained by imposing that the interior field $\left(\mathbf{E}'_-, \mathbf{H}'_-\right)$ and the actual field have equal tangential components on Σ_R [10–12]. Since external equivalence must be preserved in this case, too, the absence of either equivalent current results in zero jump in the tangential component of the respective field, thus ensuring the equality just mentioned. The economy achieved in this case, with half the number of unknowns and the possibility of use the free-space Green's function, is offset by the fact that the equivalent current is related to the field in a more complicated fashion than in Love's form of equivalence: the electric (magnetic) current is not simply $\hat{\mathbf{n}} \times \mathbf{H}(-\hat{\mathbf{n}} \times \mathbf{E})$ on the reconstruction surface; in this case, source and field are instead related via the radiation integral (2.2).

This formulation, in the special case of spherical surfaces, was solved explicitly in [10] and employed in a source-reconstruction problem in [13]. Besides the economy mentioned above, it will be shown that the single-current formulation provides interesting perspectives in practical applications.

2.2.2 Uniqueness of the solution to the reconstruction problem

From the previous discussion, with the exception of Love's equivalence with material filling, which will not be discussed further, fields radiated by the equivalent currents in the region *external* to Σ_R (including Σ_M) are obtained via the free-space radiation operators (see (2.2) and Figure 2.1 (right)). Specifically, on Σ_M, the relation between the equivalent sources and the tangential component of $\mathbf{E}$ is as follows:

$$\hat{\mathbf{n}} \times \mathbf{E}(\mathbf{r}) = \hat{\mathbf{n}} \times \left[-\eta_0 \mathscr{L}\left(\mathbf{J}_{eq}; \mathbf{r}\right) + \mathscr{K}\left(\mathbf{M}_{eq}; \mathbf{r}\right)\right] \quad \mathbf{r} \in \Sigma_M \tag{2.5}$$

This expression, when used to enforce that the equivalent currents lying on Σ_R radiate the measured field on Σ_M, constitutes an integral equation for the two unknown currents $\left(\mathbf{J}_{eq}, \mathbf{M}_{eq}\right)$. At this point, it should be clear that (2.5) by itself does not impose any particular choice in the internal fields $\left(\mathbf{E}_-^{'}, \mathbf{H}_-^{'}\right)$, since all the equivalent forms discussed above result in equivalent currents that satisfy it (i.e., equivalent currents reproduce the actual fields at Σ_M), yet resulting in different internal fields $\left(\mathbf{E}_-^{'}, \mathbf{H}_-^{'}\right)$. Hence, the solution to (2.5) is not unique, and if it is solved as given, the meaning of the resulting equivalent currents is unclear.

In solving (2.5), one may set $\mathbf{M}_{eq}$ to zero, which corresponds to a form of equivalence with continuous tangential E-field across Σ_R; an analog situation results when setting $\mathbf{J}_{eq} = 0$. It is worth noting that with no additional conditions imposed, there are already three different equivalent current distributions that rigorously solve (2.5) (i.e., radiate the field given at Σ_M). Furthermore, the field continuity implicit in the $\mathbf{J}$-only or $\mathbf{M}$-only formulations just mentioned confirms that at least these radiate nonzero fields inside Σ_R, which is a reminder that field equivalence applies only outside Σ_R.

In order to further demonstrate the nonuniqueness of solution to (2.5), and the fact that when it is solved with the two currents present there is no warranty that these are the Love's currents, we resort to the classic reference [14, eq. 17], where it is proven that Love's equivalent sources corresponding to sources completely on one side of Σ_R (either inside or outside) are mapped through the radiation operator in (2.5) to a field that is identically zero in the volume where original sources were located and equal to the original one elsewhere. Based on this, we could set up a source distribution equivalent to an arbitrary arrangement of sources outside Σ_R by computing Love's equivalent currents $\mathbf{J}_{eq} = \hat{\mathbf{n}} \times \mathbf{H}$ and $\mathbf{M}_{eq} = -\hat{\mathbf{n}} \times \mathbf{E}$ on Σ_R (with $\hat{\mathbf{n}}$ pointing inwards) that radiate the same fields inside Σ_R, and zero field outside. Clearly, any combination of equivalent constructed in this way can be added to the solution of our original exterior problem without affecting external equivalence (external field remains the same), while the internal fields would be different in each case.

From a numerical point of view, the discretization of (2.5) almost invariably results in a rectangular linear system, which is customarily solved resorting to direct or iterative methods that are numerically equivalent to the normal equations or the Moore–Penrose pseudo-inverse [15]. These methods implicitly impose the additional constraint of minimum-norm (in the sense of the ℓ_2 norm, i.e., the norm of the vector of unknown coefficients) and thus lead to a modified problem which has a unique solution [16, eq. 5;17]. This solution, however, is not warrantied to radiate zero internal field; indeed, the large reactive component required in the currents to achieve zero internal field seems to contradict the minimum norm condition, and this is supported by the numerical tests presented at the end of this section.

From the previous discussion we can conclude that, in order to obtain equivalent currents that radiate zero field inside Σ_R, and thus are simply related to the actual fields via $(\mathbf{J} \times \hat{\mathbf{n}} = \mathbf{H}^{tan}, \mathbf{M} \times \hat{\mathbf{n}} = -\mathbf{E}^{tan})$, the zero-field condition of Love's equivalence has to be imposed explicitly. This can be achieved by demanding that fields tangential to the interior of Σ_R are zero. It can be shown that specification of zero field inside Σ_R, which is somewhat arbitrary due to the fact that it involves defining an additional interior surface where the condition is enforced, can be put in terms of an integral boundary identity to be enforced on Σ_R [9,18,19,20: eq. 6] as follows:

$$\hat{\mathbf{n}} \times \left[-\eta_0 \mathscr{L}\left(\mathbf{J}_{eq}^{Love}; \mathbf{r}\right) + \mathscr{K}\left(\mathbf{M}_{eq}^{Love}; \mathbf{r}\right) \right] = -\frac{1}{2}\mathbf{M}_{eq}^{Love}(\mathbf{r}) \quad \mathbf{r} \in \Sigma_R \tag{2.6a}$$

$$\hat{\mathbf{n}} \times \left[-\frac{1}{\eta_0} \mathscr{L}\left(\mathbf{M}_{eq}^{Love}; \mathbf{r}\right) - \mathscr{K}\left(\mathbf{J}_{eq}^{Love}; \mathbf{r}\right) \right] = \frac{1}{2}\mathbf{J}_{eq}^{Love}(\mathbf{r}) \quad \mathbf{r} \in \Sigma_R \tag{2.6b}$$

Taking into account that in the equivalent problem (see Figure 2.1, right) we have no sources inside Σ_R (the fields are solution to the homogeneous Maxwell equations), the uniqueness theorem [1, pp. 101–102] tells us that imposing the condition $\hat{\mathbf{n}} \times \mathbf{E} = 0$ on Σ_R alone (i.e., (2.6a)), $\mathbf{E}$ is zero everywhere inside Σ_R and hence so is $\mathbf{H}$. The converse is true when $\hat{\mathbf{n}} \times \mathbf{H} = 0$ is specified (i.e., (2.6b)) so that it is enough to enforce any one of (2.6a) and (2.6b). The only exception to this is when the cavity defined by Σ_R is resonant at the operating frequency; in this case, it becomes necessary to enforce the two constraints simultaneously or use a dual-surface approach as that proposed in [21].

With the addition of Love's condition to the conventional integral equation (2.5), we obtain a system of two equations for two unknowns, in contrast to formulations with two unknown currents being computed from a single equation or only one of the currents being solved for, as in [3–9,22–26], which are all particular cases of the general formulation presented here. Possibly, the first work where the Love's condition was imposed in order to obtain physical currents in a source reconstruction problem was presented in [27] for the study of Body of Revolution radomes, while the 3D version of this, along with a comparison with simplified formulations used before, was presented in [18,19].

For ease of reference in comparing the various formulations, they will be labeled as follows: Single-Equation Formulations (SEqFs) refer to the sole use of (2.5) and encompass two-current SEqF (**JM**-SEqF), which includes both **J** and **M** currents, and the two one-current formulations: **J**-current SEqF (**J**-SEqF), where only **J** current is present and **M**-current SEqF (**M**-SEqF), that considers only **M** current. Finally, the term Dual-Equation Formulation (DEqF) refers to the addition of Love's condition with both **J** and **M** currents, i.e., based on the solution of the system composed by (2.5) and (2.6a) as follows:

$$\hat{\mathbf{n}} \times \mathbf{E}(\mathbf{r}) = \hat{\mathbf{n}} \times \left[-\eta_0 \mathscr{L}\left(\mathbf{J}_{eq}; \mathbf{r}\right) + \mathscr{K}\left(\mathbf{M}_{eq}; \mathbf{r}\right)\right] \quad \mathbf{r} \in \Sigma_M \tag{2.7a}$$

$$\hat{\mathbf{n}} \times \left[-\eta_0 \mathscr{L}\left(\mathbf{J}_{eq}^{Love}; \mathbf{r}\right) + \mathscr{K}\left(\mathbf{M}_{eq}^{Love}; \mathbf{r}\right)\right] = -\frac{1}{2}\mathbf{M}_{eq}^{Love}(\mathbf{r}) \quad \mathbf{r} \in \Sigma_R \tag{2.7b}$$

According to the labeling scheme proposed, every formulation considered can be trivially obtained from DEqF, the most general one. For additional discussions into the differences between the various formulations, the reader is referred to [19].

Finally, we observe that in principle, the actual fields and Love's currents could be recovered in a trivial way from the single-equation solution; to do so, it is enough to apply the radiation operators and/or field-boundary integral identities (e.g., [20]). However, only the DEqF provides the accuracy required for good estimates of the field close to Σ_R as shown in [28].

We conclude this section by illustrating the solution of the source-reconstruction problem using the formulations just described. For completeness, the test includes also a comparison between the field radiated by each of the resulting equivalent sources. The reference structure is depicted in Figure 2.2. It consists of a strip dipole fed with a transmission line of the type paired strips that emerges from a coaxial cable; the coaxial cable is grounded on a metallic rectangular box. The reconstruction surface Σ_R is an enclosing sphere with radius 9.4 cm (see Figure 2.2, right).

Full-wave MoM simulation was used to compute the field generated by the reference structure on the measurement surface Σ_M, which is a sphere with radius 1.5 m (compatible with commercial antenna measurement ranges), and is 6.7 cm

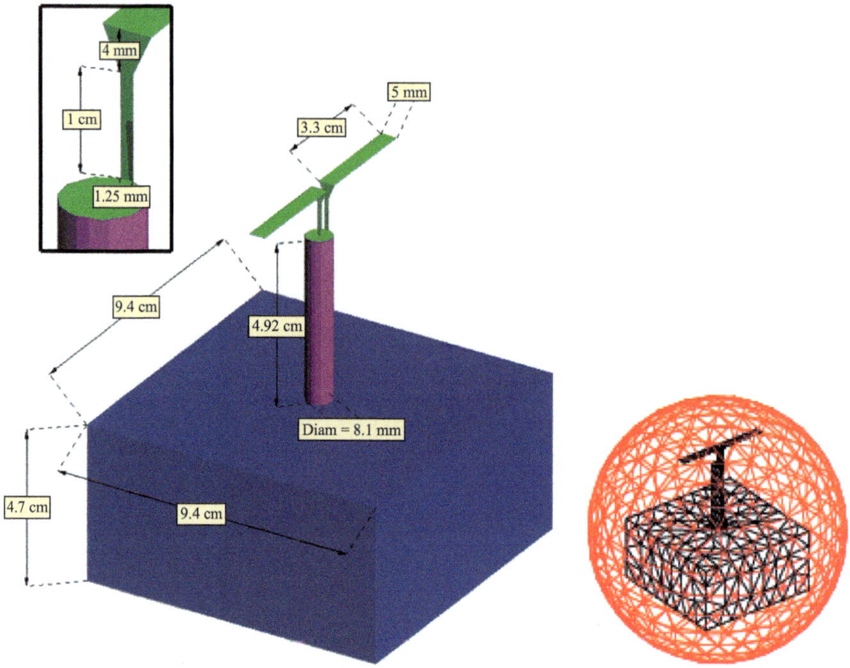

Figure 2.2 Left: structure used for the numerical assessment of the various equivalence formulations: dipole on a metallic box fed by a coaxial cable. Right: discretization of the reference structure and the enclosing surface Σ_R. Left: Reproduced, with permission, from Reference [19]

off-center with respect to Σ_R in the *z* axis. Then, equivalent currents were computed on Σ_R using all the formulations discussed above. The results of this test are shown in Figure 2.3. For reference, we include the exact solution, which can be computed in this case from the MoM analysis of the structure shown in Figure 2.2 just as the "measured" fields on Σ_M were obtained. For completeness, alongside the equivalent currents, we show the electric field modulus due to each set of equivalent currents on three coordinate planes including points inside and outside Σ_R (the contours of Σ_R's intersection with these planes are shown in dashed lines). It can be seen that only the DEqF is able to reproduce the reference Love's currents and radiate zero-field inside Σ_R. Finally, we note that all the formulations reproduce the reference field outside Σ_R, supporting the nonuniqueness of the solution as discussed previously.

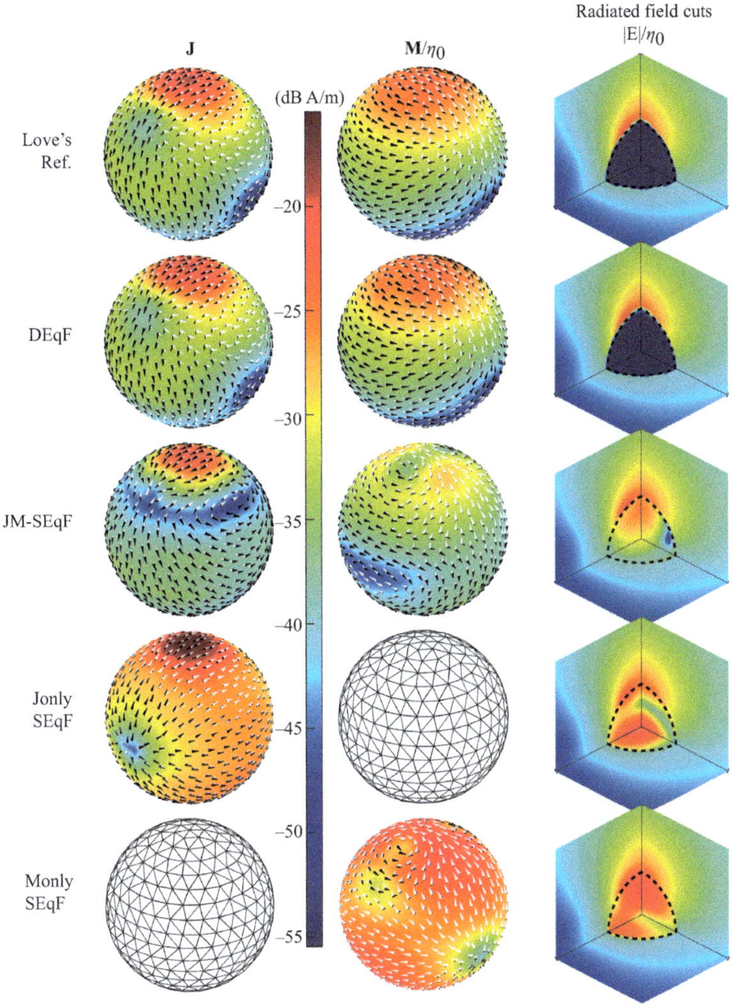

Figure 2.3 Solution of the source reconstruction problem using all the formulations presented. In the first two columns, electric and magnetic equivalent currents. In the third column, the modulus of the total electric field produced by these currents. The exact Love's equivalent currents and the electric field they produce are included in the top row for reference

2.3 Numerical solution of the source reconstruction problem

2.3.1 Discretization

Equations (2.5) and (2.6a) are cast into two systems of linear equations by expressing the unknown currents as sum of basis functions (e.g., Rao–Wilton–Glisson (RWG), [29]) and testing the two equations against appropriate test functions. When considering the field-matching equation (2.5), constant functions defined on a tessellation of Σ_M along the measured polarizations can be used with a one-point quadrature rule. When considering the Love's conditions, MoM test proves to be an efficient way to handle the singularities.

2.3.2 Setup of the linear system

Writing equivalent electric and magnetic currents as

$$\mathbf{J}_{eq} = \sum_{n=1}^{N} C_n^J \mathbf{f}_n, \quad \mathbf{M}_{eq} = \eta_0 \sum_{n=1}^{N} C_n^M \mathbf{f}_n \tag{2.8}$$

and denoting with $\mathbf{h}_\ell, \ell = 1, \ldots, L$, the set of test functions defined on the tessellation of Σ_M considering all the measured polarizations in each point $\mathbf{r}_\ell$, the discrete version of (2.5) and (2.6a) reads as, respectively:

$$\begin{aligned} &-\eta_0 \sum_{n=1}^{N} \left(\int_{\Sigma_M} (\hat{\mathbf{n}} \times \mathscr{L}(\mathbf{f}_n; \mathbf{r})) \cdot \mathbf{h}_\ell \mathrm{ds} \right) C_n^J \\ &+ \eta_0 \sum_{n=1}^{N} \left(\int_{\Sigma_M} (\hat{\mathbf{n}} \times \mathscr{K}(\mathbf{f}_n; \mathbf{r})) \cdot \mathbf{h}_\ell \mathrm{ds} \right) C_n^M \\ &= \int_{\Sigma_M} (\hat{\mathbf{n}} \times \mathbf{E}(\mathbf{r})) \cdot \mathbf{h}_\ell \mathrm{ds} \quad \ell = 1, \ldots, L \end{aligned} \tag{2.9}$$

and

$$\begin{aligned} &-\eta_0 \sum_{n=1}^{N} \left(\int_{\Sigma_R} (\mathscr{L}(\mathbf{f}_n; \mathbf{r})) \cdot \mathbf{f}_m \mathrm{ds} \right) C_n^J \\ &+ \eta_0 \sum_{n=1}^{N} \left(\int_{\Sigma_R} (\mathscr{K}(\mathbf{f}_n; \mathbf{r})) \cdot \mathbf{f}_m \mathrm{ds} + \frac{1}{2} \int_{\Sigma_R} \mathbf{f}_n \cdot (\hat{\mathbf{n}} \times \mathbf{f}_m) \right) C_n^M \\ &= 0, \quad m = 1, \ldots, N \end{aligned} \tag{2.10}$$

The two sets of linear equations (2.9) and (2.10) are combined in a unique system of linear equations in the unknowns C_n^J and C_n^M—so equality at the measurement points and null-field constraints are treated in the same way—and solved by means of an iterative solver [30].

Iterative solvers allow the use of fast algorithm (as the multilevel fast multipole method (ML-FMM) [31] employed in [26] or more algebraic oriented as [32]) to perform the matrix–vector products (MVPs) needed in the solution, a mandatory request when electrically large radiators are considered. Iterative solvers also perform an intrinsic regularization, an important characteristic when dealing with inverse problems based on real data which are affected by unavoidable noise.

2.3.3 Application of existing Method of Moments codes in source reconstruction

One can borrow from the extensive literature on fast algorithms for MoM to devise fast algorithms for the MVP needed in the inversion algorithm, bearing in mind that the two sets of testing functions $\mathbf{f}_n$ and $\mathbf{h}_\ell$ are on different regimes, with $\mathbf{f}_n$ behaving similarly to the classical MoM, whereas the functions $\mathbf{h}_\ell$ are generally on an intermediate distance and distributed at an almost-Nyquist rate.

When both field matching and Love's conditions are imposed, there are no references which show an efficient dynamical scaling in broad frequency bands, comparable to the $N \log N$ of standard MoM codes, and which account for both sparse-measurement points (almost-Nyquist rate) and for a dense mesh on the reconstruction surface. However, the algorithms derived from MoM literature have been profitably applied to the treatment of electrically large cases [26,33–35].

2.4 Probe correction

As already discussed, an input for source-reconstruction problems are complex vector field samples at given points in space, which could be at large distances, as in FF data. In practical applications, measurements proceed by capturing the response, often in terms of complex voltage/current, of a sensing antenna ("field probe") with specific location and orientation. When the probe is located in the FF of the antenna under test (AUT), the electric field is proportional to the probe's response, and the proportionality factors are as specified in the complex form of the Friis' transmission formula (see below). However, in general, the FF distance cannot be achieved in practice due to the large dimensions of the AUT or the wavelength involved, or because measurements in the NF are desired. In these situations, the probe's response is not simply related to the field measured due to the finite extent of the probe and its directive properties; the field-response link can actually be expressed in terms of an integral relation, as it will be discussed below, and some processing is required to recover the desired field (probe correction) or to obtain correct results without explicitly recovering it (probe-corrected source reconstruction, probe-corrected NF to FF (NF–FF) transformation). For the above, and in view of the practical applications in measurement processing to be discussed subsequently, it is desirable to generalize the source reconstruction technique so that it can be directly applied in cases when the input is given as probe response values, instead of complex vector-field samples.

The problem of correcting electromagnetic measurements for probe effects has a long history, and the results presented here are based on earlier works. Arguably, the earliest works related to probe correction (or probe compensation) of field measurements are the following: [36] introduced the scattering matrix representation of antennas, [37] presented a correction procedure based on the plane wave expansion, [38,39] presented a correction procedure in terms of wave expansion in cylindrical coordinates, and [40] described the procedure using spherical waves. A comprehensive treatment of spherical NF (SNF) measurements including probe correction is given in [41]. A more recent work [33] uses the plane wave expansion and the translation operator applied in the FMM [42] to perform probe correction; unlike [37], this work does not include the evanescent portion of the plane wave spectrum, therefore the integration domain is finite. Furthermore, it can be easily integrated in large-scale ML-FMM solvers, so the presentation that follows is based on it.

2.4.1 Reciprocity and the reception formula

The reciprocity theorem is a classical result that has been proven in many well-known textbooks for the case of antennas [1,43]. For an antenna composed by reciprocal materials, it may be used to relate its FF $\mathbf{e}$ in the direction $\hat{\mathbf{r}}$ with input current I_{in} (Tx mode), and the open-circuit voltage V_{oc} at its terminals in response to an impinging plane wave with electric field E_{inc} and propagation direction $-\hat{\mathbf{r}}$ (Rx mode), via the complex vector effective length ℓ_e as follows, assuming that the antenna is immersed in free space:

$$\mathbf{e}(\hat{\mathbf{r}}) = jk_0\eta_0 I_{in}\ell_e(\hat{\mathbf{r}}) \tag{2.11a}$$

$$V_{oc} = \ell_e(\hat{\mathbf{r}}) \cdot \mathbf{E}_{inc}^{-\hat{\mathbf{r}}}(0) \tag{2.11b}$$

These expressions refer to the transmission mode (2.11a) and reception mode (2.11b), and state the fact that the antenna vector effective length ℓ_e determines both the vector FF (i.e., the electric field with the spherical wave factor $e^{-jk_0 r}/4\pi r$ removed), and the open-circuit voltage at the antenna terminals when excited by a plane wave with arbitrary direction of incidence and polarization. For non-reciprocal antennas (e.g., antennas including magnetized ferrite elements), the effective vector length is different in transmission and reception, and the appropriate one must be used in the formulas that follow according to the measurement setup employed (AUT as transmitter or receiver).

The vector effective length ℓ_e is simply related to the radiation vector $\mathbf{N}$ used by Schelkunoff, which is the principal term for the magnetic vector potential $\mathbf{A}$ in the FF [43,44]:

$$\mathbf{E} = -j\eta_0 k_0 \mathbf{N}\frac{e^{-jk_0 r}}{4\pi r} \tag{2.12a}$$

$$\ell_e = \frac{-\mathbf{N}}{I_{in}} \tag{2.12b}$$

Expression (2.11a) and (2.11b) predicts the antenna's response to a plane wave. This is not a limitation at all, since quite general fields can be expressed in terms of plane wave expansions [45]. Indeed, considering the situation illustrated in Figure 2.4, we can use the Gegenbauer's addition theorem [42] to provide a plane wave expansion of the field due to the AUT on the probe's support:

$$\mathbf{E}(\mathbf{r}_2) = jk_0\eta_0 \oiint \mathbf{e}_{aut}(\hat{\mathbf{r}})T_L(k_0, \hat{\mathbf{r}}, \mathbf{D})e^{-jk_0\hat{\mathbf{r}}\cdot\mathbf{r}_2}\mathrm{d}\Omega \tag{2.13a}$$

$$T_L(k_0, \hat{\mathbf{r}}, \mathbf{D}) = \frac{k_0}{4\pi}\sum_{l=0}^{L}(-j)^{l+1}(2l+1)h_l^{(2)}(k_0D)P_l(\hat{\mathbf{r}}\cdot\hat{\mathbf{D}}) \tag{2.13b}$$

$$\mathrm{d}\Omega = \sin(\theta)\mathrm{d}\phi\mathrm{d}\theta \tag{2.13c}$$

Note that (2.13a) expresses the field at $\mathbf{r}_2$ as the superposition of a continuous spectrum of plane waves with propagation directions spanning the whole solid angle around the probe's local origin. This continuous spectrum is actually the angular spectrum (FF) of the AUT $\mathbf{e}_{aut}$ weighted by the so-called translation operator T_L, which depends on the relative location of the local coordinate systems $\mathbf{D}$ and the wave vector for each wave $\mathbf{k} = k_0\hat{\mathbf{r}}$, assuming propagation in free space, which is the case in practical measurement conditions. An important point is that the number of multipoles L to be included in the sum in the translation operator (2.13b) is chosen based on a trade-off between accuracy and stability [46]: for β digits accuracy and a probe with radius A referred to its local coordinate system, $L = k_0A + \beta(k_0A)^{1/3}$ is suggested; on the other hand, the spherical Hankel's function of the second kind $h_l^{(2)}(k_0D)$ grows exponentially with l for $k_0D < l$ [41]; therefore, L cannot be significantly larger than k_0D, which imposes an inherent limitation in the accuracy that can be attained for reduced probe

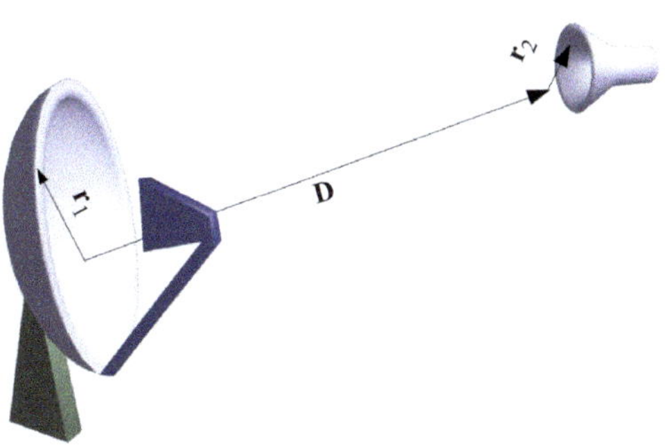

Figure 2.4 Measurement setup with indication of coordinates to compute. $\mathbf{r}_1$ and $\mathbf{r}_2$ are local position vectors referred to local coordinate systems in the domain of the AUT and the probe respectively, and $\mathbf{D}$ is the relative position vector between their origins. The AUT's and probe's FFs are referred to their respective local coordinate systems

distance D. The condition required to achieve the desired accuracy can be stated as follows:

$$k_0 D > k_0 A + \beta (k_0 A)^{1/3} \tag{2.14}$$

which is sometimes roughly stated as $D > A$.

Given such expansion, superposition can be used to obtain the probe's response to the arbitrary field from (2.11b) and (2.13a)–(2.13c):

$$V_{oc} = \oiint \ell_e(-\hat{\mathbf{r}}) \cdot \mathbf{e}_{aut}(\hat{\mathbf{r}}) T_L(k_0, \hat{\mathbf{r}}, \mathbf{D})\, \mathrm{d}\Omega \tag{2.15}$$

This is the reception formula sought, which provides the antenna response when measuring a given AUT. As given, this formula along with the comments following (2.13a)–(2.13c) is valid provided that superposition applies, and that the AUT is in transmit mode; then, $\mathbf{e}_{aut}$ is its FF in transmission and the probe responds via its vector effective length ℓ_e in reception. When the probe is reciprocal, ℓ_e can be obtained from the probe's FF as given in (2.11a); otherwise, it must be measured from the probe's response to plane waves as given in (2.11b).

2.4.2 Formulation of the source reconstruction problem with probe correction

The formulation presented in (2.7a) and (2.7b) must be modified in order to account for the probe's effect, where the theoretical basis is the reception formula presented in the previous section. In the DEqF presented, which consists of two equations, we note that the equation concerning the Love's condition on the surface, i.e., (2.7b), remains the same, while the equation referring to measurements, i.e., (2.7a), is to be modified. The modification consists in inserting the unknown equivalent sources in (2.15) and the FF counterpart of the operators $\mathscr{L}$ and $\mathscr{K}$, and consider that at each measurement location, the probe can be rotated in order to point its maximum at the AUT according to the rotation matrix $\mathscr{R}$, expressed in global coordinates:

$$\begin{aligned} V_{oc}(\mathbf{D}, \mathscr{R}) = \oiint \big(\ell_e^{\mathscr{R}}(-\hat{\mathbf{r}}) \cdot \left[-\eta_0 \mathscr{L}_f(\mathbf{J}_{eq}; \hat{\mathbf{r}}) + \mathscr{K}_f(\mathbf{M}_{eq}; \hat{\mathbf{r}})\right] \\ T_L(k_0, \hat{\mathbf{r}}, \mathbf{D})\big)\, \mathrm{d}\Omega \end{aligned} \tag{2.16}$$

$$\mathscr{L}_f(\mathbf{J}_{eq}; \hat{\mathbf{r}}) = jk_0 \oiint_{\Sigma_R} e^{-jk_0 \hat{\mathbf{r}} \cdot \mathbf{r}'} \hat{\mathbf{r}} \times \mathbf{J}_{eq} \times \hat{\mathbf{r}}\, \mathrm{d}s' \tag{2.17}$$

$$\mathscr{K}_f(\mathbf{M}_{eq}; \hat{\mathbf{r}}) = -jk_0 \oiint_{\Sigma_R} e^{-jk_0 \hat{\mathbf{r}} \cdot \mathbf{r}'} \mathbf{M}_{eq} \times \hat{\mathbf{r}}\, \mathrm{d}s' \tag{2.18}$$

$$\mathscr{R} = \begin{bmatrix} x_x^{loc} & x_y^{loc} & x_z^{loc} \\ y_x^{loc} & y_y^{loc} & y_z^{loc} \\ z_x^{loc} & z_y^{loc} & z_z^{loc} \end{bmatrix} \tag{2.19}$$

In the rotation matrix $\mathscr{R}$, each row is a basis vector of the probe's local coordinate system expressed in the global coordinate system. When FF is expressed in Cartesian coordinates, the rotated probe's FF can be obtained as follows:

$$\ell_{xyz}^{\mathscr{R}}(-\hat{\mathbf{r}}) = \mathscr{R}^{-1} \cdot \ell_{xyz}(-\mathscr{R} \cdot \hat{\mathbf{r}}) \tag{2.20}$$

where $\hat{\mathbf{r}}$ and other vectors are assumed as *column* vectors and denote the conventional MVP. It should be noted that only rotations are considered in the transformation (not displacements) since the probe's FF remains referred to the origin of its coordinate system, as required by the reception formula in (2.15). If the FFs both in local and global coordinates are expressed in terms of the spherical unit vectors, additional coordinate transformations are required:

$$\ell_{r\theta\phi}^{\mathscr{R}}(-\hat{\mathbf{r}}) = \mathscr{C}^{-1} \cdot \mathscr{R}^{-1} \cdot \mathscr{C}_{\mathbf{loc}} \cdot \ell_{r\theta\phi}(-\mathscr{R} \cdot \hat{\mathbf{r}}) \tag{2.21}$$

$$\mathscr{C} = \begin{bmatrix} \sin(\theta)\cos(\phi) & \cos(\theta)\cos(\phi) & -\sin(\phi) \\ \sin(\theta)\sin(\phi) & \cos(\theta)\sin(\phi) & \cos(\phi) \\ \cos(\theta) & -\sin(\theta) & 0 \end{bmatrix} \tag{2.22}$$

In (2.21), $\mathscr{C}$ is a matrix that transforms vectors expressed in the spherical base $(\hat{\mathbf{r}}, \hat{\theta}, \hat{\phi})$ into the Cartesian base $(\hat{\mathbf{x}}, \hat{\mathbf{y}}, \hat{\mathbf{z}})$. In $\mathscr{C}_{loc}$, these angles are measured in the probe's local coordinate system.

To summarize, the DEqF of the inverse source problem with probe correction consists of the following system of integral equations:

$$\begin{aligned} V_{oc}(\mathbf{D}, \mathscr{R}) = \oiint (\ell_e^{\mathscr{R}}(-\hat{\mathbf{r}}) \cdot \left[-\eta_0 \mathscr{L}_f(\mathbf{J}_{eq}; \hat{\mathbf{r}}) + \mathscr{K}_f(\mathbf{M}_{eq}; \hat{\mathbf{r}})\right] \\ T_L(k_0, \hat{\mathbf{r}}, \mathbf{D}))d\Omega \quad \mathbf{D} \in \Sigma_M \end{aligned} \tag{2.23a}$$

$$\hat{\mathbf{n}} \times \left[-\eta_0 \mathscr{L}\left(\mathbf{J}_{eq}^{Love}; \mathbf{r}\right) + \mathscr{K}\left(\mathbf{M}_{eq}^{Love}; \mathbf{r}\right)\right] = -\frac{1}{2}\mathbf{M}_{eq}^{Love}(r) \quad \mathbf{r} \in \Sigma_R \tag{2.23b}$$

2.5 Application examples

2.5.1 Application as an investigation and diagnostic tool

We show how equivalent currents can be used to investigate the radiation mechanism of an antenna and diagnose possible causes of malfunction starting from NF (or FF) measurement [47,48].

Data of this section, courtesy of the European Space Agency, refer to the EDAS-CASA Espacio array antenna intended for the European Navigation System GALILEO. The measured antenna is the predevelopment model flying on the In-Orbit Validation Element GIOVE-B satellite. A two-level beam-forming network (BFN) feeds 42 patch elements divided into six sectors and provides the complex excitations needed to obtain the desired iso-flux pattern. The measurements have been performed in the NF and compact range facility in the ESA ESTEC CPTR, Figure 2.5. From these NF data, the FF radiation pattern can be

Figure 2.5 The GIOVE-B array antenna measured in the ESA ESTEC CPTR. Reproduced, with permission, from Reference [47]

obtained using the classical spherical wave expansion (SWE); the FF pattern on cuts at fixed ϕ coordinate is shown in Figure 2.6, separating the right-hand circular polarization (co-polar) from the left-hand circular polarization (cross-polar) component.

In Figures 2.6 and 2.7, we show the FF pattern obtained from SNF data. We note a small axial asymmetry.

Equivalent currents are computed from NF data on a surface enclosing the antenna and part of the supporting stand-off (Figure 2.8). This computation uses no *a priori* information about the antenna except its spatial occupancy and position with respect to the measurement points, as described in Section 2.2.

The low levels of the currents supported on the stand-off show the low interaction between the antenna and its support. Using current filtering [49], we can highlight the contribution of the currents around the stand-off to the FF radiation pattern: we compute the FF radiation pattern due to *filtered* equivalent currents, i.e., excluding radiation of those on the stand-off and comparing it against the FF pattern due to the original currents (Figure 2.9). As expected, the difference is minimal.

The co-polar and cross-polar components of the electric and magnetic currents on the top of the reconstruction surface Σ_R, Figure 2.10, give very interesting information about the feeding mechanism and about a possible explanation for the axial asymmetry noted in the FF pattern.

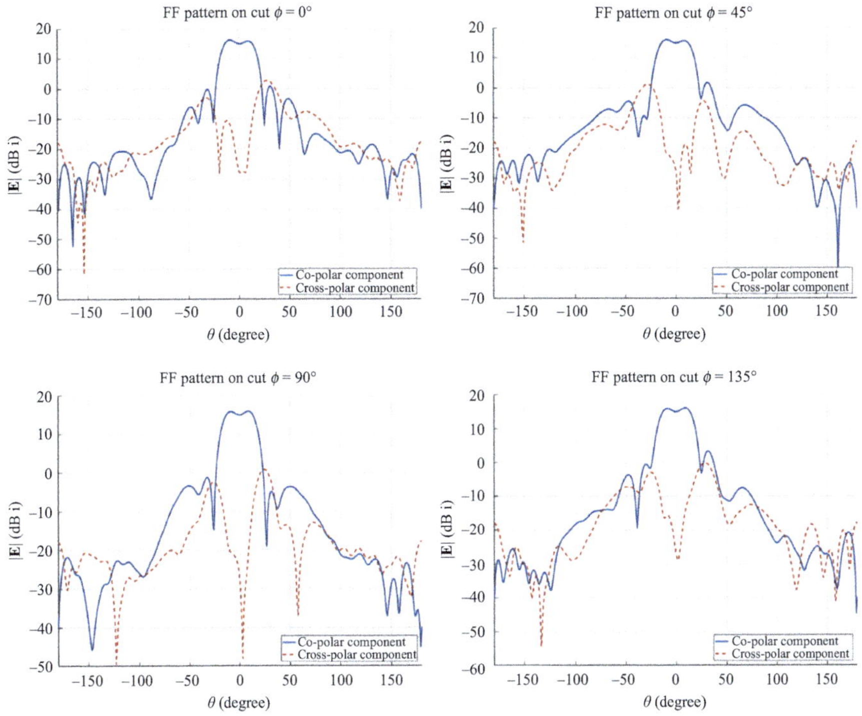

Figure 2.6 FF of the GIOVE-B array antenna on 4 cuts at constant ϕ coordinate (data courtesy of the European Space Agency)

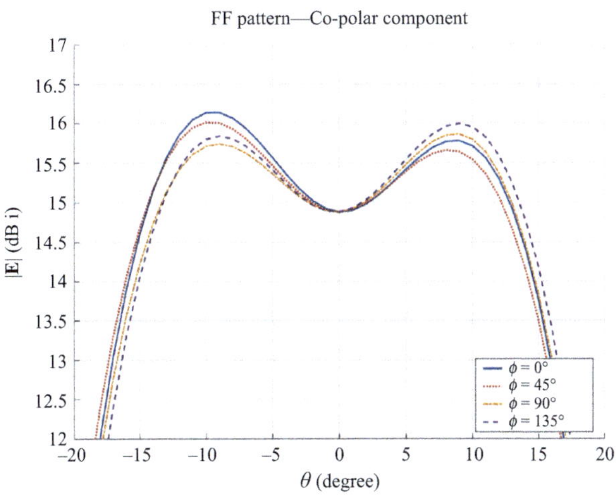

Figure 2.7 FF of the GIOVE-B array antenna on 4 cuts at constant ϕ coordinate: zoom on the main beam of the co-polar component (data courtesy of the European Space Agency)

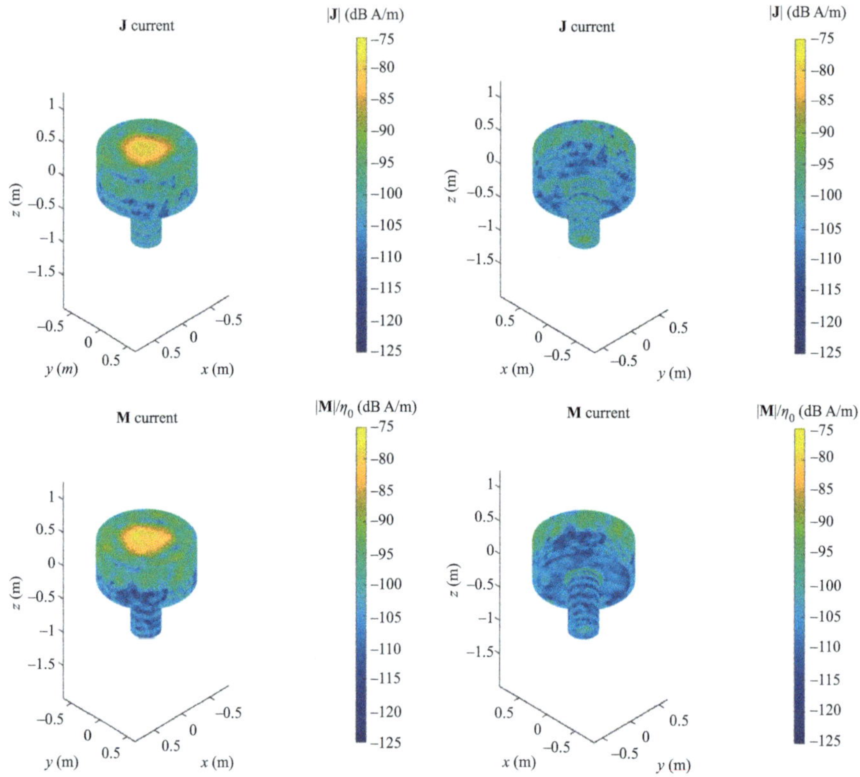

Figure 2.8 Equivalent electric currents on a surface enclosing the GIOVE-B array antenna and the stand-off

The second row from the top of Figure 2.10 shows the cross-polar amplitude, with levels normalized to the co-polar peak value. The cross-polar excitation at the center of the array is approximately −12 dB below the co-polar peak. Despite this remarkably high value, the on-axis cross-polar FF radiation pattern is very low. A careful examination of the phase of the cross-polar component of the currents (bottom row of Figure 2.10) highlights the three-element sequential rotation mechanism used to obtain such performance. Three different phase shifts are implemented in the BFN to create the desired null in the boresight direction while keeping the in-phase co-polar excitation. Such complex feeding mechanism could give rise to some residual error on the phase of the co-polar part. This could explain the co-polar phase dispersion observed in the third row of Figure 2.10 which, in turn, could explain the axial asymmetry of the FF pattern.

In summary, the equivalent currents, computed starting from measured samples of the field radiated by the antenna and without any *a priori* information about the antenna itself but its spatial occupancy, uncover the complex inner working mechanism of the array.

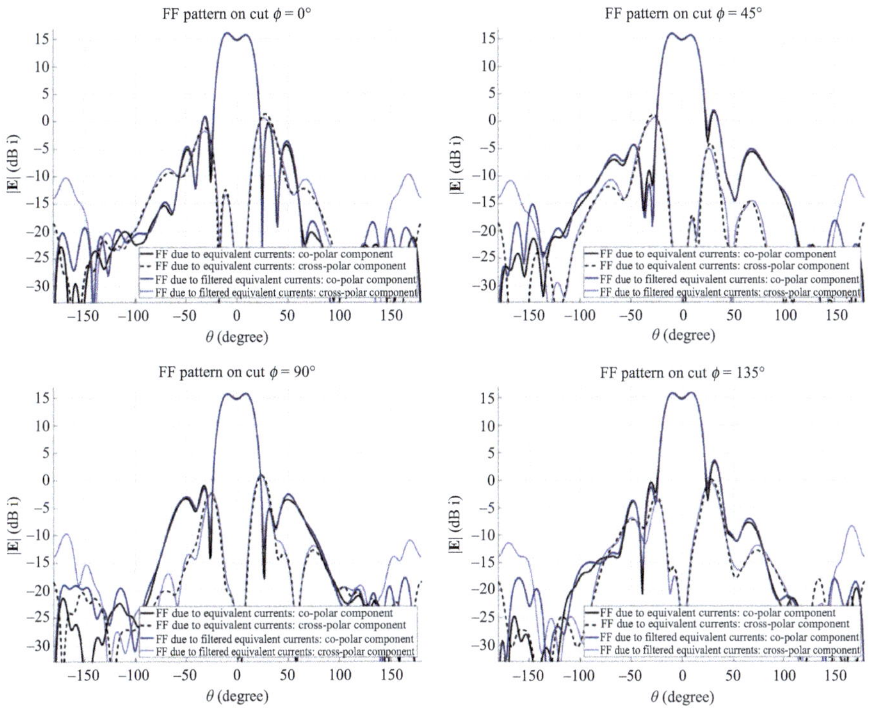

Figure 2.9 Comparison between FF pattern due to all equivalent currents and FF pattern due to filtered equivalent, i.e., currents not supported on the stand-off. These curves should be compared with the curves in Figures 2.6 and 2.7

2.5.2 Application in field transformations

Equivalent currents can be used as a flexible tool to perform field transformation, as NF–FF and NF–NF.

Indeed, we can evaluate the field radiated by the equivalent currents on any surface exterior to Σ_R and the equivalence theorem at the basis of the formulation guarantees that this field is the same as the field radiated by the original sources (Section 2.2.1).

In particular, the generality of the surface Σ_M, where field samples are acquired, allows us to evaluate the FF pattern of a radiating structure starting from NF samples acquired on noncanonical surfaces, where standard technique based on Fourier transform and SWE cannot be used efficiently. When working with equivalent currents, instead, we can leverage on fast methods as mentioned in Section 2.3.3.

To show the performance of the method, we consider two radiating structures, one in Section 2.5.2.1 and the other in Section 2.5.2.2.

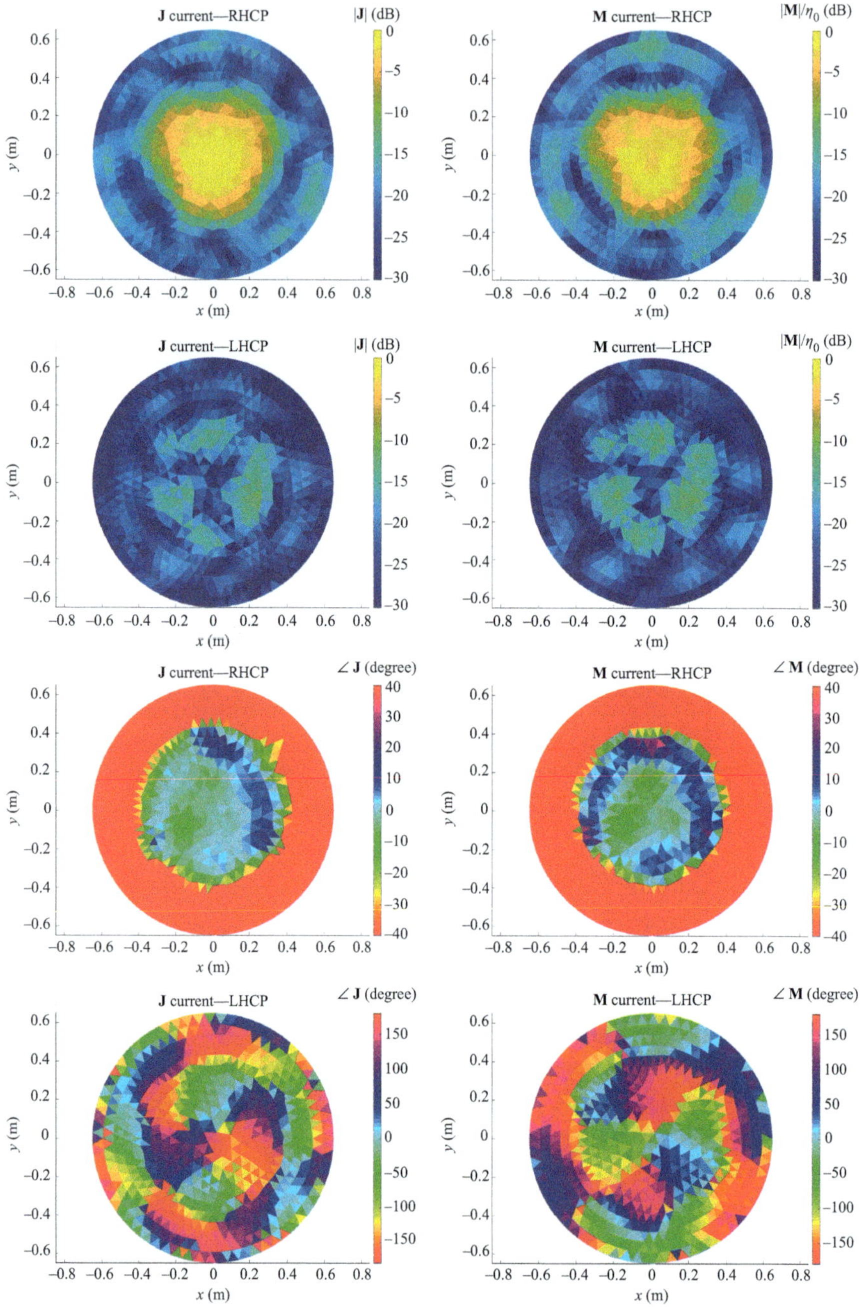

Figure 2.10 Amplitude and phase of co-polar and cross-polar components of the electric and magnetic equivalent currents on the top of the equivalent surface. Equivalent currents are evaluated at the barycenter of each mesh triangle and plotted with no interpolation (piecewise constant)

2.5.2.1 Near-field to near-field

The first structure is a linearly polarized patch antenna working at 7.5 GHz. We consider three sets of measurements (Figure 2.11):

1. In the first set, the field samples are acquired on a sphere centered with radiator (Σ_1 in Figure 2.12, top panel).
2. In the second set, the field samples are acquired on a sphere whose center is offset with respect to the radiator position (Σ_2 in Figure 2.12, top panel).
3. In the third set, the field samples are acquired on a sphere centered with the radiator, with radius larger than in point 1.

Figure 2.11 Patch antenna in the anechoic chamber

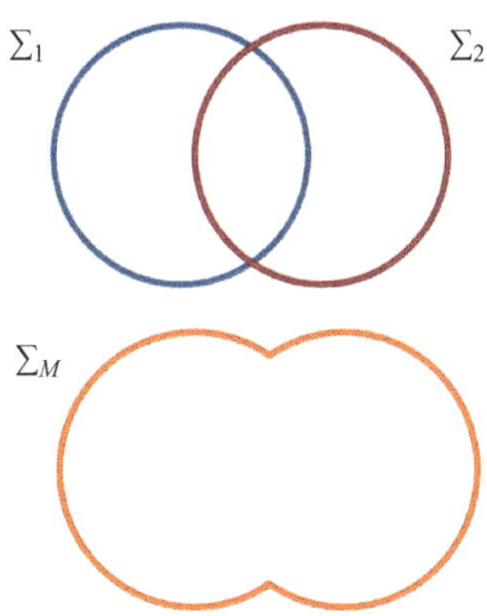

Figure 2.12 Schematic illustration of the procedure applied to obtain a noncanonical surface Σ_M from the disjoint union of two spheres Σ_1 and Σ_2

Data from points 1 and 2, namely, surfaces Σ_1 and Σ_2, are combined to obtain what we consider as Σ_M: we keep the measurement points which correspond to the boundary of the disjoint union of the two balls which have the spheres in points 1 and 2 as boundaries. The resulting surface is such that standard field transformation from this surface to the surface of point 3 cannot be applied efficiently (Σ_M in Figure 2.12, bottom panel).

Once the equivalent currents are obtained on a surface enclosing the radiator, Σ_R, we evaluate the field radiated on the points where samples of 3 are acquired, performing then an NF–NF, and we compare the results. In Figure 2.13, we present the two fields, obtained from NF–NF and measured on 4 cuts at constant ϕ coordinate. The transformation using equivalent currents has a clear smoothing effect with respect to the actual measurement; however, the low difference between the measured field and the transformed field highlights the good performance of the procedure.

2.5.2.2 Near-field to far-field

The second radiator we consider is composed of the patch antenna of Section 2.5.2.1 mounted on a mock-up of a vehicle, Figure 2.14. The overall largest dimension of the radiator is 30 wavelengths. We consider two sets of measurements:

1. In the first set, the field samples are acquired on a sphere centered with radiator.
2. In the second set, the field samples are acquired on a sphere whose center is offset with respect to the radiator position.

Data from 1 and 2 are combined to obtain what we consider as Σ_M as explained in Section 2.5.2.1. We look for the FF pattern of the radiator, evaluated starting from these samples distributed on a surface where standard field transformation cannot be applied.

To perform the NF–FF transformation, we compute the equivalent currents on a surface enclosing the radiator, Σ_R, using a fast algorithm to perform the MVP, and then we evaluate their FF pattern. We compare this reconstructed pattern against the FF pattern obtained using a commercial software starting from data in point 1. In Figure 2.15, we present the two fields on 4 cuts for constant ϕ coordinate. Again, the low difference between the pattern obtained from data on a noncanonical surface and the NF–FF based on equivalent currents, and the pattern obtained from data on a canonical surface and NF–FF based on SWE highlights the good performance of the procedure.

2.5.3 Array antenna synthesis

In this section, we present a technique introduced in [50], which is based on the observation of fields radiated by equivalent sources that represent real radiators when using any of the SEqFs. In these situations, when the internal field is *not* forced to zero, it has been (numerically) observed that non-Love equivalent sources generate a field in the interior of Σ_R that provides a clear indication of the original

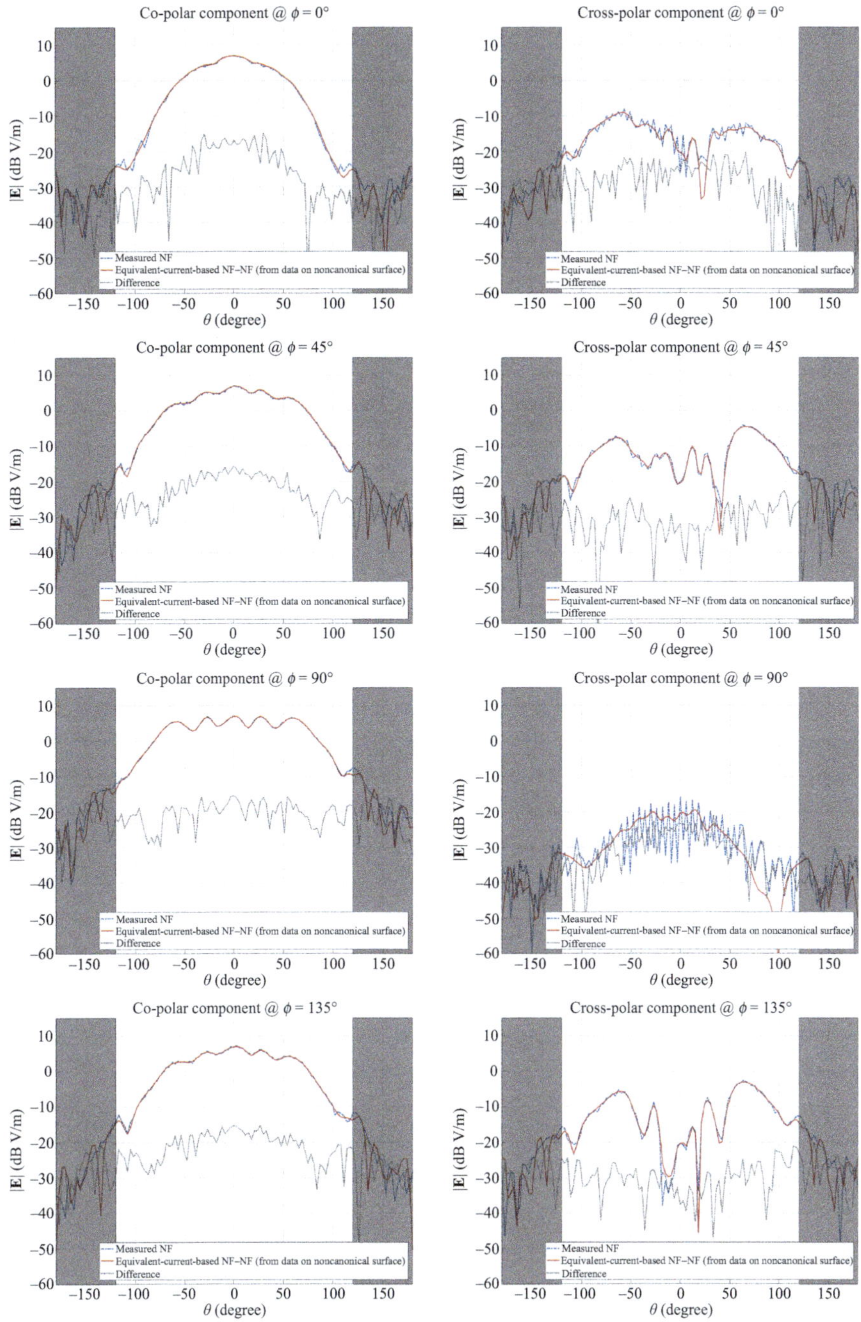

Figure 2.13 NF–NF for patch antenna. Left column: co-polar component and right column: cross-polar component. Each row corresponds to a constant ϕ cut. Gray areas shade the angular regions where the blockage of the supporting structure and rotor make the data less reliable (see Figure 2.11)

Figure 2.14 Patch antenna mounted on a vehicle mock-up in the anechoic chamber

sources. To illustrate this, consider a pair of monocone antennas located above a circular ground plane with radius $r = 3\lambda$. Full-wave simulation results are shown in Figure 2.16, which were obtained with our MoM solver.

The full-wave currents depicted in Figure 2.16 can be used to obtain numerical "field measurements" via the radiation operators in (2.7a) and (2.7b). We thus set up a source reconstruction problem where Σ_R is the minimum sphere enclosing the structure ($r = 3\lambda$) and both electric and magnetic currents are computed to reproduce the numerical field measurements without imposing the Love's condition (i.e., we solve (2.7a) and (2.7b) alone). As discussed above and in [19], these radiate nonzero fields inside Σ_R, and it is interesting to observe that internal fields give a clear view of the original radiators, as seen in Figure 2.17.

These results indicate the potential of source reconstruction for array antenna synthesis, and especially for elements location in *sparse* array synthesis according to the following procedure:

1. Enclose the desired array's volume within Σ_R.
2. Use radiation of any source distribution on Σ_R to generate a field that matches the pattern requirements (e.g., an amplitude mask); as the sources are on a given mesh, this can be done with any source-based synthesis procedure, e.g., [51].
3. Consider this field (amplitude and phase) as the "measured" field for a source reconstruction and obtain the equivalent sources that generate this "measured" field using one of the SEqFs.
4. Compute the field radiated by these equivalent surface currents through the volume inside Σ_R.
5. Determine array elements locations (and number) by local maxima of the internal field.

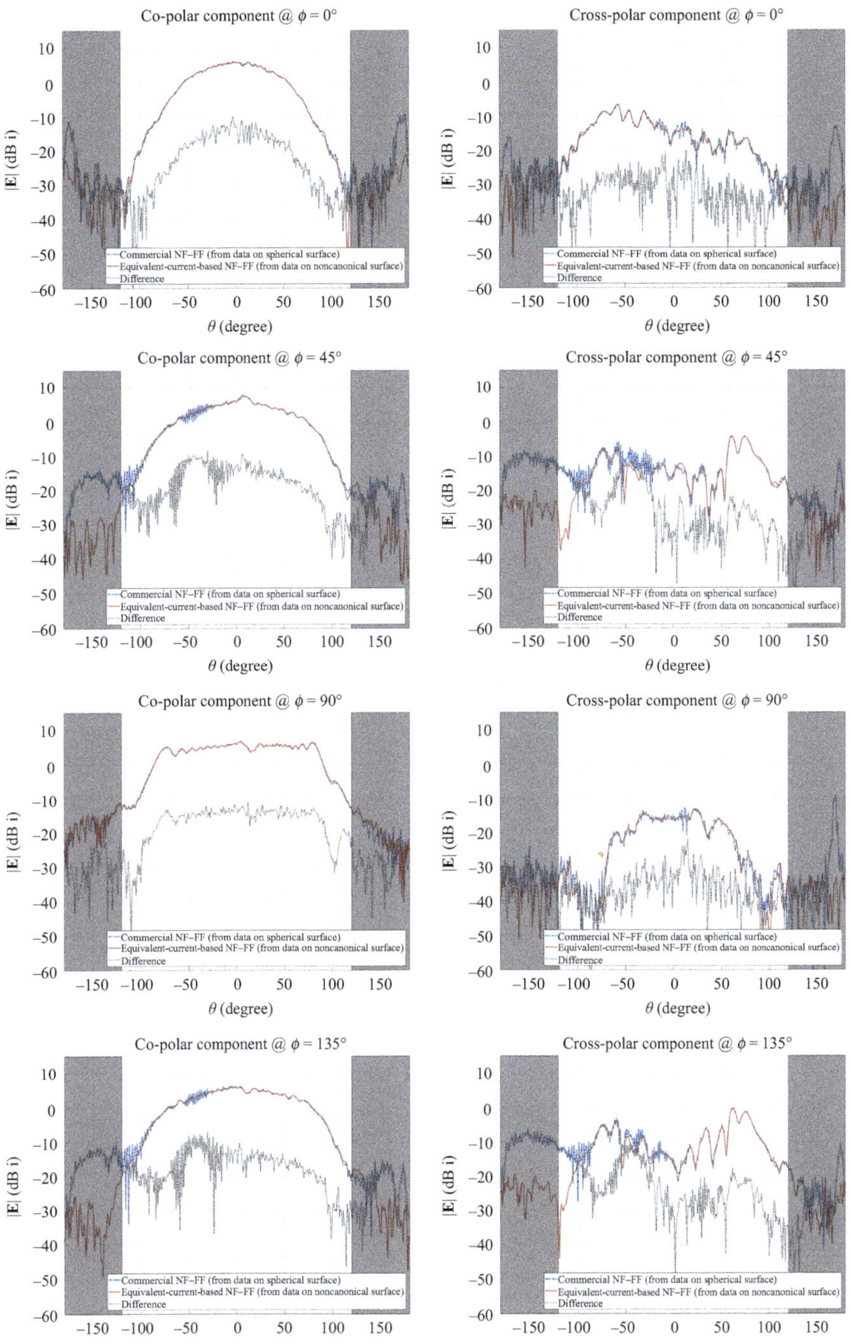

Figure 2.15 NF–FF for patch antenna mounted on a mock-up. Left column: co-polar component; right column: cross-polar component. Each row corresponds to a constant ϕ cut. Gray areas shade the angular regions where the blockage of the supporting structure and rotor make the data less reliable (see Figure 2.14)

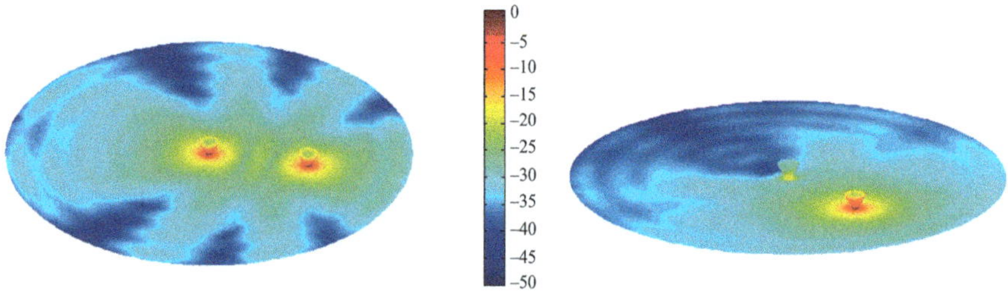

Figure 2.16 Full-wave simulation currents for the test structure with both monocones excited (left), and only the off-center monocone excited (right)

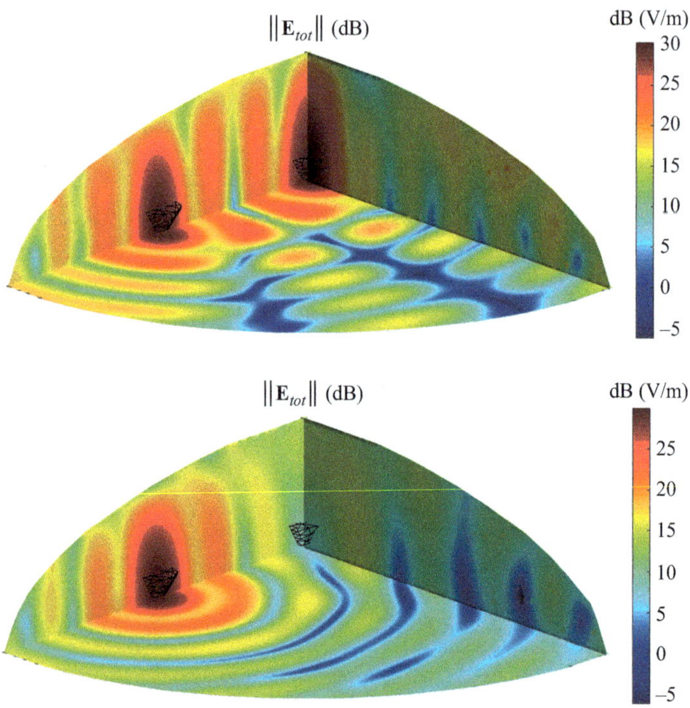

Figure 2.17 Cuts of the electric field produced by the equivalent sources inside the reconstruction surface Σ_R when both monocones are excited (top) and when only the external monocone is excited (bottom)

If step 2 above is unsuccessful, this can be taken as an indication that the considered volume Σ_R needs to be enlarged, as the goal field cannot be radiated by sources lying inside the volume specified. Note that this fact can be exploited to look for the smallest Σ_R that can be used, starting from a fairly large volume and shrinking it progressively, while the synthesis in step 2 is successful.

The procedure just described was used in [50] to synthesize a 3D array and a planar array that generate a given goal field.

2.6 Numerical tests on the accuracy of source reconstruction

In this section, we present detailed results on the accuracy of source reconstruction for a test case already reported in literature [28]. The comparison includes NF accuracy for varying observation distance and signal-to-noise ratio (SNR), and also the accuracy of the reconstructed equivalent currents for varying SNR. All the data are based on full-wave simulation in order to have fine control over SNR and to be able to do comparisons against accurate reference data (NF and equivalent currents).

The radiator considered is a monocone antenna that operates at 3.3 GHz, and it is located above the center of a circular ground plane with radius 15 cm, which is simulated using our in-house MoM code to obtain the induced currents, which in turn can be used to compute the electromagnetic field at arbitrary locations. These simulation results are shown in Figure 2.18, where it is clear that the small ground plane contributes with significant diffraction in the radiated field.

The test case is depicted in Figure 2.19. The reconstruction surface Σ_R is an ellipsoid that encloses the reference structure tightly, and "measurements" (computed using full-wave simulation) are available on a spherical range with radius 1.6 m co-centered with the ground plane, where samples have an angular step of 5.3° along θ and ϕ. The angular portion $\theta > 164°$ is not included in these measurements to account for the effect of a supporting mast. The sampling grids for NF accuracy testing are depicted in Figure 2.19, too. These are semi-ellipses whose semiaxes are those of Σ_R plus values Δs that are approximately the distance from Σ_R at which NF is checked, and there are ten of these in the range $0.1\lambda_0 \leq \Delta s \leq 4\lambda_0$.

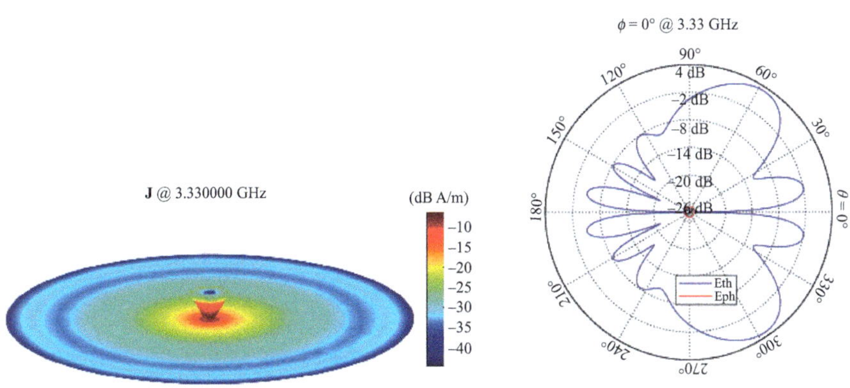

Figure 2.18 MoM simulation results for the monocone structure used in the accuracy tests: induced currents (left) and FF (right)

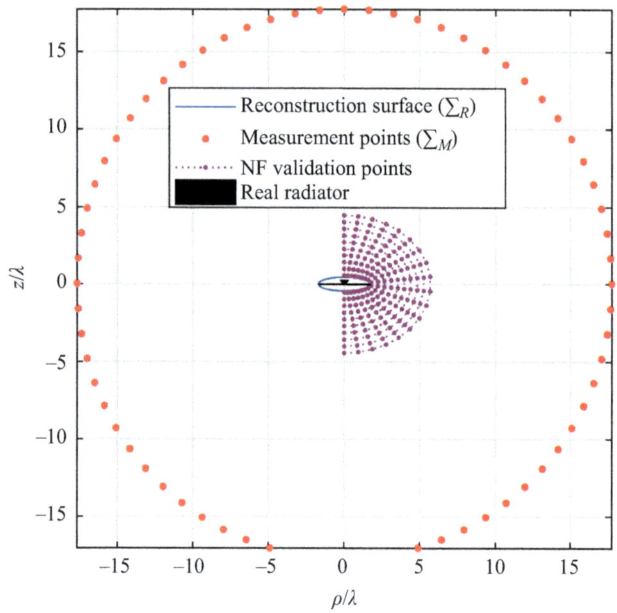

Figure 2.19 Depiction of the synthetic measurement and source reconstruction scenario: measurement points cover a full sphere with the angular sampling shown both in θ and ϕ, with the exception of an angular sector in the south pole which is modeled as unavailable due to the supporting mast; on the other hand, the NF validation points include only the semi-plane $\phi = 0$

Besides varying the observation distance d, the SNR is swept in the range $20\text{ dB} \leq \text{SNR} \leq 70\text{ dB}$. In addition to these, the case with no noise (i.e., $\text{SNR} = \infty$) is included for reference. The noise applied is additive Gaussian, and the exact SNR is guaranteed in each case.

Initially, we present the error on the current reconstruction, which is computed as follows:

$$\varepsilon_R = \frac{\mathscr{M}(\mathbf{J}_{ex} - \mathbf{J}_R, \mathbf{M}_{ex} - \mathbf{M}_R)}{\mathscr{M}(\mathbf{J}_{ex}, \mathbf{M}_{ex})}$$
$$\mathscr{M}(\mathbf{J}, \mathbf{M}) = \sqrt{\iint_{\Sigma_R} |\,\mathbf{J}|^2 + |\,(\mathbf{M})/\eta_0|^2\, ds} \tag{2.24}$$

where the integral involved in the total current norm $\mathscr{M}$ can be computed exactly (our MoM discretization uses the RWG basis), $(\mathbf{J}_{ex}, \mathbf{M}_{ex})$ are the reference equivalent currents (computed from full-wave simulation), and $(\mathbf{J}_R, \mathbf{M}_R)$ are the currents obtained in the reconstruction process.

Since only the DEqF aims at reconstructing Love's currents, it is the only formulation tested with respect to the accuracy of reconstructed currents.

As already discussed in [19], the SEqF present large errors even with no noise present, of the order resulting with the DEqF at SNR $=$ 20 dB. The results in Figure 2.20 indicate that in order to obtain an acceptable current reconstruction in this case, one must ensure that SNR $\geq$ 30 dB. This threshold depends on the particular test case and the measurement distance (radius of Σ_M when it is a sphere) and turns out to be particularly demanding in the present scenario since the enclosing surface is almost touching hot zones of the reference antenna, and measurements are made close to the FF boundary $\left(R_{far} = 1.98\text{ m}\right)$; the limit would be relaxed should the measurements be available at a smaller distance, since in such case, the higher order (evanescent) modes would have been less attenuated and could be recovered with less noise amplification [41,52]; a similar effect would result if a somewhat larger Σ_R was used. Finally, we note that in general, the accuracy of the reconstructed currents depends heavily on the particularities of the case considered. The results presented in [19] demonstrate a halved equivalent current error for the noiseless case when using a looser spherical enclosure and the same reference radiator.

Another metric of interest is the error in the NF reconstructed at arbitrary observation points. For the present test, given a fixed value Δs (recall that it is the enlargement of the semiaxes of the NF semiellipses with respect to Σ_R), the error is computed by treating the set of measurements at the sampling points of each semiellipse as a vector as seen in the following expression:

$$\varepsilon_{NF}(\Delta s, \text{SNR}) = 10\log_{10}\frac{\Sigma_n\left|E_{rec}\left(\mathbf{r}_n^{\Delta s}, \text{SNR}\right) - E_{ref}\left(\mathbf{r}_n^{\Delta s}\right)\right|^2}{\Sigma_n\left|E_{ref}\left(\mathbf{r}_n^{\Delta s}\right)\right|^2} \tag{2.25}$$

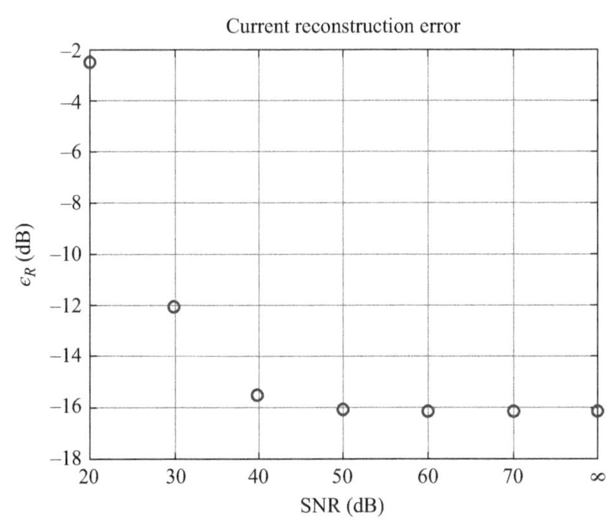

Figure 2.20 Current reconstruction error as a function of SNR

where $E_{rec}\left(\mathbf{r}_n^{\Delta s}, \mathrm{SNR}\right)$ is the field reconstructed from radiation of the equivalent currents, which in turn were computed using noisy data and $E_{ref}\left(\mathbf{r}_n^{\Delta s}\right)$ is the noiseless data. In this sense, the technique is being tested not only on the accuracy with which it approaches a given field but also the filtering capability to discard the noise and ideally reconstruct sources that radiate the "clean" measurement data from noisy data.

The results for NF accuracy are shown in Figure 2.21 in the form of level curves, separately for each of the formulations considered: **J**-SEqF, **M**-SEqF, **JM**-SEqF, and DEqF. Based on these results, one can make the following observations:

- The **M**-SEqF is the lowest performer in all situations, arguably due to the nature of the only integral operator used in that case magnetic field integral equation, and hence must be avoided in practice.
- For $(\mathrm{SNR}) \leq 60$ dB, the DEqF provides superior results to all the formulations, and hence it must be used in diagnostics and in general for solutions that need robustness to noise. Considering that most practical measurement

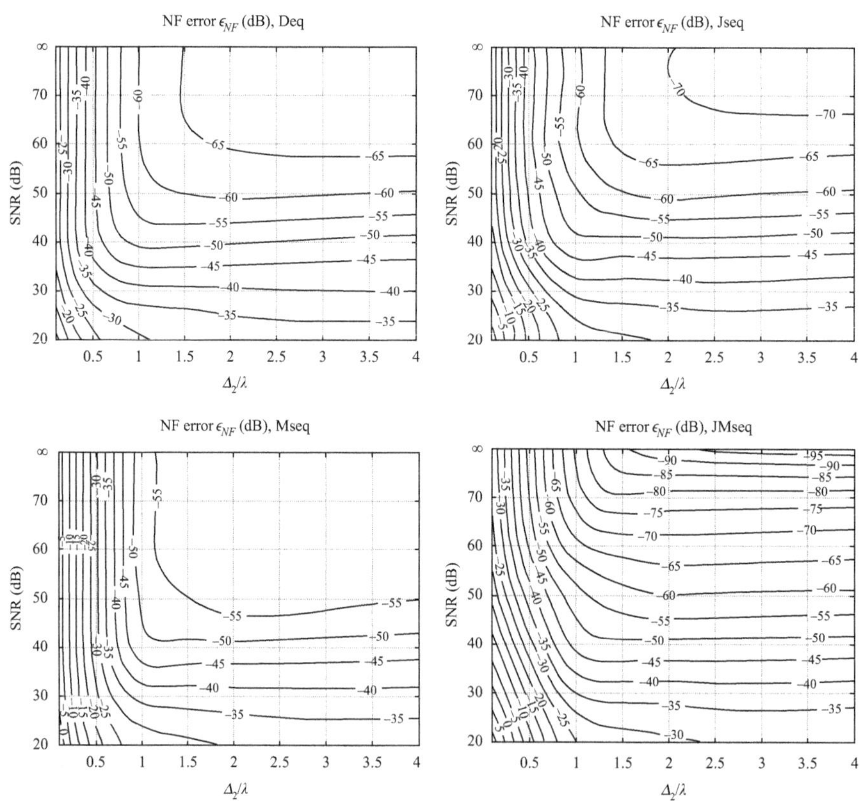

Figure 2.21 NF error as a function of distance and SNR

situations conform to the threshold mentioned, this formulation is suggested for general purpose.

- The **JM**-SEqF presents the best performance for large SNR ($\geq$ 60 dB), so it is appropriate only for tests with very accurate numerical data or when the NF observation distance is larger than λ_0.
- The formulation **J**-SEqF provides a performance similar to **JM**-SEqF, so it is recommended for large SNR and observation distance $\Delta s > \lambda_0$ in view of the memory and time savings resulting from the halved number of unknowns.

2.7 Conclusions and outlook

Source reconstruction using equivalent currents is a mature technique with a number of interesting applications such as field transformation (NF–FF/NF–NF), electromagnetic diagnostics, design optimization, general source modeling, and antenna synthesis. Its formulation and numerical solution are based on the widely popular MoM and therefore inherits its well-known properties such as accuracy and flexibility and profits directly from advanced techniques developed around MoM such as low-frequency, large-scale, and multi-resolution solvers to mention some.

It has been found that the DEqF provides robust results and the best accuracy in real-world situations both for diagnostics and for field transformations, while the simpler SEqFs could be of interest in order to reduce memory and time requirements when the loss in accuracy is acceptable, or when dealing with data from simulation.

Present topics of interest in the application of the technique are as follows:

1. Optimization of large-scale solvers to obtain a truly efficient scaling of complexity even in broad frequency bands, and considering the multi-resolution nature of the problem (sparse in the measurement points, dense in the reconstruction surface).
2. The integration of constraints into the solution process that accounts for *a priori* information such as total radiated power, maximum directivity, and the input impedance of the AUT.

References

[1] R. F. Harrington. *Time Harmonic Electromagnetic Fields*. New York: IEEE Press, 2001.

[2] C. A. Balanis. *Antenna Theory: Analysis and Design*. Hoboken, NJ: Wiley-Interscience, 3 edition, April 2005.

[3] Y. Alvarez, F. Las-Heras, and M. R. Pino. Reconstruction of equivalent currents distribution over arbitrary three-dimensional surfaces based on integral equation algorithms. *IEEE Transactions on Antennas and Propagation*, 55(12):3460–3468, 2007.

[4] P. Petre and T. K. Sarkar. Planar near-field to far-field transformation using an equivalent magnetic current approach. *IEEE Transactions on Antennas and Propagation*, 40(11):1348–1356, 1992.

[5] T. K. Sarkar and A. Taaghol. Near-field to near/far-field transformation for arbitrary near-field geometry, utilizing an equivalent magnetic current. *IEEE Transactions on Electromagnetic Compatibility*, 38(3):536–542, 1996.
[6] B. Nadeau and J. J. Laurin. Extrapolations using vectorial planar near-field measurements for EMC. In *IEEE Int. Symp. Electromag. Compat.*, pages 924–928. Denver CO, August 1998.
[7] T. K. Sarkar and A. Taaghol. Near-field to near/far-field transformation for arbitrary near-field geometry utilizing an equivalent electric current and MoM. *IEEE Transactions on Antennas and Propagation*, 47(3):566–573, 1999.
[8] F. Las-Heras, M. R. Pino, S. Loredo, Y. Alvarez, and T. K. Sarkar. Evaluating near-field radiation patterns of commercial antennas. *IEEE Transactions on Antennas and Propagation*, 54(8):2198–2207, 2006.
[9] J. J. Laurin, J. F. Zurcher, and F. Gardiol. Near-field diagnostics of small printed antennas using the equivalent magnetic current approach. *IEEE Transactions on Antennas and Propagation*, 49(5):814–828, 2001.
[10] E. Martini, G. Carli, and S. Maci. An equivalence theorem based on the use of electric currents radiating in free space. *IEEE Antennas and Wireless Propagation Letters*, 7:421–424, 2008.
[11] E. Marx. Integral equation for scattering by a dielectric. *IEEE Transactions on Antennas and Propagation*, 32(2):166–172, 1984.
[12] A. Glisson. An integral equation for electromagnetic scattering from homogeneous dielectric bodies. *IEEE Transactions on Antennas and Propagation*, 32(2):173–175, 1984.
[13] M. Mohajer, S. Safavi-Naeini, and S. K. Chaudhuri. Surface current source reconstruction for given radiated electromagnetic fields. *IEEE Transactions on Antennas and Propagation*, 58(2):432–439, 2010.
[14] J. A. Stratton and L. J. Chu. Diffraction theory of electromagnetic waves. *Physical Review*, 56(1):99–107, 1939.
[15] G. H. Golub and C. F. Van Loan. *Matrix Computations (Johns Hopkins Studies in Mathematical Sciences)(3rd Edition)*. Baltimore, MD: The Johns Hopkins University Press, 3rd edition, October 1996.
[16] W. C. Chew, Y. M. Wang, G. Otto, D. Lesselier, and J. C. Bolomey. On the inverse source method of solving inverse scattering problems. *Inverse Problems*, 10(3):547–553, 1994.
[17] E. A. Marengo and R. W. Ziolkowski. Nonradiating and minimum energy sources and their fields: Generalized source inversion theory and applications. *IEEE Transactions on Antennas and Propagation*, 48(10):1553–1562, 2000.
[18] J. Araque and G. Vecchi. Improved-accuracy source reconstruction on arbitrary 3-D surfaces. *IEEE Antennas and Wireless Propagation Letters*, 8:1046–1049, 2009.
[19] J. L. A. Quijano and G. Vecchi. Field and source equivalence in source reconstruction on 3D surfaces. *Progress in electromagnetics Research, PIER*, 103:67–100, 2010.

[20] G. C. Hsiao and R. E. Kleinman. Mathematical foundations for error estimation in numerical solutions of integral equations in electromagnetics. *IEEE Transactions on Antennas and Propagation*, 45(3):316–328, 1997.

[21] M. B. Woodworth and A. D. Yaghjian. Derivation, application, and conjugate gradient solution of dual-surface integral equations for three-dimensional, multi-wavelength perfect conductors. *Progress in Electromagnetics Research, PIER*, 05:103–129, 1991.

[22] F. Las-Heras, Y. Alvarez, M. R. Pino, and M. Alvarez. Sources reconstruction techniques for the diagnosis and characterization of antennas of complex geometry. In *Proceedings of ICONIC*, pages 182–187, June 2007.

[23] Y. Alvarez, T. Sarkar, and F. Las-Heras. Improvement of the sources reconstruction techniques: Analysis of the SVD algorithm and the RWG basis functions. In *Antennas and Propagation International Symposium*, pages 5644–5647, June 2007.

[24] Y. Alvarez, F. Las-Heras, M. R. Pino, and J. A. Lopez. Acceleration of the sources reconstruction method via the fast multipole method. In *Antennas and Propagation Society International Symposium, 2008. AP-S 2008. IEEE*, pages 1–4, July 2008.

[25] M. M. Leibfritz, F. M. Landstorfer, and T. F. Eibert. An equivalent source method to determine complex excitation levels of antenna arrays from near-field measurements. In *Antennas and Propagation, 2007. EuCAP 2007. The Second European Conference on*, pages 1–7, November 2007.

[26] T. F. Eibert and C. H. Schmidt. Multilevel fast multipole accelerated inverse equivalent current method employing Rao–Wilton–Glisson discretization of electric and magnetic surface currents. *IEEE Transactions on Antennas and Propagation*, 57(4):1178–1185, 2009.

[27] K. Persson and M. Gustafsson. Reconstruction of equivalent currents using a near-field data transformation – with radome applications. *Progress in Electromagnetics Research, PIER*, 54:179–198, 2005.

[28] J. L. A. Quijano and G. Vecchi. Near- and very near-field accuracy in 3-D source reconstruction. *IEEE Antennas and Wireless Propagation Letters*, 9:634–637, 2010.

[29] S. Rao, D. Wilton, and A. Glisson. Electromagnetic scattering by surfaces of arbitrary shape. *IEEE Transactions on Antennas and Propagation*, 30(3): 409–418, 1982.

[30] C. Paige and M. Saunders. LSQR: An algorithm for sparse linear equations and sparse least squares. *ACM Transactions on Mathematical Software*, 8(1):43–71, 1982.

[31] R. Coifman, V. Rokhlin, and S. Wandzura. The fast multipole method for the wave equation: A pedestrian prescription. *IEEE Antennas and Propagation Magazine*, 35(3):7–12, 1993.

[32] M. A. E. Bautista, M. A. Francavilla, P. Martinsson, and F. Vipiana. $O(n)$ nested skeletonization scheme for the analysis of multiscale structures using the method of MoMents. *IEEE Journal on Multiscale and Multiphysics Computational Techniques*, 1:139–150, 2016.

[33] C. H. Schmidt, M. M. Leibfritz, and T. F. Eibert. Fully probe-corrected near-field far-field transformation employing plane wave expansion and diagonal translation operators. *IEEE Transactions on Antennas and Propagation*, 56(3):737–746, 2008.

[34] L. J. Foged, L. Scialacqua, F. Saccardi, J. L. Araque Quijano, and G. Vecchi. Diagnostics on electrically large structures by fast multipole method enhancement of the equivalent current technique. In *Proceedings of the 35th European Space Agency (ESA) Antenna Workshop on Antenna and Free Space RF Measurements*. ESTEC Noordwijk, The Netherlands, 10–13 September 2013.

[35] L. J. Foged, L. Scialacqua, F. Saccardi, J. L. Araque Quijano, and G. Vecchi. Application of the dual-equation equivalent-current reconstruction to electrically large structures by fast multipole method enhancement [AMTA corner]. *IEEE Antennas and Propagation Magazine*, 56(5):264–273, 2014.

[36] C. G. Montgomery, R. H. Dicke, and E. M. Purcell. *MIT Radiation Laboratory Series: Principles of Microwave Circuits*. New York: McGraw-Hill, 1948.

[37] D. M. Kerns. Correction of near-field antenna measurements made with an arbitrary but known measuring antenna. *Electronics Letters*, 6(11):346–347, 1970.

[38] W. Leach and D. Paris. Probe compensated near-field measurements on a cylinder. *IEEE Transactions on Antennas and Propagation*, 21(4):435–445, 1973.

[39] G. Borgiotti. Integral equation formulation for probe corrected far-field reconstruction from measurements on a cylinder. *IEEE Transactions on Antennas and Propagation*, 26(4):572–578, 1978.

[40] F. Jensen. *Electromagnetic Near-field-far-field correlations*. Ph.D. Thesis, Laboratory of Electromagnetic Theory, Technical University of Denmark, July 1970.

[41] J. E. Hansen. Spherical near-field antenna measurements. *IEE Electromagnetic Waves Series*, 26. Stevenage, Herts, England: Peter Peregrinus Ltd, 1988.

[42] R. Coifman, V. Rokhlin, and S. Wandzura. The fast multipole method for the wave equation: A pedestrian prescription. *IEEE Antennas and Propagation Magazine*, 35(3):7–12, 1993.

[43] S. A. Schelkunoff. *Electromagnetic Waves*. New York: D. Van Nostrand Company, 1943.

[44] G. Sinclair. The transmission and reception of elliptically polarized waves. *Proceedings of the IRE*, 38(2):148–151, 1950.

[45] P. C. Clemmow. *The Plane Wave Spectrum Representation of Electromagnetic Fields*. New York: Wiley-IEEE Press, 1997.

[46] W. C. Gibson. *The Method of MoMents in Electromagnetics*. Boca Raton, FL: Chapman & Hall/CRC, 2008.

[47] L. J. Foged, F. Saccardi, L. Scialacqua, L. Salghetti Drioli, J. L. A. Quijano, and G. Vecchi. Diagnostic and investigation of sub-array radiation of large array antenna by inverse source technique. In *Proceedings of the Antennas*

and Propagation Society International Symposium (APSURSI), 2014 IEEE, pages 303–304. Memphis, Tenessee, USA, 6–11 July 2014.

[48] L. J. Foged, L. Scialacqua, F. Saccardi, L. Salghetti Drioli, J. L. A. Quijano, and G. Vecchi. Investigation of complex excitation errors at single element and sub-array level in large array antennas by inverse source technique. In *Proceedings of the Antennas and Propagation (EuCAP), 2014 8th European Conference on*, pages 2655–2658. The Hague, The Netherlands, 6–11 April 2014.

[49] J. L. A. Quijano, L. Scialacqua, J. Zackrisson, L. J. Foged, M. Sabbadini, and G. Vecchi. Suppression of undesired radiated fields based on equivalent currents reconstruction from measured data. *IEEE Antennas and Wireless Propagation Letters*, 10:314–317, 2011.

[50] J. L. A. Quijano and G. Vecchi. Source reconstruction in antenna synthesis. In *Proceedings of European Conference on Antennas and Propagation (EuCAP)*, Barcelona, Spain, April 2010.

[51] J. L. A. Quijano and G. Vecchi. Alternating adaptive projections in antenna synthesis. *IEEE Transactions on Antennas and Propagation*, 58(3):727–737, 2010.

[52] P. Koivisto. Analytical solution for characteristic modal power distribution and truncation limit for spherical wave expansion of antenna radiation pattern. *Journal of Electromagnetic Waves and Applications*, 16:1307–1328, 2002.

Chapter 3

Antenna and measurement improvements based on equivalent currents reconstruction

Maria Alberica Saporetti[1], *Andrea Giacomini*[1], *Francesco Saccardi*[1], *and Lars J. Foged*[1]

3.1 Introduction

By post-processing of measured data in near field (NF) or far field (FF), the equivalent current (EQC) technique can reconstruct the extreme NFs and radiating currents on an arbitrary 3D surface enclosing a test object. In Chapter 2, the theoretical foundation for the EQC technique has been presented and illustrated by applicative examples concerning antenna diagnostic, field transformation and array synthesis. The precise knowledge of the radiating currents distribution on a measured antenna gives access to a wide range of advanced post-processing which can be summarized as follows:

- Antenna diagnostics
- Measurement diagnostics and filtering
- Echo suppression/Spatial filtering
- Truncation error mitigation
- Electromagnetic compatibility (EMC) investigation
- NF to FF (NF/FF) and NF/FF transformation (back transformation)
- Data interpolation/extrapolation
- Source for numerical computation (see Chapter 7).

In this chapter, different applicative examples of the EQC technique will be considered.

In particular, in Section 3.2, we focus on diagnostic and filtering capabilities of the EQC technique and how such investigative techniques can facilitate design improvement from antenna prototyping. The design advancements are achieved through a better understanding of the cause of both desired and undesired radiation phenomenon on the measured prototype. The inspection of the EQC determined on

[1]Microwave Vision Italy (MVI), Pomezia, Italy

a surface conformal to the measured prototype is in many ways similar to what is achieved from numerical modeling of the antenna with some very important differences. The EQCs are a result of the measured radiation, whereas currents determined by numerical modeling include both radiating and non-radiating sources. A further distinction of the EQC is that the radiation from a distinct physical feature, like a connector, can be isolated and its contribution to the overall radiation determined. The spatial filtering capabilities of the EQC is a very important feature to achieve design enhancement [1,2].

In Section 3.3, we will show how the spatial filtering properties of the EQC can be used to isolate the antenna radiation from disturbances from its measurement environment and thus improve the measurement accuracy. In particular, we will show how scattering from special features of the measurement setup such as standoff, mast and rails can be removed in post-processing by isolating the EQC only from the antenna. Moreover, we will see how the EQC has intrinsic spatial filtering properties that can be exploited for echo suppression purposes, similarly to what has been investigated in Chapter 5 in the case of the spherical and cylindrical modal expansions. In the same section, we will also see how the EQC can be exploited as NF/FF transformation tool capable of reducing the truncation errors in the case of partial spherical NF (SNF) scanning.

The capability of the EQC technique to reconstruct the extreme NF from the determined EQC distribution with sufficient accuracy can also be used in areas such as radio frequency (RF) safety and EMC. This will be briefly discussed in Section 3.4.

Finally, the advantages and limitations of the EQC technique will be summarized in Section 3.5.

3.2 Application of EQC in the improvement of antenna system design

In this section, we will see examples on how knowledge of the radiating currents by the EQC technique can be used to improve the antenna design. Different types of antennas will be taken into account and diagnosed with the EQC technique in order to show how important information regarding the radiation mechanism of the devices can be retrieved directly from the measurement and used to improve the design of the antenna. The antennas that will be considered are

- a wideband dual-ridge horn,
- a 32×8 patch array working as X-band,
- an array for space application and
- an NF first-order probe.

In addition to these antennas, a more complex system such as satellite will be also considered in order to assess the coupling between the installed antennas. Finally, an example regarding the application of the EQC technique in the EMC field will also be shown.

3.2.1 Advanced diagnostic of a wideband dual-ridge horn

A wideband dual-ridge horn is considered in this section in order to show the advanced diagnostic capabilities of the EQC technique. The horn is the MVG SH4000[*] shown in Figure 3.1. Such device works in the 4–40 GHz frequency range and is typically used as reference gain antenna and as feed for parabolic antennas. As can be seen, the dual-ridge is on the top part of the device. The lower part of the device is composed of a metallic skirt which allows to limit the scattering due to the back-radiation.

The SH4000 has been measured in the MVG SL18GHz system from 4 to 18 GHz. The higher frequencies of the device have been measured in the MVG SL50GHz shown in Figure 3.1 (right).

The EQC has been computed on a geometry conformal to the test object as shown in Figure 3.2. As can be seen, the defined equivalent geometry includes the details of the dual-ridge, the two side walls, the small supporting box of the dual-ridge and the skirt. The EQCs have computed at 8 and 28 GHz considering a triangular mesh size of $\lambda/15$ and $\lambda/4$, respectively. The computed equivalent electrical currents are shown for different views in Figures 3.3 and 3.4, respectively, for 8 and 28 GHz.

The two processing highlight how the currents are distributed on the device. As expected, at both frequencies the peak of the currents is clearly identifiable on the dual ridge. At 8 GHz, it can be noticed that the illumination of the skirt is quite

Figure 3.1 MVG SH4000 dual-ridge horn during measurement in the MVG StarLab 50 GHz multi-probe system

[*]https://www.mvg-world.com/en/products/field_product_family/antenna-1/dual-ridge-horns.

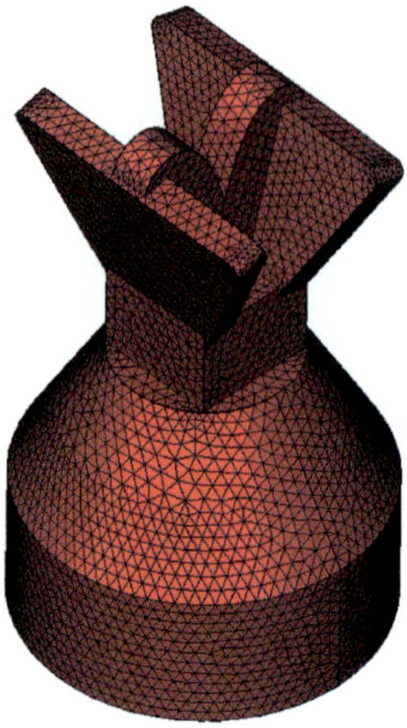

Figure 3.2 Conformal geometry used to compute the EQC of the SH4000

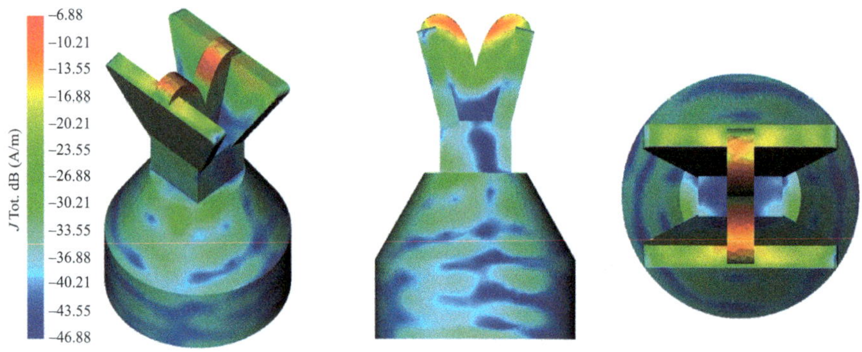

Figure 3.3 Electric EQC at 8 GHz computed from SNF measurements on an equivalent geometry conformal to the shape of the SH4000: perspective view (left); side view (center); top view (right)

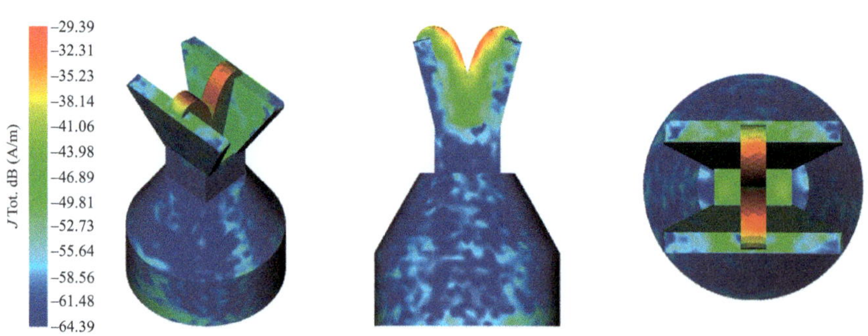

Figure 3.4 Electric EQC at 28 GHz computed from SNF measurements on an equivalent geometry conformal to the shape of the SH4000: perspective view (left); side view (center); top view (right)

high while it is much less intense at 28 GHz. This is due to the back-radiation of the SH4000 which is much higher at lower frequencies.

The effect of the skirt on the radiation pattern at 8 GHz can be estimated by switching off the currents flowing on the skirt (i.e., spatial filtering [1,2]) as shown in Figure 3.5 (top right). The 3D pattern on the bottom-left side of Figure 3.5 is obtained from all the computed EQC while the one on the bottom-right side is obtained by switching off the currents on the skirt. As can be seen, the filtering of the skirt affects the directivity and the back-radiation of pattern.

The presented analysis is an example of how unique information regarding the radiation mechanism can be gained directly from measurements using the EQC technique. The visual inspection of the amplitude and phase of the EQC and the spatial filtering allow for advanced diagnostic of the device under test which can be exploited to find out possible malfunctioning of the manufactured device and/or improve the design of the device.

In the examples reported in the later sections, we will understand how the EQC can be effectively used to improve the design of the antennas/antenna system.

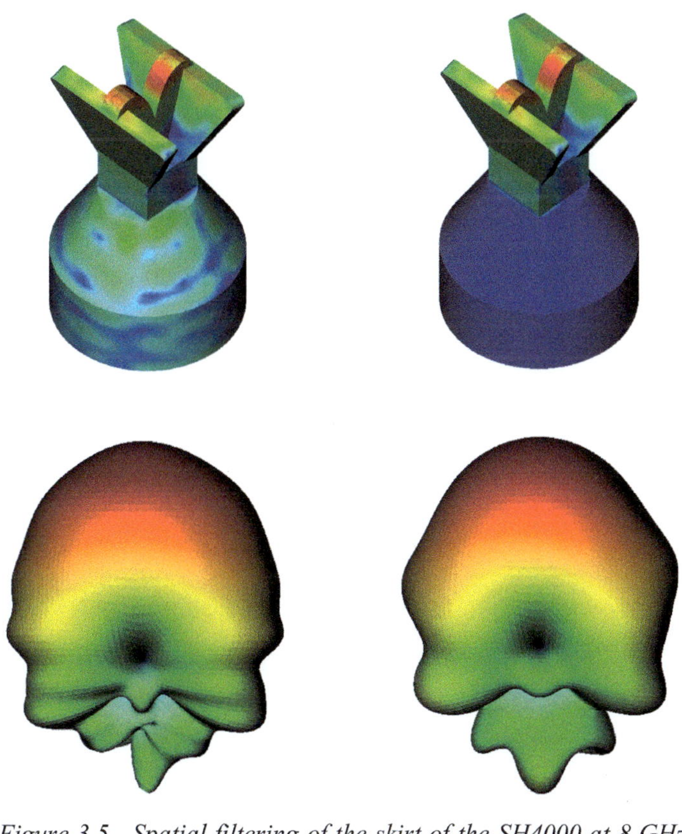

Figure 3.5 Spatial filtering of the skirt of the SH4000 at 8 GHz

3.2.2 Design improvement of an X-band 32 × 8 array

The possibility to determine extreme NF and currents on arbitrary 3D surfaces by the EQC technique makes it highly suitable to investigate excitation errors in large antenna arrays. In this example, an X-band patch array antenna with electrically large dimensions is investigated. The antenna is intended for large-scale low-cost production. The low-cost array materials are likely to differ from their nominal values, leading to performance errors. The antenna array of 32 × 8 elements is designed to illuminate a target at 1 m distance with constant field strength. The physical dimensions of the array are $L \times W \times H = 666 \times 206 \times 14$ mm. The electrical dimensions of the array are approximately $20\lambda \times 6\lambda \times 1\lambda$ at 9.4 GHz with equivalent antenna volume of $120\lambda^3$. The dimensions of each patch element are very small compared to wavelength, less than $\lambda/10$. The initial prototype is shown in Figure 3.6.

The stringent radiation requirements put particular attention on the factual array excitation as provided by the beam-forming network (BFN). In particular, a 15 dB excitation taper is desired in the vertical plane, and an equal amplitude/phase distribution is needed in the horizontal plane. The initial prototype of this antenna has been measured in a multi-probe SNF system MVG SL18GHz. The FF patterns of the antenna in the vertical and horizontal planes are shown in Figure 3.7.

The measured pattern shows an unexpected behavior as seen in Figure 3.7. In the horizontal plane of the array, the desired uniform amplitude and phase distribution of the array elements should result in a symmetric main lobe and side lobes distribution. However, it can be clearly observed that this is not the case, and some of the first side lobes are higher than expected, and the nulls are not well defined. The vertical plane pattern seems in agreement with the design and the desired amplitude distribution.

In order to diagnose the source of discrepancy, the EQC reconstruction technique has been applied to the antenna measured data. The reconstruction surface has been chosen as a rectangular box slightly larger than the array dimensions in order to enclose entirely the antenna array. The geometry has been triangulated using a mesh step of $\lambda/4$ at 9,400 MHz. The reconstructed equivalent J total currents and the measured radiated field are shown in Figure 3.8.

The amplitude of the reconstructed electric currents (J) on the upper face of the reconstruction surface with co-polar and cross-polar components, respectively, is shown in Figure 3.9(a). From the EQCs, it can be observed that the tapered excitation

Figure 3.6 First prototype of printed 32 × 8 patch array antenna

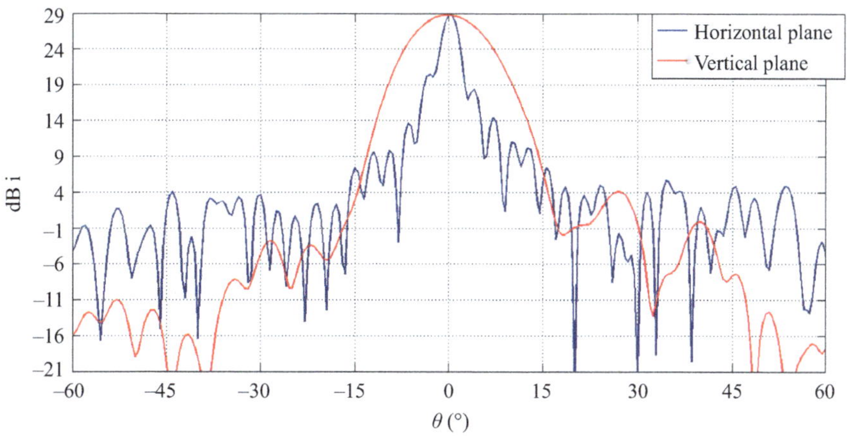

Figure 3.7 Measured directivity pattern of the first array prototype. Horizontal and vertical planes at 9,400 MHz

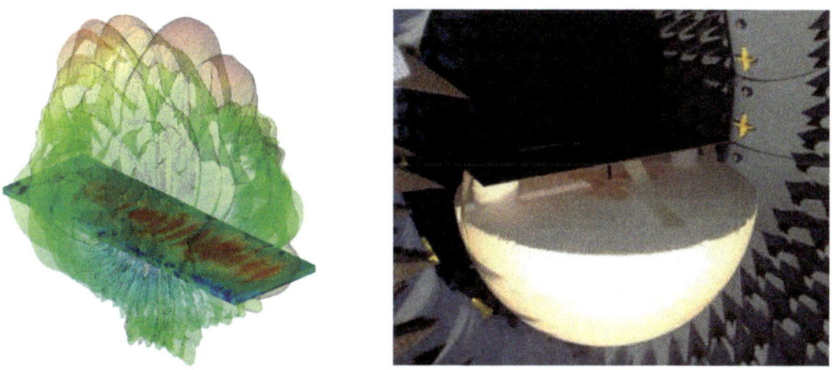

Figure 3.8 First array prototype reconstructed equivalent electric currents and measured radiated field (Left). Measurement of the array in SL18GHz spherical NF system (right)

in the vertical plane has been obtained successfully. On the other hand, it is evident that the correct power distribution has not been achieved in the horizontal plane. The entire left half of the array is clearly not excited at the same level as the right half. From this information, it is easy to conclude that the BFN does not work correctly.

The instantaneous phase of the radiating currents can be investigated by phase animation. The relative phase of the radiating currents is illustrated by an arrow. The pointing of the arrow from 0° to 360° indicates the instantaneous phase. By observing the phase of the right part of the array, the causes of the incorrect radiation are examined in further detail as shown in Figure 3.9(b). The correct phasing of the currents has not been obtained in the horizontal plane but differs with as much as 180° in some points. With this information, it is easy to deduce that part of the problem is due to the miscalculation of the electrical length of the BFN feeding lines that are not correctly dimensioned to ensure a constant phase

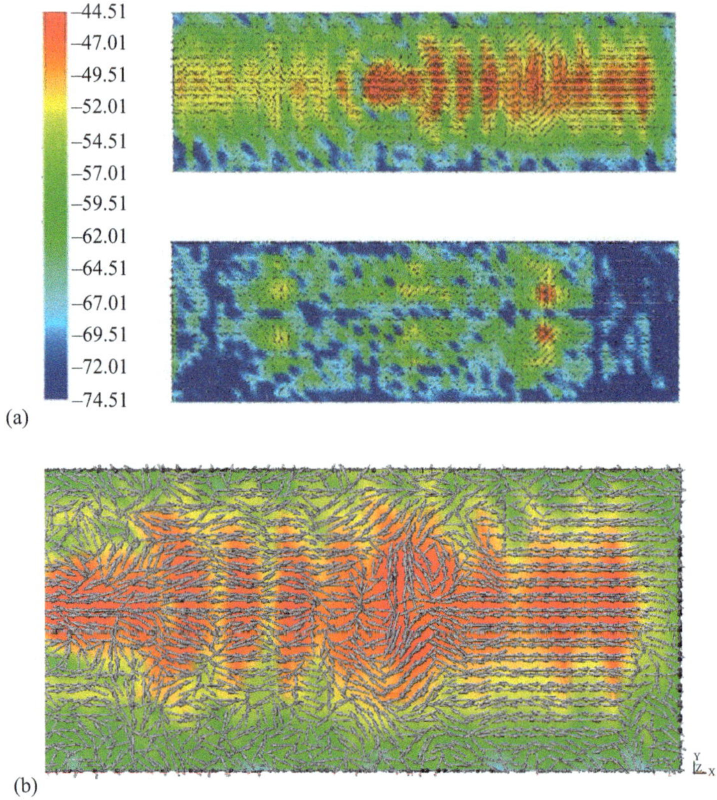

Figure 3.9 Reconstructed equivalent electric currents for the first array version: (a) co-polar currents (top) and cross-polar currents (bottom) and (b) phase distribution of the excited part (part on the right) array antenna

Figure 3.10 Picture of the new array with improved BFN

excitation along the horizontal plane. The same inaccuracy is likely the root cause in the nonsymmetric power division between the left and right halves of the array from offsets in the resulting impedance of the different branches of the BFN.

Based on the indication from the EQC diagnostics on the prototype array, a new array has been designed and manufactured in order to achieve the proper excitation of each array and sub-array element. The location of the BFN has been moved to the back of the array, and each patch is excited through via-holes in the ground plane. The new array with improved BFN is shown in Figure 3.10.

The new array has been measured in the SL18GHz. The measured patterns of the array are closer to the expected performance as seen in Figure 3.11. The elevation plane is more symmetric and with well-defined, symmetric side lobes and nulls indicating that the phase excitation error in the horizontal plane has been reduced considerably. The vertical plane also shows an increased symmetry.

The normalized NF comparison at 1 m from the two devices, obtained with back-propagation of the field, is shown in Figure 3.12. The blue trace is the

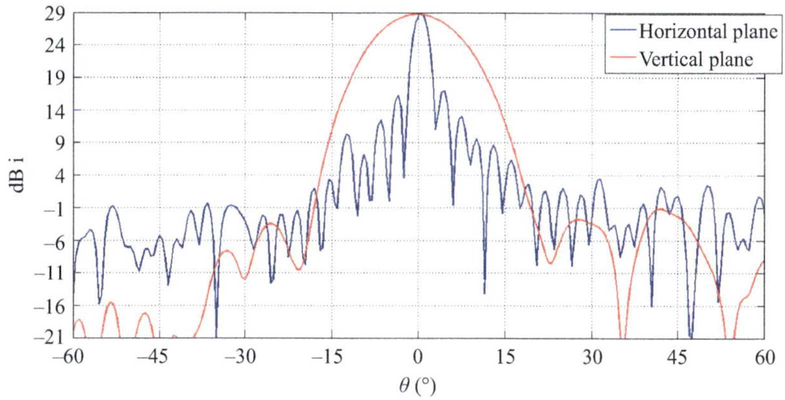

Figure 3.11 Measured directivity pattern of the improved array. Horizontal and vertical planes at 9,400 MHz

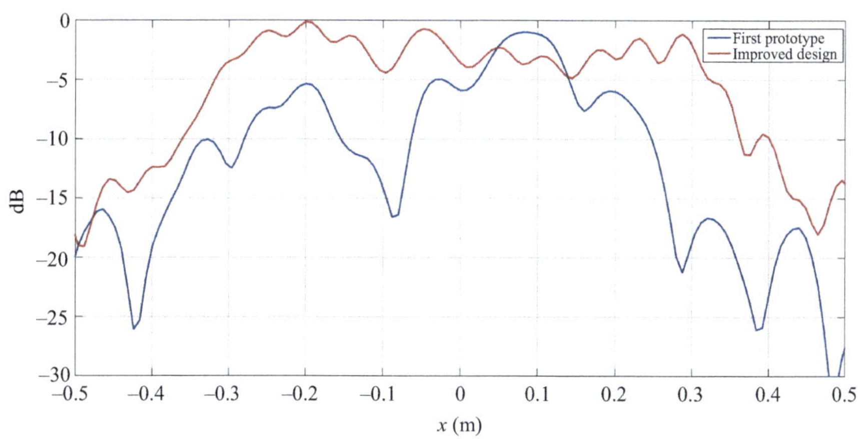

Figure 3.12 Normalized near field at 1 m from the array at 9,400 MHz. Cut along the longest dimension of the array. Co-polar component of the first prototype (blue trace); improved design (orange trace)

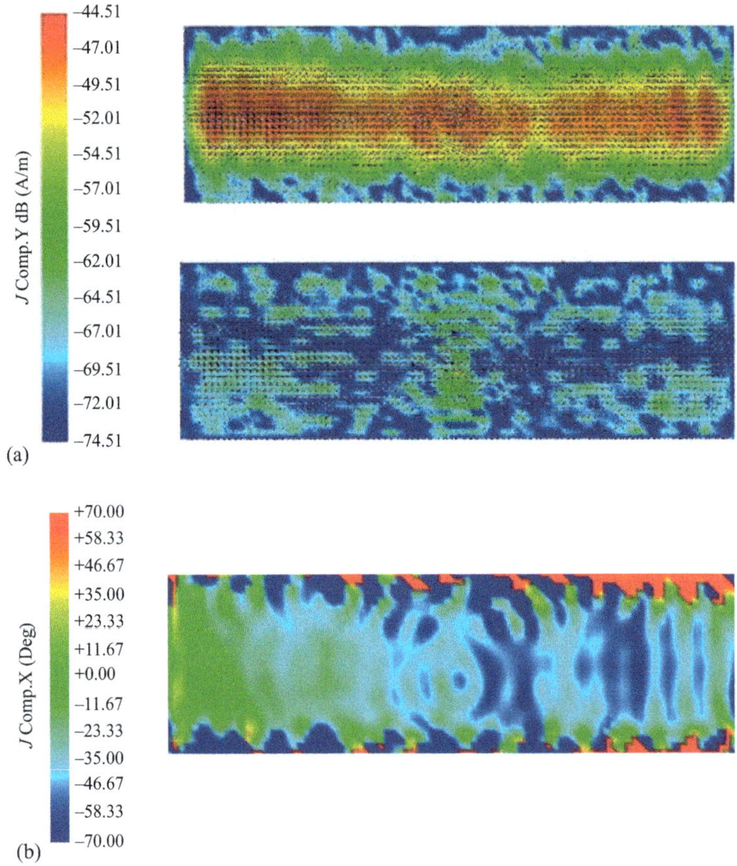

Figure 3.13 Reconstructed equivalent electric currents for the improved array version: (a) co-polar currents (top) and cross-polar currents (bottom) and (b) phase distribution of the reconstructed equivalent electric currents co-polar currents

co-polar NF along the longest dimension of the first prototype (Figure 3.6), while the orange trace is the corresponding cut of the improved device (Figure 3.10). As can be seen, approximately between −30 and +30 cm, the desired field uniformity at 1 m distance from the device has been sensibly improved in the second version of the array.

Despite the satisfactory improvements of the array performance, the EQC reconstruction technique has been also applied to the measurement of the second antenna. The equivalent electrical currents (*J*) computed from the measurement are reported in Figure 3.13(a). The overall amplitude distribution confirms the successful outcome of the design improvement of the array, although the distribution is less uniform than intended. Even in this case, the ECQ method

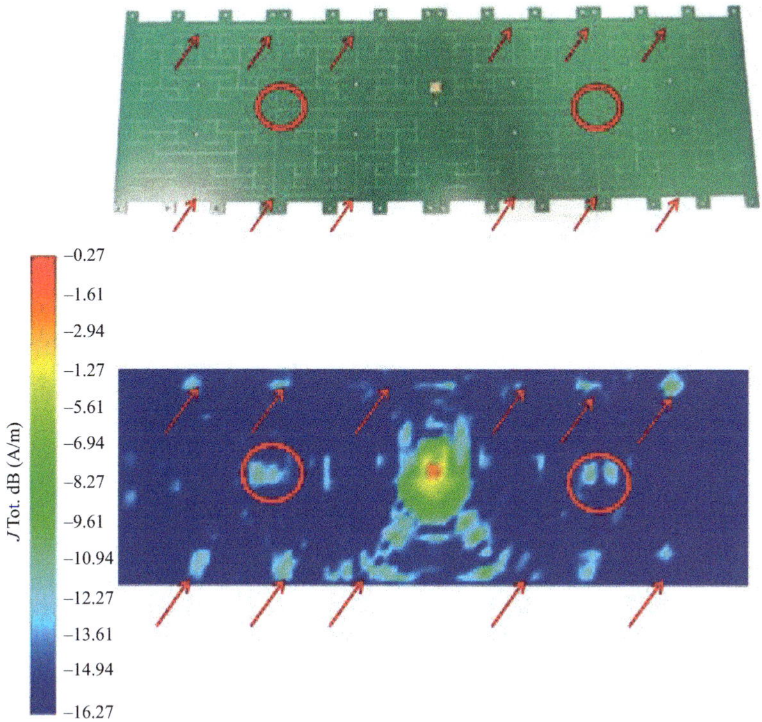

Figure 3.14 Backside of the improved array (top); electrical current distribution on the backside of the improved array (bottom)

indicates minor asymmetries in the array excitation. This becomes more evident by the analysis of the phase distribution of the co-polar currents on the antenna as shown in Figure 3.13(b). Indeed, it can be noted that the phase distribution is highly uniform in the vertical plane but shows up to 70° deviations in the horizontal plane.

By examining the radiating EQCs on the backside of the array (see Figure 3.14, bottom) a possible explanation to the excitation differences becomes clear. As seen in Figure 3.14 (top), the connector is soldered orthogonal to the array but has a small fixture to allow room for the input microstrip line of the array. This fixture seems to launch a surface wave that couples to the BFN structure of the array and thus through the coupling changes the effective amplitude and phase distribution of the array. The EQC expansion only shows where the surface wave couples with the BFN structure so the actual coupling cannot be determined with any given accuracy. The coupling clearly affects the upper central part of the array where the front side co-polar distribution is altered as shown in Figure 3.13(a). The error due to this coupling is considered acceptable and does not require correction.

3.2.3 Verification of complex excitation coefficient of a space array antenna

The EQC technique can be used in the verification of the realized or actual excitation coefficients of array antennas on both single element and sub-array level and thus acts as an important tool in the array design improvement from measurements on an appropriate prototype. In this example, a satellite array antenna is considered. The array is composed of a two-level BFN feeding 42 patch elements divided into six sectors providing the complex excitations needed to obtain the desired iso-flux right-handed circularly polarized (RHCP) pattern.

The antenna, also briefly discussed in Chapter 2, has been measured in the HERTZ range at European Space Agency (ESA)/ESTEC in SNF geometry [3]. The array is developed by EADS-CASA Espacio and is intended as part of the European Navigation System GALILEO. It is a predevelopment model flying on the In-orbit validation element the GIOVE-B satellite [4,5]. The measured data is courtesy of the ESA.

In Chapter 2, the EQC analysis of this antenna aimed at finding the root cause of asymmetries in the radiation pattern at 1,592 MHz. It was shown that the asymmetries were caused by residual excitation errors in the sequential rotation, implemented on the array to achieve low on-axis cx-polar.

In the following section, the complex excitation of the array element is analyzed with more detail by isolating different parts of the antenna EQC using spatial filtering. The considered geometry where the EQC is computed is composed of two cylinders as seen in Figure 3.15: one containing the array and the other containing the stand-off used to handle the device. It should be noted that such geometry slightly differs from the one considered in Chapter 2. The top part of the cylinder enclosing the array is in fact closer to the array element allowing to better identify the different element contributions. In Figure 3.15, the currents lying on an inner and an outer ring of elements are clearly identifiable. Such working mechanism is needed in order to create the desired iso-flux pattern.

The top view of the normalized RHCP equivalent electric and magnetic currents is shown in Figures 3.16 and 3.17, respectively, for the amplitude and the phase. Figure 3.17 is an illustration of the causes of the pattern asymmetries explored in Chapter 2. Asymmetries are caused by the phase dispersion originating from residual excitation errors in the sequential rotation that has been implemented in the array to achieve low on-axis cx-polar.

The spatial filtering of the computed EQC has first been exploited in order to analyze the effect of the inner and outer rings of the antenna separately. To do that, the two filtering masks shown in Figure 3.18 have been used in order to evaluate the two separate contributions needed to create the wanted iso-flux pattern. The co-polar radiation pattern contributions associated respectively to the inner and outer ring currents are shown in Figure 3.19 for both amplitude and phase. As can be seen, the patterns associated to both rings show asymmetries, but those associated to the outer ring are more pronounced. The unexpected behavior of the

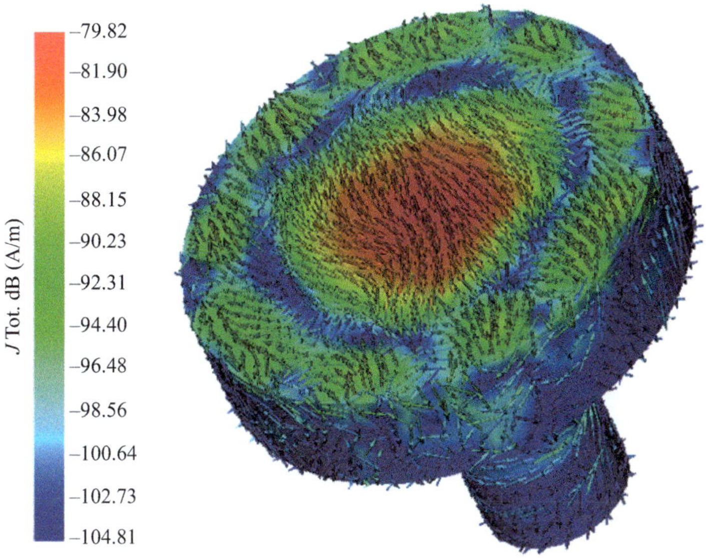

Figure 3.15 Equivalent electric currents on the whole antenna structure plus the stand-off used to perform the NF measurement

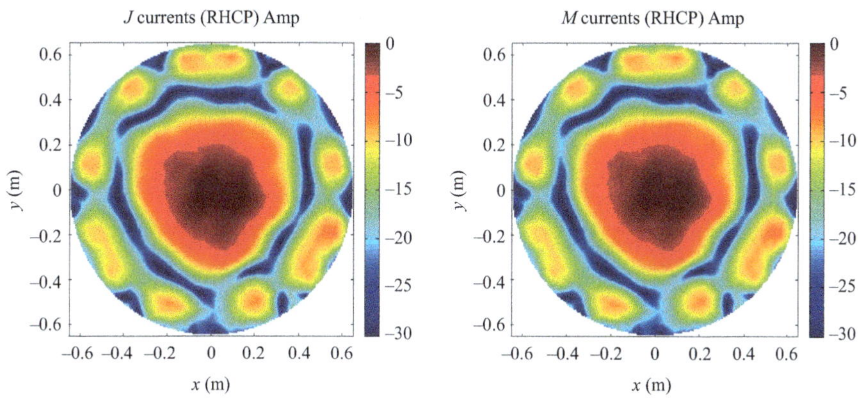

Figure 3.16 Amplitude of the co-polarized electric (left) and magnetic (right) currents of the GIOVE-B array determined by the EQC technique

iso-flux pattern is thus due to the outer ring which probably has some issue related to the BFN.

A further step in the analysis of the excitation of the array has been performed, again exploiting the spatial filtering properties of the EQC, by

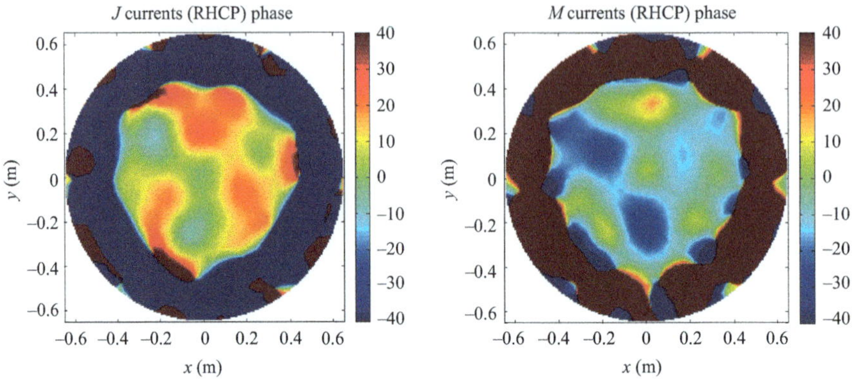

Figure 3.17 Phase of the co-polarized electric (left) and magnetic (right) currents of the GIOVE-B array determined by the EQC technique

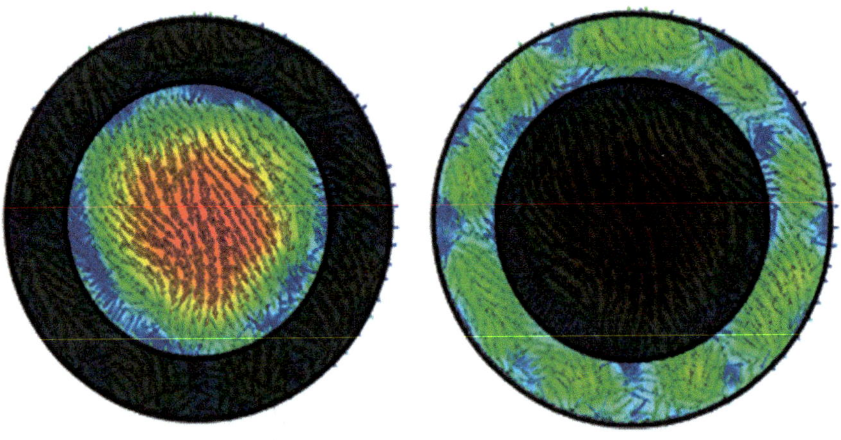

Figure 3.18 Equivalent currents associated to the inner ring (left) and associated to the outer ring (right)

isolating three sub-array sections located on the outer annular ring from the inner section. The four different sets of spatial filtered currents are illustrated in Figure 3.20 (first column). The 3D radiation patterns associated to each EQC dataset are also shown in the same figure (second column). The complex excitation coefficients of the four sub-array sections have been obtained computing for each section the total radiated power over the full sphere and taking the phase value on the boresight direction. The obtained complex coefficients normalized with respect to the central sub-array section are reported in the third column of Figure 3.20.

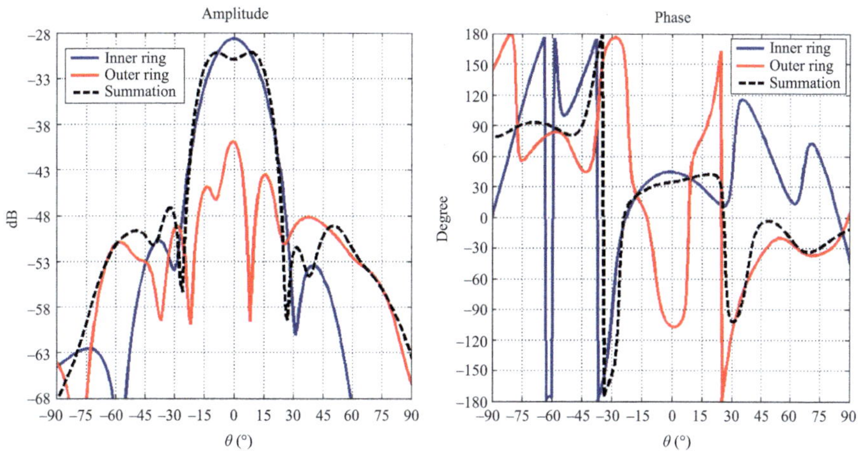

Figure 3.19 Separation of different contributions to the co-polar radiation pattern along vertical cut: inner ring pattern (blue), outer ring pattern (red)

Current isolation	Radiation pattern	Complex excitation
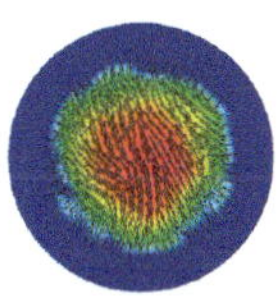		Amp: 0.00 dB Phase: 0.00°
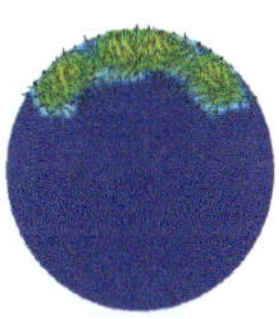	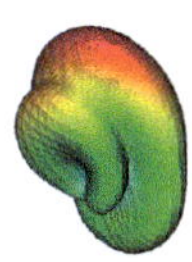	Amp: –18.76 dB Phase: –184.06°
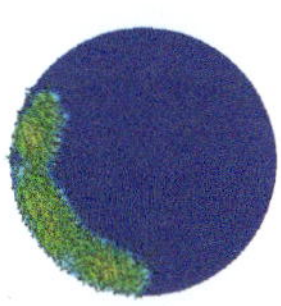		Amp: –18.45 dB Phase: –166.89°
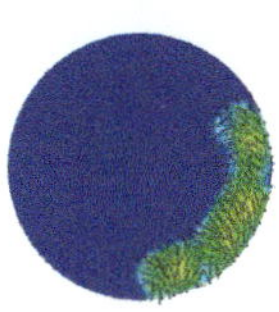	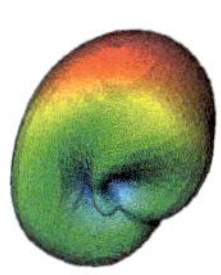	Amp: –17.85 dB Phase: –161.52°

Figure 3.20 Verification of sub-array complex excitation coefficients

The determined complex values are reasonably in line with the expected values especially regarding the amplitude. The obtained excitation phase is also quite good. In fact, the wanted phase shifts of 0°/180° between central elements and outer elements have been reached within less than 20° error in the worst case. Nevertheless, this excitation error can also contribute to the observed co-polar axial asymmetry of the radiation pattern.

3.2.4 Analysis of coupling phenomena between spacecraft antennas

In this example, the EQC technique is applied to analyze the coupling phenomena between antennas installed on a complex structure. The considered complex structure is the GALILEO FOC satellite produced by OHB shown in Figure 3.21 [3–7]. The measured data and pictures are courtesy of the ESA. The satellite is composed by several antennas of different types used for different purposes, which are as follows:

- UHF Search and Rescue antenna (Sarant UHF), the array of helixes located on the top part of the satellite;
- L-band Search and Rescue antenna (Sarant L-band) located within the Sarant UHF;
- navigation antenna (Navant), the big circular antenna located in the central part of the satellite;
- mission antenna (Misant), the horn antenna located in the bottom-right part of the satellite and
- omnidirectional antenna located in the bottom-left part of the satellite.

Figure 3.21 GALILEO FOC Satellite antenna farm testing at ESA/ESTEC HERTZ facility

The satellite has been fully measured in the same NF system at ESA/ESTEC [3] (see HERTZ range shown in Figure 5.1 of Chapter 5) also used to measure the GIOVE-B antenna presented in previous sections. Each antenna mounted on the satellite has been measured using a Planar NF (PNF) or SNF geometry depending on the characteristics of antenna.

The measurements of the Sarant UHF antenna installed on the top part of the satellite have been taken into account for the analysis because its measured radiation pattern on the satellite showed some differences with respect to the one measured in a stand-alone configuration. A non-negligible coupling with other antennas was thus expected and thus it is analyzed.

For comparison, the Sarant L-band antenna, located in the same area of the satellite where the UHF antenna is placed, has also been analyzed even if its measured radiation pattern on the satellite was in good agreement with the corresponding stand-alone measured behavior.

The two Search and Rescue antennas have been measured in PNF configuration. The size of the planar scanner used for the measurement of the Sarant UHF is $(X \times Y) = (5.78 \times 5.78)$ m, and the distance from the satellite is 2.25 m. Instead, concerning the Sarant L-band, the size of the planar scanner is $(X \times Y) = (5.9 \times 5.9)$ m, and the distance from the satellite is 1.38 m. For both measurements, a standard half-wavelength sampling has been considered along both scanning axes. Furthermore, the center of the scanning plane has been aligned with the center of the fed antennas.

To analyze the coupling between the different satellite antennas, the EQC technique has been applied considering an equivalent geometry conformal to the spacecraft (see Figure 3.22). The equivalent geometry is mainly composed by a large box fitting the body of the satellite. Furthermore, on the top part of such box, the following four additional equivalent surfaces enclosing the different antennas have been considered:

- a cross-shaped surface on the top part enclosing the two Search and Rescue antennas,
- a big cylindrical surface on the central part enclosing the Navant,
- a small cylindrical surface on the bottom-right part enclosing the Misant and
- a small box on the bottom-left part enclosing the omnidirectional antenna.

The computed equivalent electric currents from the PNF measurement of the two antennas are shown in Figure 3.22. The currents of the Sarant UHF are shown on the left side of the figure while those of the Sarant L-band are shown on the right side.

In both cases, the currents are of course concentrated on the equivalent surface associated to the Sarant antennas. As expected, the illumination of the whole satellite is much stronger when the UHF antenna is fed.

When the Sarant UHF is turned on, a non-negligible portion of currents flows on the upper part of the Navigation antenna. Minor contributes of induced EQCs

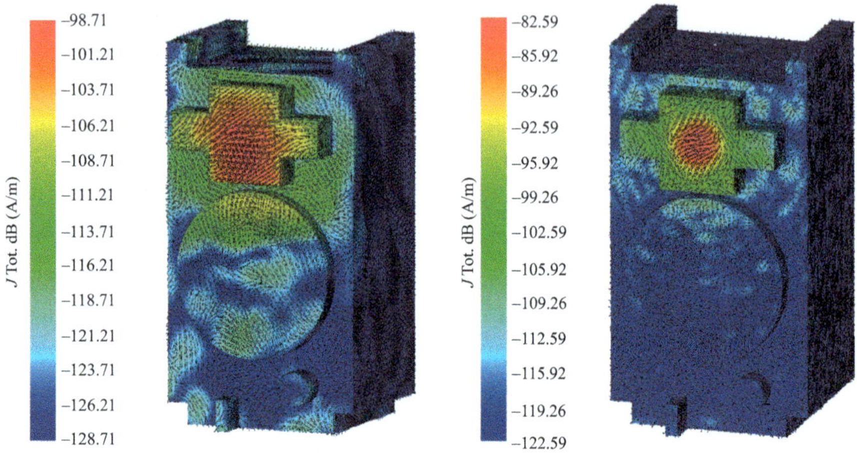

Figure 3.22 Equivalent electric current on the satellite when the Sarant UHF is fed (left) and when the Sarant L-band is fed (right)

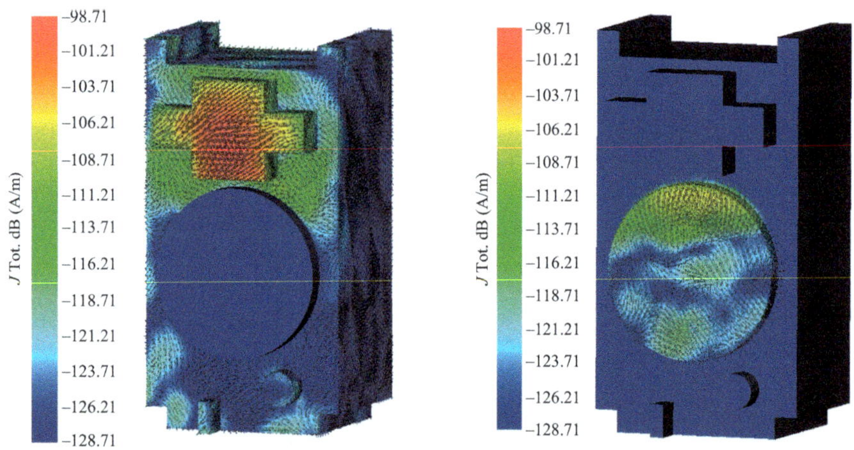

Figure 3.23 Spatial filtering of the Navigation Antenna when the Sarant UHF antenna is fed

are instead identified on the Misant and omnidirectional antennas. In order to better understand the influence of the discovered induced currents on the radiation pattern, a spatial filtering has been applied on the satellite EQCs. In particular, as illustrated in Figure 3.23, the induced EQCs on the Navant have been first switched off (see Figure 3.23, left), and then only the induced currents on the Navant have been turned on (see Figure 3.23, right).

The normalized radiation patterns along the horizontal cut computed from the two current distributions are shown in Figure 3.24 together with the pattern radiated from the currents distribution without any spatial filtering applied. In such a plot, solid lines are the RHCP components while dashed lines are the LHCP components. The RHCP pattern radiated only by the EQCs induced on the Navant (solid green trace) is relatively high (up to approximately −17.5 dB in the main beam area). As a consequence, the radiation pattern of the Sarant UHF is sensibly modified as observable comparing the solid-blue and solid-red traces.

The same spatial filtering processing has been also applied to the Sarant L-band. The corresponding radiation patterns comparison along the horizontal cut shown in Figure 3.25 where solid lines are the LHCP components while dashed lines are the RHCP components. The pattern radiated by the currents induced on the Navant is in this case relatively low (maximum value is around −35 dB). Consequently, the pattern radiated by the full set of currents (blue traces) and the pattern obtained by switching off the Navant (red traces) are in good agreement, confirming that the interaction between the Sarant L-band and the Navant is very low.

It is highlighted that the information derived from these analyses could be exploited in the design of the next generation of GALILEO satellite in order to optimize the position of antennas on the farm.

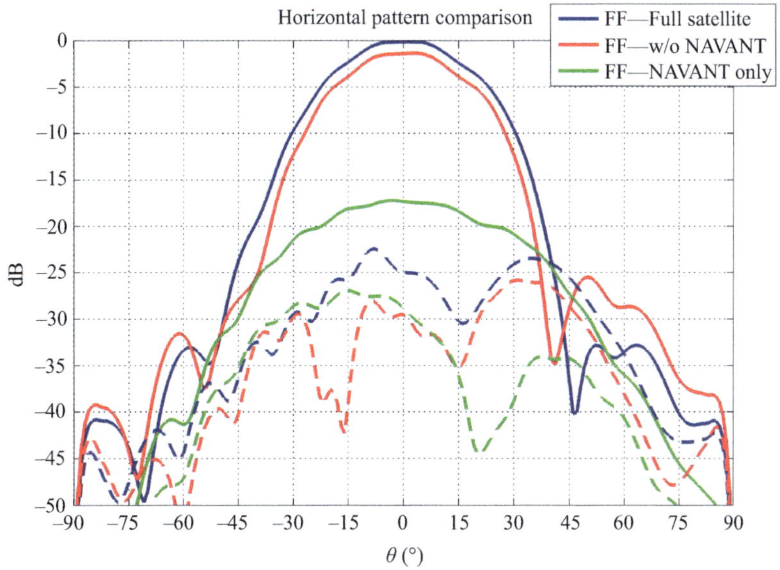

Figure 3.24 Sarant UHF normalized pattern comparison: FF from full EQC (blue traces); FF with Navant switched off (red traces); FF with only Navant turned on (green traces). Solid lines are RHCP components; dashed lines are LHCP components

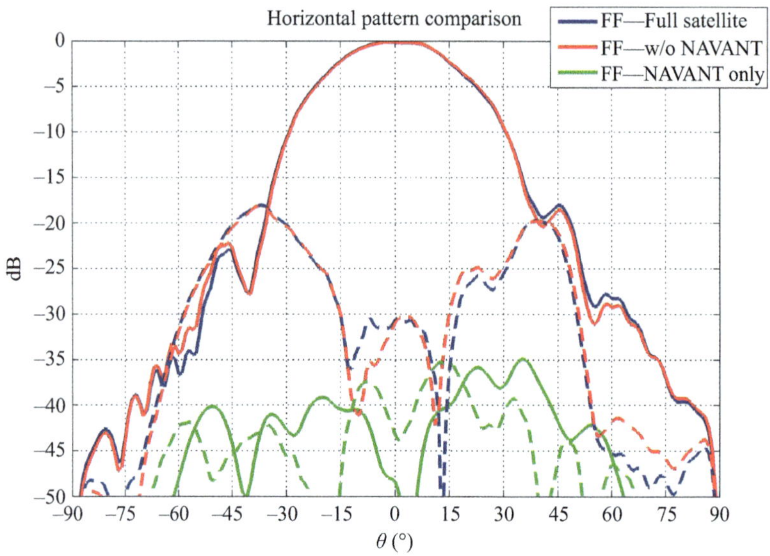

Figure 3.25 Sarant L-band normalized pattern comparison: FF from full EQC (blue traces); FF with Navant switched off (red traces); FF with only Navant turned on (green traces). Solid lines are LHCP components; dashed lines are RHCP components

3.2.5 Investigation of spherical higher order modes in the design of NF probes

The EQC technique is here exploited to understand the source of higher order spherical mode radiated by a probe designed for SNF antenna measurements.

Classical probe–corrected SNF measurements assume probes with limited $|\mu| = 1$ spherical wave spectrum† [8] (first-order probe). This requirement is needed to fully compensate the effect of the probe during the NF/FF transformation discussed in Chapter 5 when the first-order probe correction technique is used [8]. If the probe spectrum contains higher order modes, a residual error will affect the measured pattern (probe modal truncation), unless a suitable higher order probe correction technique is adopted [9–11]. This requirement leads to challenging probe designs, especially if the required bandwidth is wide and/or there is a limitation on the acceptable physical dimension. Consequently, in many practical cases, higher order spherical modes are unavoidable, and the design task is to minimize their impact on the probe quality.

The device considered for the analysis of the higher order spherical mode is the MVI SP1100 L-band probe shown in Figure 3.26 (right). Such device is a wideband and electrically small probe working in the frequency range of 1.1–2.0 GHz in dual linear polarization. The probe consists of a quad-ridge wideband dual polarized

†IEEE Std P1720™/D2 (Draft) “Recommended Practice for Near-Field Antenna Measurements”.

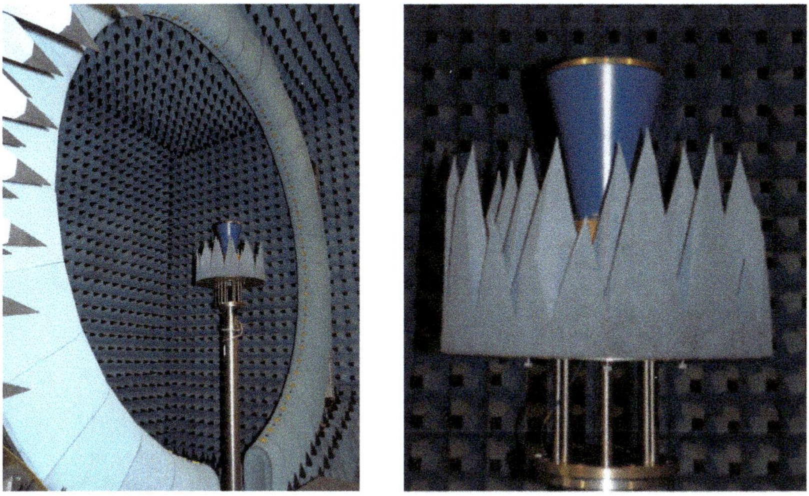

Figure 3.26 MVI SP1100 probe during calibration in SNF in MVG SG64 spherical multi-probe system (left); zoom on the SP1100 probe (right)

orthomode junction (OMJ) [12] feeding a conical aperture. The OMJ is based on a quad-ridge circular waveguide with external BFN.

In [13], it has been shown that excitation errors in the BFN generate higher order spherical modes. Another interesting aspect to be analyzed is the influence of the absorber collar that in practical SNF measurement scenario is used in order to limit the reflections from the supporting structure (see Figure 3.26, right). As shown in [14], the absorber collar has an impact on the radiated field which, due to the uncertainties and repeatability of the absorber material, is very difficult to predict.

The SP1100 has been measured in the MVG SG64 spherical multi-probe system as shown in Figure 3.26 (left). As can be seen, the circular absorber collar has been included as integral part of the probe. The probe was characterized in terms of radiation pattern and spherical mode spectrum. The measured directivity patterns are reported in Figure 3.27 for the main cuts at 1.25 GHz. The corresponding spherical modes represented by the P_n and P_m power spectrum as defined in Chapter 5 are illustrated in Figure 3.28. From the P_m curve, it can be seen that spherical modes of higher order ($|\mu| \neq 1$) are well below the -35 dB threshold. Despite these low levels, it is interesting to understand the source of such higher order modes. For example, in order to further optimize the probe performance in future designs, it is important to know if such modes are directly radiated by the aperture or generated by the interaction between the probe itself and other probe parts such as the absorber collar.

The EQC of the measured SP1100 probe has first been computed using the NF data as measured. The EQC has been determined on a geometry conformal to the

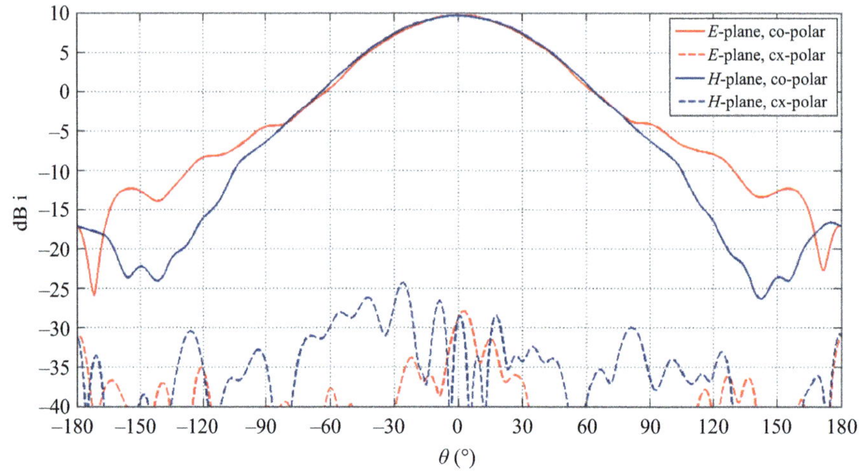

Figure 3.27 Measured directivity pattern of the MVI SP1100 probe at 1,250 MHz. E-plane pattern cut (red traces); H-plane pattern cut (blue traces). Solid traces are co-polar components; dashed traces are cx-polar components

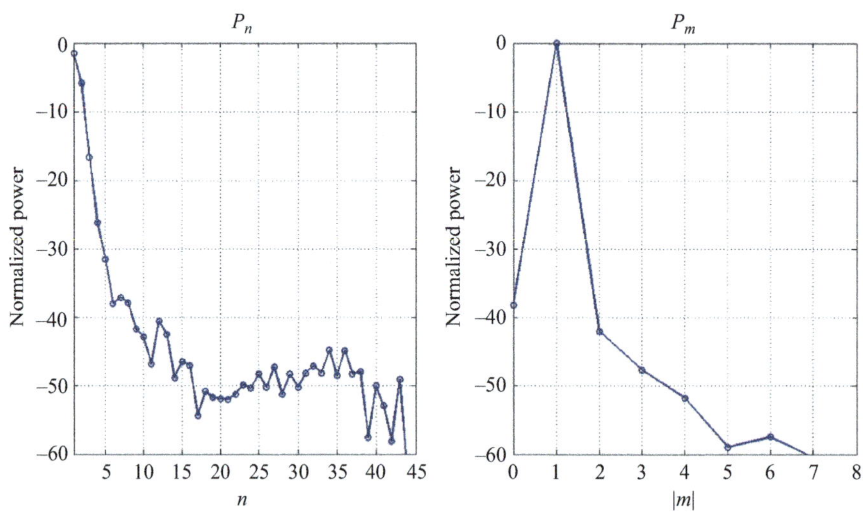

Figure 3.28 Measured spherical mode spectrum at 1,250 MHz of the MVI SP1100 probe. P_n-modes (left); P_m modes (right)

probe shape, including the absorber collar and stand-off as illustrated in Figure 3.29 (top). The top view of the co-polar and cx-polar equivalent electrical currents is also shown in Figure 3.29. As expected, most of the radiation originates from the probe aperture. A much lower, but not negligible, radiation originates from the

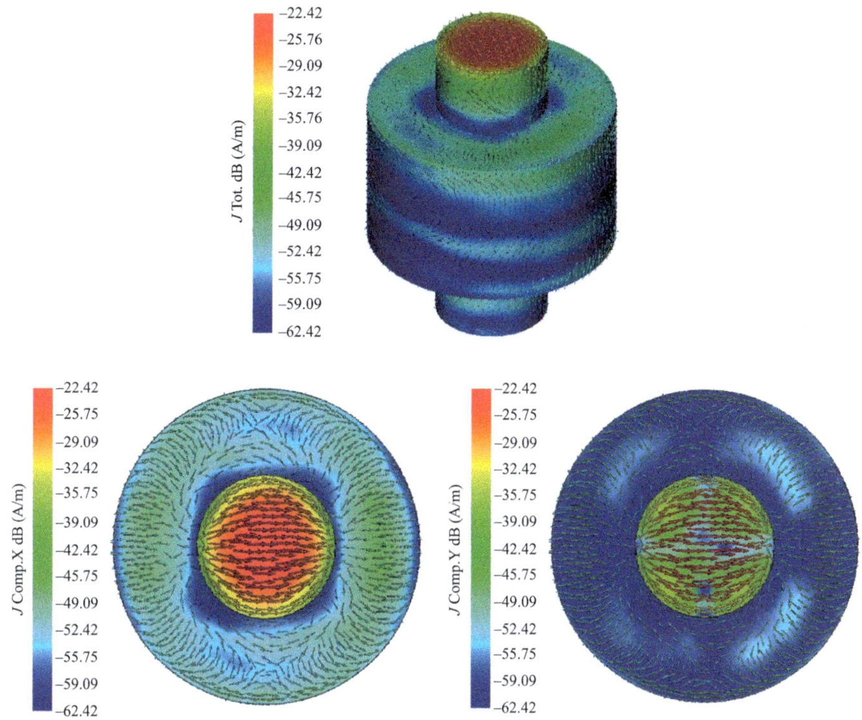

Figure 3.29 Equivalent electric currents at 1,250 MHz evaluated on an equivalent geometry encompassing the MVI SP1100 probe: total electric EQC (top); top view of the co-polar EQC (left); top view of the cx-polar EQC (right)

surface of the absorber collar. It can also be deduced that the majority of the cx-polar radiation is due to the probe aperture itself.

From the spectrum shown in Figure 3.28, different spherical mode subsets to be analyzed have been isolated from the full probe spectrum using band-pass filtering. The reduced spectrum is used as input to the EQC processing which is performed considering the same equivalent reconstruction surface previously shown. The obtained currents illustrate the origin of the considered spherical mode.

As a first step, the $|\mu| = 1$ modes have been isolated from the $|\mu| \neq 1$ modes. The EQCs associated to the two subsets of spherical modes are shown in Figure 3.30. In agreement with the P_m modes shown in Figure 3.28, the EQCs associated to the $|\mu| \neq 1$ modes are well below the dominant currents associated to the $|\mu| = 1$ modes.

Successively, the analysis has been repeated considering spherical modes with index $\mu = 0$, $|\mu| = 2$ and $|\mu| > 2$. The obtained EQC of each spherical mode subset is shown in Figure 3.31.

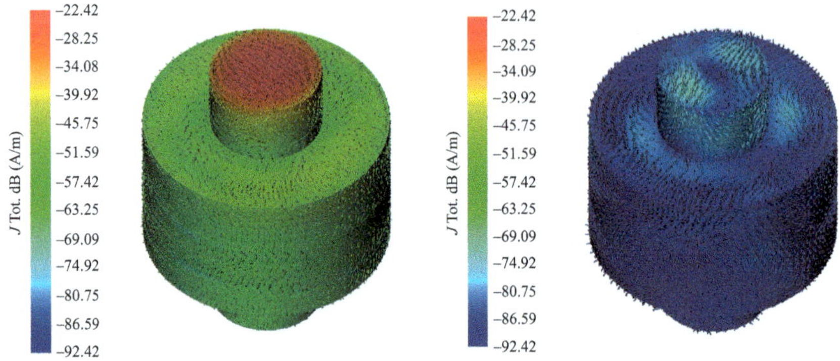

Figure 3.30 Equivalent currents at 1,250 MHz associated to the $|\mu| = 1$ spherical modes (left) and to the $|\mu| \neq 1$ (right)

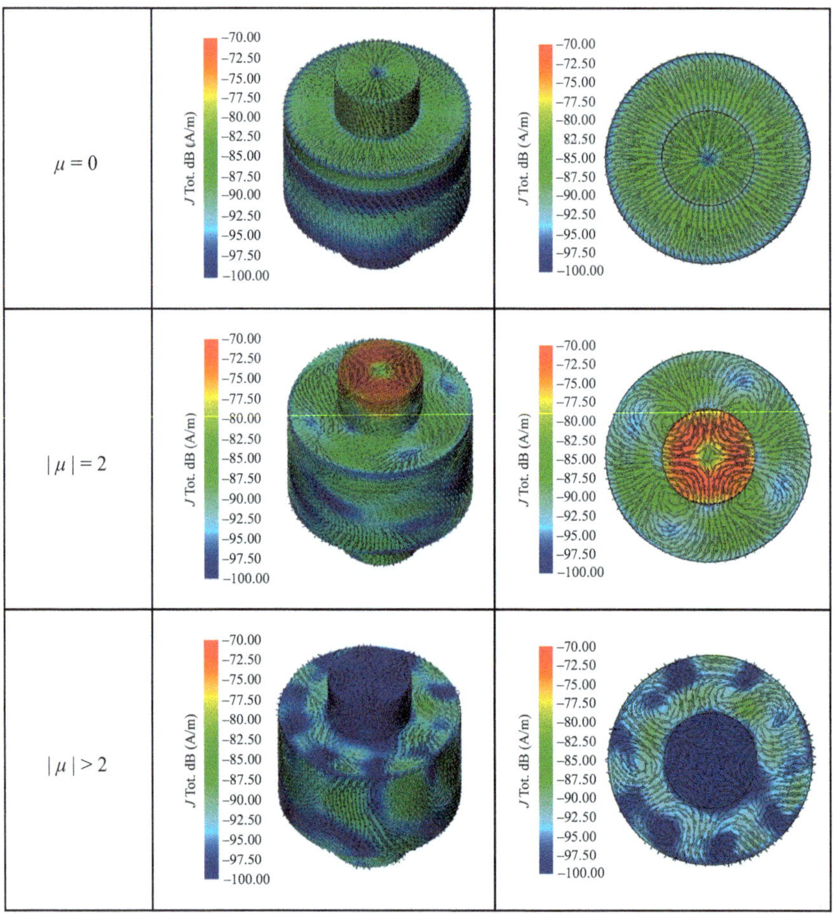

Figure 3.31 Equivalent currents at 1,250 MHz associated to different azimuthal spherical modes. 3D view (second column); top view (third column)

The amplitude of the currents associated to the $\mu = 0$ modes (first row of Figure 3.31) is more than 55 dB below the peak (peak is at -22.42 dB as shown in Figure 3.29). Nevertheless, it is possible to locate their main contribution on the main radiating part of the probe. Observing the top view of the probe aperture it is possible to recognize the shape of the undesired TM_{01} circular waveguide mode [15]. The very low TM_{01} modal excitation is compatible with expected excitation errors in the BFN. The short probe and inverted ridge structure are unable to attenuate significantly the propagation of this mode which is therefore radiated by the aperture and then reflected/diffracted by the absorber collar.

The amplitude of the $|\mu| = 2$ currents (second row of Figure 3.31) is also very low (more than 47 dB below the peak). From the current distribution, it is possible to recognize the shape of the undesired TE_{21} circular waveguide mode [15] which is compatible with expected excitation errors in the BFN. The origin of these modes is therefore likely excitation errors more than interaction between probe aperture and absorbers.

The currents associated to the $|\mu| > 2$ spherical modes (third row of Figure 3.31) are instead distributed as scattered hot spots only on the absorber collar, hence, they are not due to the aperture radiation.

To summarize, it has been found out that the source of spherical modes with indices $|\mu| = 0$ and $|\mu| = 2$ is on the probe aperture, and they must be due to BFN errors which generate unwanted waveguide modes (TM_{01} and TE_{21}), which in turn originates the higher order spherical modes. It is observed that $|\mu| > 2$ spherical modes are instead due to the field reradiated by the absorber collar rather than to a direct radiation from the probe aperture.

3.3 Application of EQC in the improvement of antenna measurements accuracy

In this section, we will show how the EQC technique can be used as a tool to improve the measurement accuracy in NF measurements. In particular, the spatial filtering properties of the EQC will be exploited as

- filtering of supporting measurement structures,
- echo suppression and
- NF/FF transformation for partial spherical NF scanning (reduction of truncation errors).

3.3.1 Diagnostic and spatial filtering of supporting structure

In antenna measurements, close interactions by objects, like a support structure or the feeding cable of the antenna, can significantly alter the measured performance with respect to the desired free-space performance. This issue is particularly relevant in the measurement of small antennas in which the wide-angle radiation is prone to couple with structures in the close vicinity. Some measurement tricks employed to cope with this situation involve the use of additional hardware during measurement, such as baluns, electro-optic transducers, chokes and ferrites on the cable. The EQC technique has shown to be very efficient to detect such interaction and provide a mean to isolate the antenna radiation from the scattered radiation in the surrounding structure by spatial filtering [1,16–18].

In the following example, the EQC technique has been applied to investigate possible interferences and coupling between a linear array antenna, BTS1940 at 1,940 MHz, shown in Figure 3.32, and nearby objects that are part of the measurement setup, such as positioners and mounting support. The antenna is an array of eight dual slant +45°/−45° polarized radiating elements, fed with the same amplitude and phase input signals.

During measurements in a cylindrical scan range the BTS1940 antenna has been mounted on a horizontal scanner as shown in Figure 3.33. It is suspected that interaction between the antenna and the scanner could alter the measured result. EQCs have been reconstructed on a geometry which encloses the BTS1940 antenna with the linear scanner with the scope to investigate the EQC amplitude level and distribution on the scanner at 1,930 MHz.

Maximum values of the J electrical equivalence currents are located where the BTS1940 antenna is positioned, see Figure 3.34. The level of the currents on the linear scanner is below −30 dB with respect to the peak of the currents on the antenna. The interaction between the antenna and the linear scanner can therefore be considered negligible. No spatial filtering of the currents on the scanner is needed for further investigations.

The second example involves a base station antenna with operative frequency band centered at 824 MHz. The antenna is mounted on a vertical support to be measured in the SNF range, see Figure 3.35. At lower frequencies, the interaction

Figure 3.32 Linear array antenna BTS1940 measured in the MVG spherical near field range SG64

Figure 3.33 BTS1940 antenna in the StarLab system with the reconstruction geometry shown overlaid

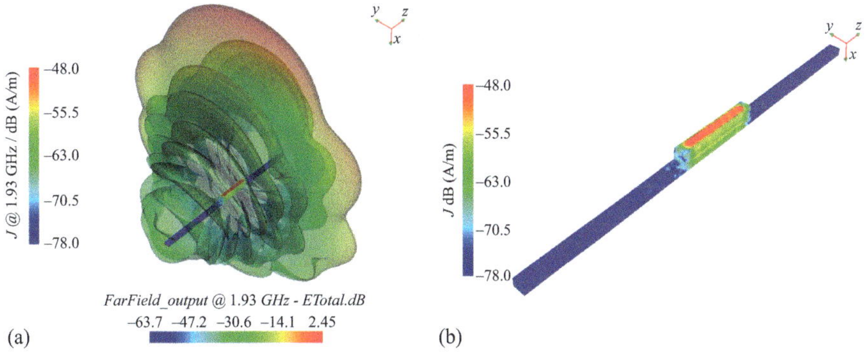

Figure 3.34 (a) Electrical J equivalent currents on a geometry enclosing the BTS1940 antenna and the linear scanner; left: EQC and measured FF at 1,930 MHz and (b) zoom on EQC

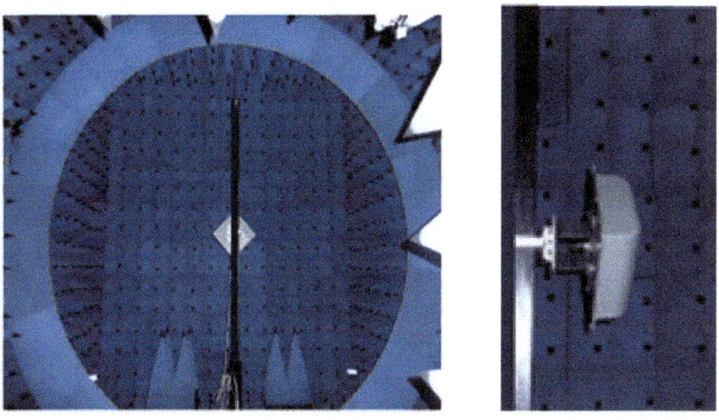

Figure 3.35 Base station antenna in the StarGate128 measurement system

between the antenna and the mounting structure is generally more relevant with respect to the frequency band previously investigated.

From the radiation patterns, a slight ripple is detected mainly on the elevation pattern. Reconstruction of the EQCs has been performed on a geometry enclosing the antenna and the mounting support to investigate eventual interaction between the antenna and the surrounding structures.

Maximum values of the J electrical and M magnetic EQCs are located where the base station antenna is positioned, see Figure 3.36, but level of the currents on the linear scanner cannot be considered negligible.

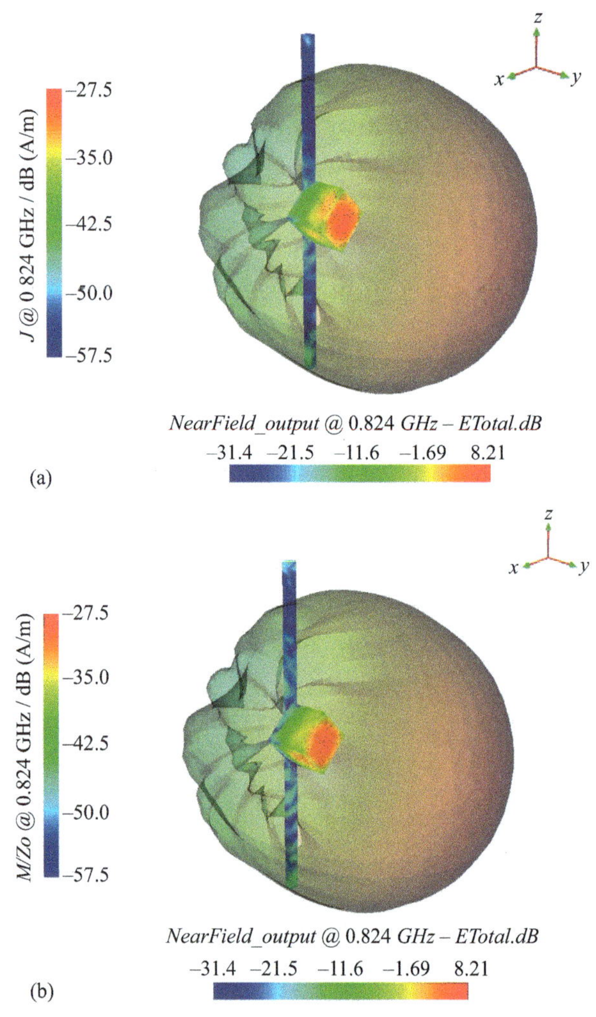

Figure 3.36 Base station antenna—(a) J and (b) M equivalent currents and measured FF at 842 MHz

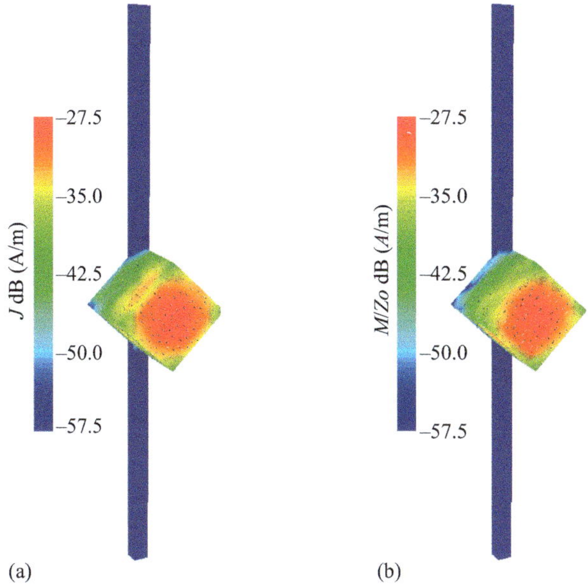

Figure 3.37 Base station antenna—Filtered (a) electrical J and (b) magnetic M equivalent currents at 842 MHz

This means that an interaction between the antenna and the measurement setup occurs. Spatial filtering of the currents on the vertical mounting support has been performed in order to remove the undesired scattering effect, see Figure 3.37. As can be seen in Figure 3.38, after filtering the ripple on the elevation patterns is reduced, while on the azimuth patterns, no relevant differences are present. This is due to the fact that the support where currents are removed is vertically oriented.

3.3.2 Echo suppression

In the majority of measurement scenario with well-designed anechoic chambers, the measurement uncertainty contribution from echoes or chamber scattering can be considered negligible. However, as explained in Chapter 5, in special measurement conditions and/or when very high precision measurements are needed, significant echo contributions can occur in the case of large angle of incidence even on absorber covered surfaces. Also, when the anechoic chamber is used toward or below the lower frequency limit of the absorbers, the quality of the test region is compromised by strong echos due to diminishing absorption properties with frequency of the absorbing material and the limited spatial attenuation of the echo fields.

In Chapter 5, it has been shown how the spherical and cylindrical modal expansions can be exploited as echo suppression tools (i.e., spatial/modal filtering). The EQC expansion applies an intrinsic spatial filtering defined by the selected

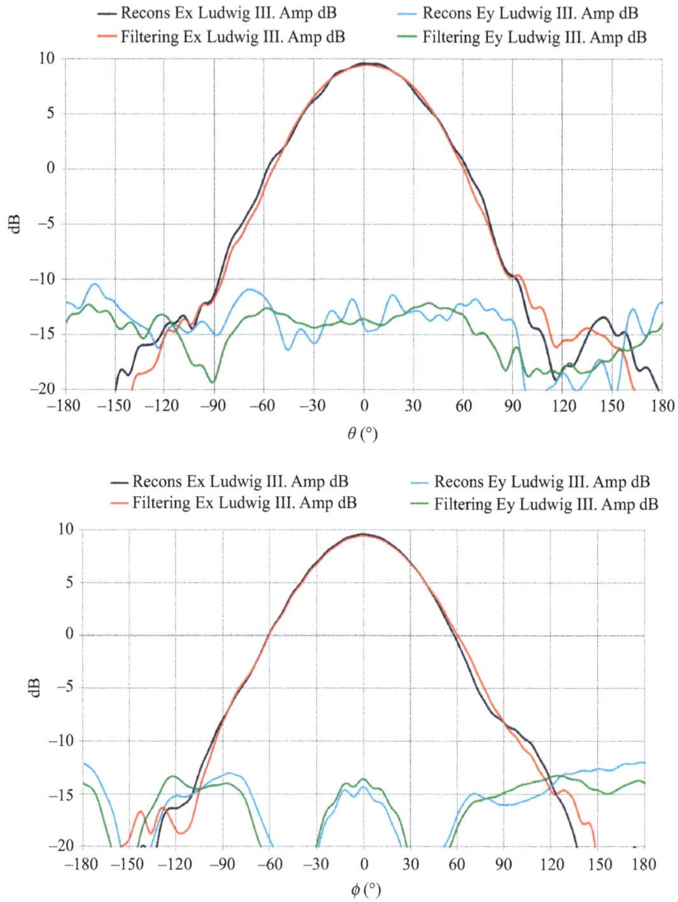

Figure 3.38 Directivity elevation (top) and azimuth pattern (bottom) at 824 MHz before (recons) and after filtering

reconstruction surface which filters out any contribution lying outside that surface. The EQC technique can thus also be used as echo suppression tool [2].

The following examples show the application of the EQC technique as echo suppression tool in PNF and SNF ranges considering two different MVG reference horns.

The PNF example is the same one considered in Section 5.1.2.3 of Chapter 5, where an X-band standard gain horn, the SGH820 is measured. A scattering plate, with dimensions 1.25 m × 1.25 m, approximately $41\lambda \times 41\lambda$ at 10 GHz has been intentionally placed behind the scanner, parallel to the wall, in order to create an echoic scenario. Reference data has taken by performing the measurement without the scattering plate.

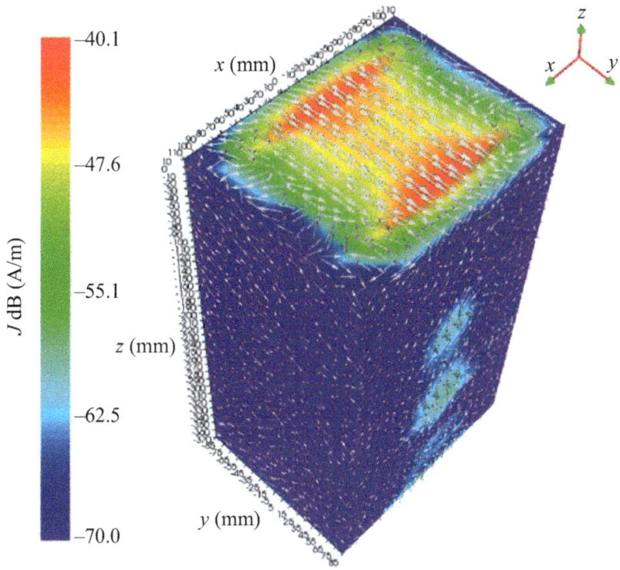

Figure 3.39 SGH820: Reconstructed electric EQC on the equivalent surface enclosing the AUT

PNF data measured in the echoic environment has been expanded with the EQC technique. The reconstructed EQCs on the defined equivalent surface fully enclosing the antenna are illustrated in Figure 3.39. As reconstruction surface enclosing the antenna under test (AUT) a box of size (0.195 × 0.170 × 0.380) m with a mesh step of $\lambda/2$ at 10 GHz has been considered.

The equivalent surface is slightly larger than the volume of the AUT. The computed currents are therefore associated only to the antenna and unwanted sources of echo signals are thus automatically filtered out in the recomputation of the FF from the EQCs.

The comparison of E-plane and the diagonal plane ($\varphi = 135°$) patterns are reported in Figure 3.40. The comparison is between reference FF (blue curve), perturbed (black curve) and echo suppression with the EQC (red curve). For comparison, the echo suppression with the spherical wave expansion (SWE), already shown in Chapter 5, is also reported (green curve) [2]. The improvements coming from the application of both echo suppression techniques are noticeable in both the co-polar (solid traces) and cx-polar (dashed traces) components.

In the second example, the dual-ridge MVG SH2000 is measured SNF setup illustrated in Figure 3.41(a). A scattering plate of size 0.5 m × 0.5 m ($10\lambda \times 10\lambda$ at 6 GHz) has been placed outside the measurement sphere at 0.9 m from the origin of the coordinate system. The measurement has been performed considering a ten-times oversampling with respect to the minimum needed have been adopted.

The measurement has been expanded with EQC considering as reconstruction a box of size (0.12 × 0.07 × 0.12) m meshed with a step of $\lambda/4$ at 6 GHz.

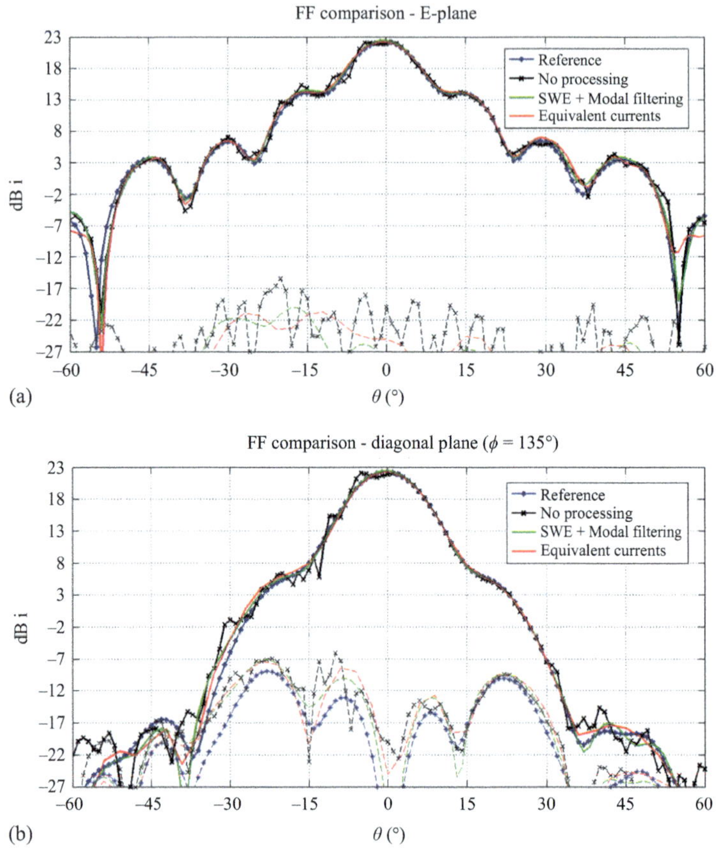

Figure 3.40 Directivity pattern comparison for the SGH820 measured in the PNF system. Solid lines show co-polar components, dashed lines show cx-polar components. E-Plane (a), diagonal plane, $\varphi = 135°$ (b)

The computed EQCs are shown in Figure 3.41(c). The computed currents are associated only to the antenna. The unwanted sources of echo signal lying outside the volume occupied by the antenna are thus automatically filtered out in the recomputation of the FF from the EQCs.

Comparison of *E*-plane and *H*-Plane pattern cuts is shown in Figure 3.42 for the reference (i.e., measurement without the scattering plate—blue curve), perturbed (black curve), echo reduction with the SWE (green curve) and with the EQC (red curve). The improvements coming from the application of the echo suppression techniques are easily appreciated also in this case. In particular, the FF ripple identifiable in the co-polar *E*-plane around $\theta = -75°$, and caused by the interaction between the reflecting plate and the probe arch, is filtered out by both techniques. Furthermore, as can be seen by the two cuts, also the overall cx-polar performances have been improved.

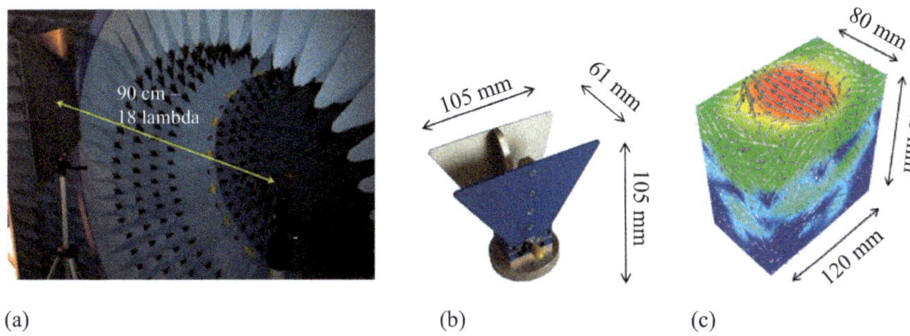

Figure 3.41 SNF measurement setup (a), details of the SH2000 measured antenna (b), reconstructed equivalent currents of the SH2000 (c)

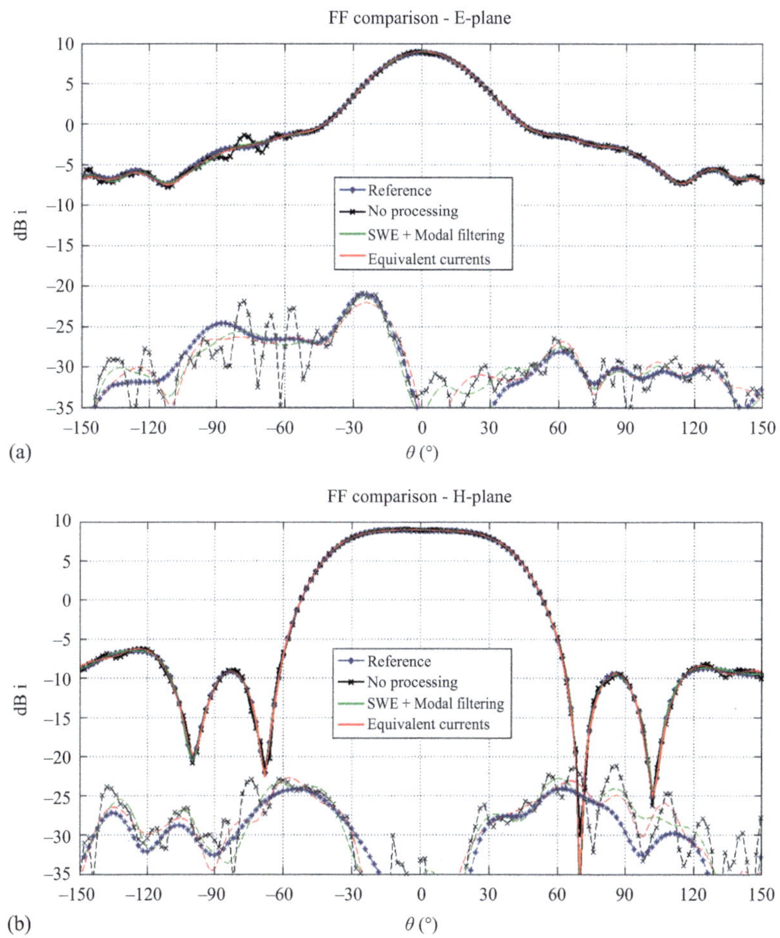

Figure 3.42 Directivity comparison. Solid lines show co-polar components, dashed lines show cx-polar components: E-plane (a) and H-plane (b)

3.3.3 Reduction of truncation errors

In the example presented in this section, the equivalent technique is applied for mitigation of truncation errors in automotive testing scenarios. Antennas installed in modern cars are often highly integrated. In such cases, the entire vehicle is contributing to the radiated field, in particular at frequencies below 400 MHz. The complete characterization of the full vehicle is thus typically required. For frequencies down to 70 MHz, a widely accepted and cost-effective solution is a multi-probe SNF system in which the scanning area is truncated below the horizon to minimize system dimensions. In order to emulate a proper free-space condition in the lower hemisphere, the ground floor is covered with absorbing material (see Figure 5.32 of Chapter 5). As explained in Section 5.1.2.4 of Chapter 5, the partial NF acquisition of such systems leads to truncation errors if standard NF/FF transformation is applied [19–22]. EQC expansion is a minimum energy operator, providing a smooth transition between the measured and truncated region, therefore it is very suitable as post-processing technique to mitigate the effect of the truncation error.

The test case is a scaled car model (1965 Morris Minor 1000, 1:12 scaled model) with a wideband patch antenna with omnidirectional pattern installed on the rear hood of the car model and measured in the MVG SL18GHz portable multi-probe system as visible in Figure 3.43.

The considered frequency range for the test is 1.1–18 GHz, approximately corresponding to the 91.7–1,500 MHz scaled frequency range. In order to reduce the truncation errors, the EQCs technique has been applied to the scaled LF measurement in the 91.7–400.0 MHz frequency range (lower frequencies where the errors are much more evident).

A simple block diagram regarding the use of the EQC as NF/FF transformation tool for truncated SNF dataset is reported in Figure 3.44. Contrary to the NF/FF transformation performed with the SWE, the truncated SNF dataset is in this case directly used without performing any field padding which would result in a

Figure 3.43 Measurement of the scaled car model in the MVG StarLab SNF system

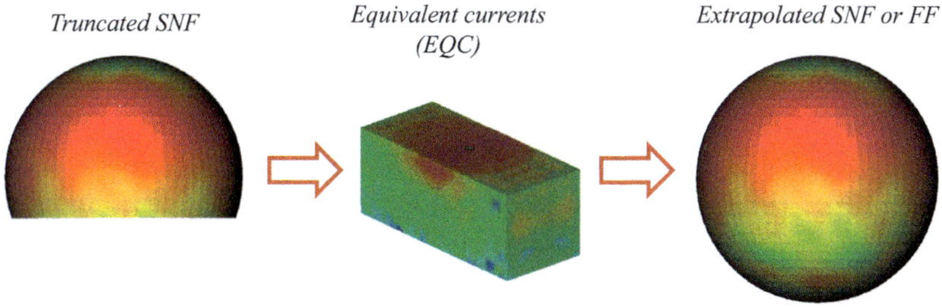

Figure 3.44 Block diagram of the EQC technique used as a tool for mitigation of truncation errors

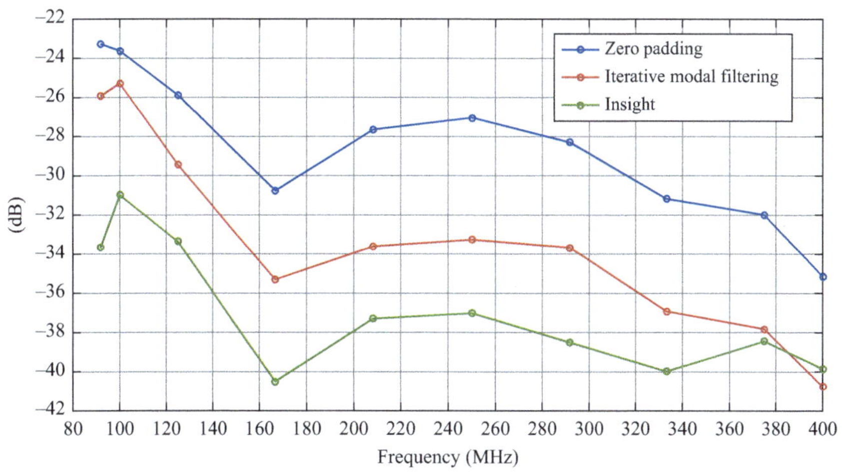

Figure 3.45 ENL with frequency obtained with zero padding and EQC

discontinuous and unphysical field. The EQC computation is in fact a minimum energy operator on the truncated region, which allows for a direct computation of the currents avoiding errors caused by the processing of a discontinuous field. From the computed EQC, the FF is finally evaluated over the full sphere completing the NF/FF transformation process.

The EQC technique has been applied considering an equivalent box for the current reconstruction of size: $L \times W \times H = 35 \times 20 \times 18$ cm, fully enclosing the scaled vehicle.

The comparison in terms of equivalent noise level (ENL—see eq. 5.22 of Chapter 5) between FF obtained with zero-padding (blue trace), FF obtained with the iterative modal filtering (IMF) described in Section 5.2.4.1 of Chapter 5 (orange trace) and EQC technique (green trace) is reported in Figure 3.45. The ENL formula has been applied considering only the portion of the FF radiation pattern corresponding to the un-truncated region (reliable part of the sphere, $\theta_{\max} = 107.5°$).

As can be seen, the EQC post-processing technique is able to improve the gain radiation pattern obtained with the simple zero-padding of the truncated area. In particular, at frequencies up to approx. 170 MHz, the relative high ENL obtained with the zero-padding are lowered to 7.5–10 dB with the EQC. In the central part of the band, the improvements obtained with EQC are very good, being around 10 dB. It should be noted that the performances of the EQC technique are even better than the ones obtained with IMF technique.

Gain radiation pattern comparison at 91.7 MHz is reported in Figure 3.46 for the vertical cut at $\varphi = 0°$ (a) and for the horizontal cut at $\theta = 90°$ (b). The reported traces are the total field cuts of the reference pattern (black trace), pattern obtained with zero-padding (blue trace), pattern obtained with the IMF technique (orange traces) and EQC technique (green trace). The shadowed areas drawn in the vertical cut illustrate the truncated and thus unreliable regions. The perturbation introduced by the truncation of the scanning area is observable in the zero-padded pattern in

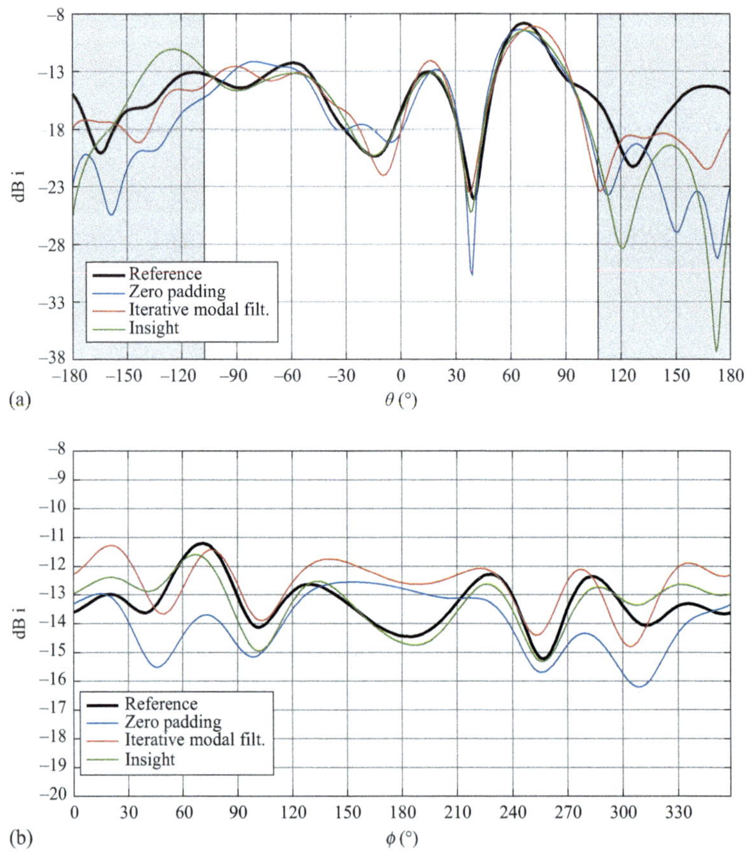

Figure 3.46 Gain pattern comparison at 91.7 MHz: (a) vertical cut at $\varphi = 0°$ and (b) horizontal cut at $\theta = 90°$

both presented cuts. It is stressed that the truncation errors are not just present in the unreliable region but all over the pattern. The improvements obtained with the EQC technique are clearly observable in the reliable region of the vertical cut. A similar comment can be applied also to the horizontal cut where the EQC technique gives a very good agreement with the reference.

3.4 Further applications of the EQC technique

In this section, we will briefly examine the application of the EQC technique in different fields.

- Human exposure to RF electromagnetic fields (EMFs) is evaluated in terms of incident power density, i.e., free space Poynting vector in close proximity to the device at frequencies higher than 10 GHz. The EQC technique is suited to determine the power density in any area in the vicinity of a measured device.
- In EMC applications, the EQC can be very efficiently used in pre-compliance, diagnostics and problem-solving.

3.4.1 Power density determination of measured devices

In the millimeter wave (mmWave) bands for 5G applications, mobile terminals become electrically large resulting in stronger effects on antenna performance, such as stronger excitation of surface waves. Such effects can be examined by the EQCs via the inverse source technique, since they allow to obtain compact numerical models of antennas that reproduce and possibly extrapolate the original measured fields. A new approach to accurately determine maximum spatial-average power density in the close vicinity to a device is calculating the radiation starting from the computed EQCs by NF/NF transformation on any surface and at any distances according to the safety normative.

In the example of this application, a 5G mobile terminal mock-up containing notch elements was fabricated by Sony Mobile, as shown in Figure 3.47, fed

Figure 3.47 Fabricated 5G mobile terminal mock-up for antenna diagnosis

through 50 microstrip lines connected with the 50 connectors. The fields were measured using the multi-probe measurement system MVG StarLab 5G,[‡] and the measured data was used for EQC reconstruction.

The size of the reconstruction surface is chosen slightly larger than the mock-up to reduce the positioning error. Figure 3.48 shows a comparison of the reconstructed electrical EQC compared to currents determined by computational electromagnetic (CEM) tools. Generally, the reconstructed EQCs agree well with the simulated EQCs. A portion of reconstructed EQCs can be observed in the corner stronger than the simulated EQCs.

This may be attributable to the fabrication discrepancy and cable effects. This could be removed by the spatial filtering allowed by EQC. As shown in Figure 3.49, the EQCs in the gray area are filtered out and suppressed cable effects resulting in lower side lobe levels can be observed.

As underlined before, the EQC can be used to calculate the incident power density. 5G user equipment, such as mobile terminals and tablets, in fact, should comply with RF safety guidelines. In mmWave frequencies, human exposure to RF EMFs is evaluated in terms of incident power density, i.e., free space Poynting vector. The obtained EQCs shown above can be used to calculate the incident power density. The power density levels are normalized to the total radiated power of 17 dB m, i.e., 50 mW. Figure 3.50 shows the maximum spatial average power density versus the distance d above the mock-up. Discrepancy can be observed between the measured power density and the simulated power density due to the positioning error. The results suggest that the misalignment could be about 3–4 mm. Such error could be improved by better alignment methods, while the maximum spatial average power density with 20 cm^2 shows more robust to the measurement error.

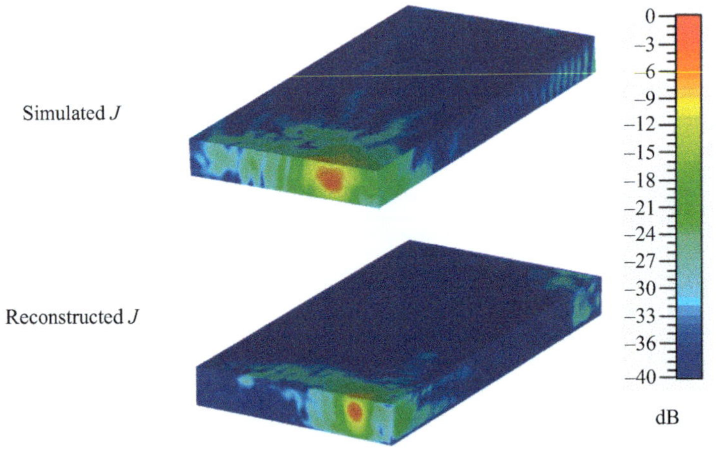

Figure 3.48 Reconstructed and simulated equivalent electric current distribution of the notch element

[‡]https://www.mvg-world.com/en/products/field_product_family/antenna-measurement-2/starlab-50-ghz.

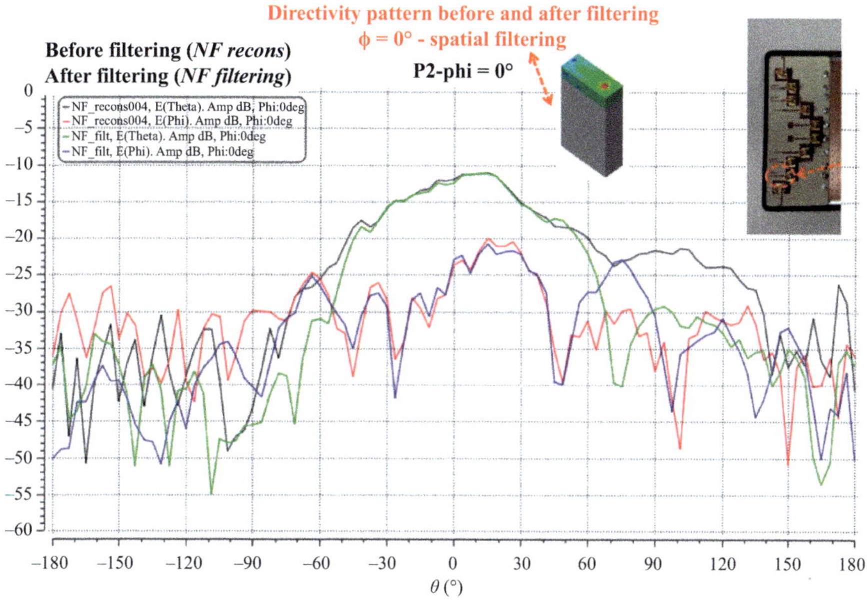

Figure 3.49 *Reconstructed and simulated equivalent electric current distribution of the notch element*

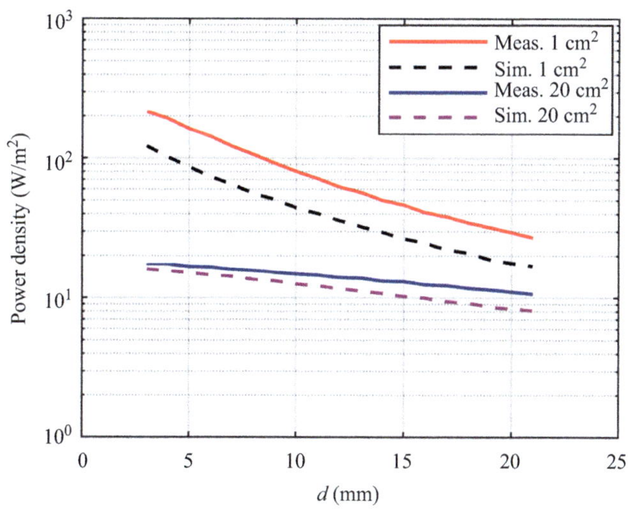

Figure 3.50 *Measured (from reconstructed EQCs) and simulated spatial-average power density for the notch element to be compliant with the FCC and ICNIRP power density limits*

3.4.2 Application of the EQC technique in EMC testing

All electronic devices, present in commercial products and integrated in their intended enclosure, need to be tested using EMC methods. The tests include pre-compliance, diagnostics and problem-solving. The goal of pre-compliance is to work out whether your device is going to pass or fail at an EMC laboratory, hence, to know whether those emissions are within the limits. The aim for diagnostics and problem-solving is to identify sources of harmonic energy and determine fixes that reduce the harmonic amplitudes, mainly looking for relative changes.

Since huge varieties of the electronic devices are produced and commercialized, these testing processes represent a special interesting activity. Hence, electronic manufacturers continuously research for new and more efficient methods to do proactive EMC work and pre-compliance test [3,6,7,23].

EQC method can be efficiently used as innovative tool for EMC studies. EQC representation of measured devices, such as PBC, is directly applicable in diagnostics/hot spot finding and in the determination of radiated emission at any distance. Additionally, the NF equivalent source, derived from measurements, can be used in the numerical simulation tools for analyzing the emission in different scenarios, when the device is located in the vicinity of different objects such as shielding and cables.

This approach is really promising since it allows to perform an EMC study by a combination of a fast NF spherical measurement and the EQC method instead of a more traditional time-consuming scan, for example, on a plane close to the PCB. Moreover, due to the capabilities of the EQC method to determine the E-field values at any distance (as post-processing of the EQC distribution), a fast EMC pre-compliance can be implemented.

In the following, the EQC method is demonstrated on a PCB with dimensions 150×225 mm shown in Figure 3.51, considering the frequency range [30–1,000] MHz.

The PCB has a substrate of 2 mm thick lossy FR4 layer, three traces on the top layer and full unbroken ground plane. Only one trace is excited and terminated. The traces are 2 mm wide, and both source and load impedances were 50 Ω. The emission from the PCB is very low, indeed the power absorbed by the load is >99%.

The measured NF radiation pattern at 640 MHz has been processed in order to create the measured equivalent model of the device. Electric J and magnetic M reconstructed currents are shown in Figure 3.52. Reconstructing box dimensions are 26 cm $\times$ 17 cm $\times$ 2 cm, while the mesh step is $\lambda/40$ at the analyzed frequency.

In order to test the accuracy of the method, E-field emission at 3 m from the measured model of the PCB is evaluated with measured EQC representation and compared with the full-wave simulated model. Results are shown in Figure 3.53. The agreement between the two curves is considered acceptable for both elevation cut planes. Differences between the two curves are due to the presence of the mounting support and the cable in the measurement setup, which are not modeled in the full-wave simulation.

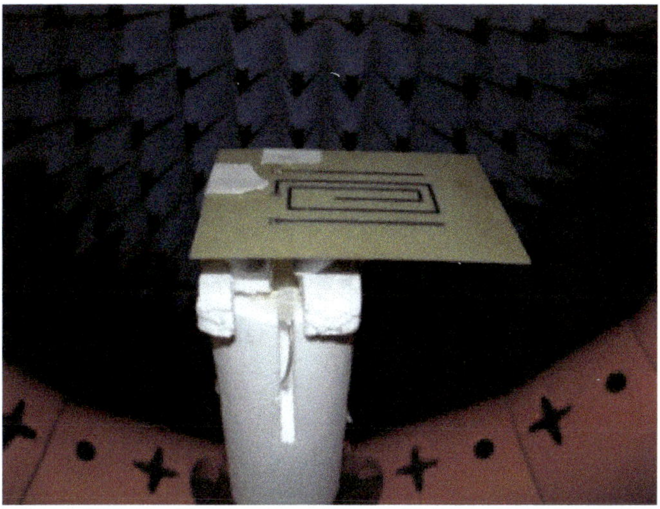

Figure 3.51 Tested PCB in the MVG multi-probe StarLab18GHz system

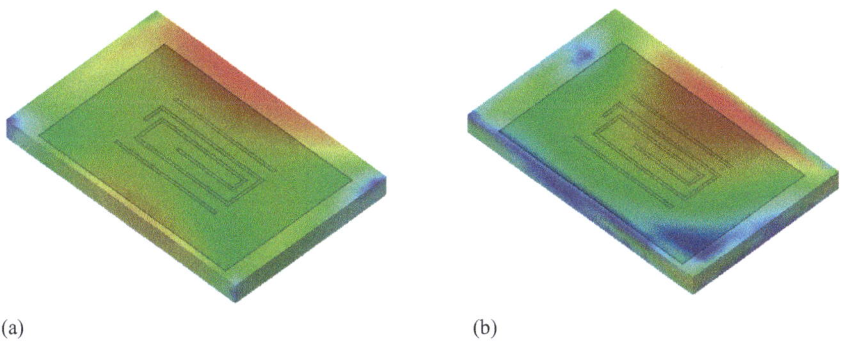

Figure 3.52 Reconstructed equivalent currents on a box enclosing the PCB at 640 MHz: (a) J currents and (b) M currents

The link between measurement and simulation tools is applied to test the device in a more complex scenario, for example, in proximity of other objects and/or scatterers. Thus, for demonstration of the 3D equivalent model of the PCB in terms of EQCs has been imported in a CEM tool[§] as NF Huygens box to be applied in computational EMC studies. Radiating emission of the stand-alone PCB and located in vicinity of a wire and inside an example of enclosure has been investigated.

[§]www.cst.com, CST STUDIO SUITE™, CST AG.

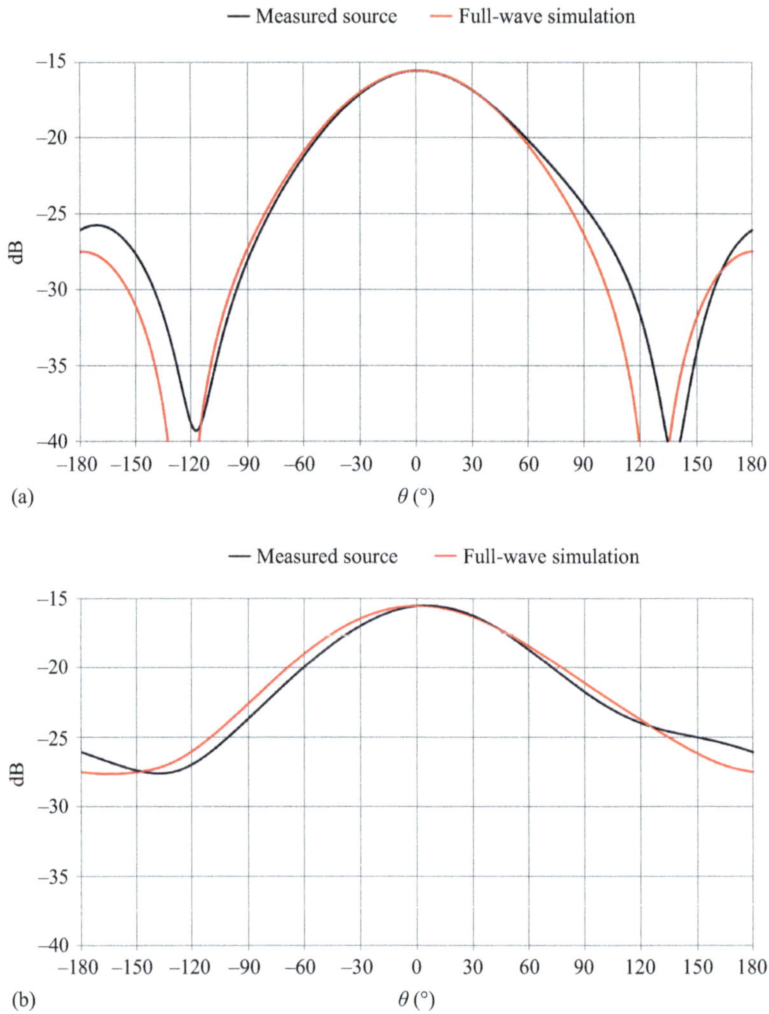

Figure 3.53 PCB E-field emission at 3 m distance at 640 MHz (a) $\phi = 0°$ elevation plane, (b) $\phi = 90°$ elevation plane; comparison between measured PBC source and full-wave simulation of the PCB

When the PCB is considered in the vicinity of objects, the Huygens box representation should include a suitable internal model of the PCB. This ensures a good accuracy of the solution during the calculation of the emitted radiation. It should be noted that due to the Love's condition (zero field inside the box), adding an object inside the equivalent surface is allowed. In Figure 3.54(a), the measured NF model has been imported in the CEM tool in order to compute the full-wave model including the NF source (computed with EQC), the wire and also the ground plane and the dielectric substrate to represent the PCB inside the box. In this case,

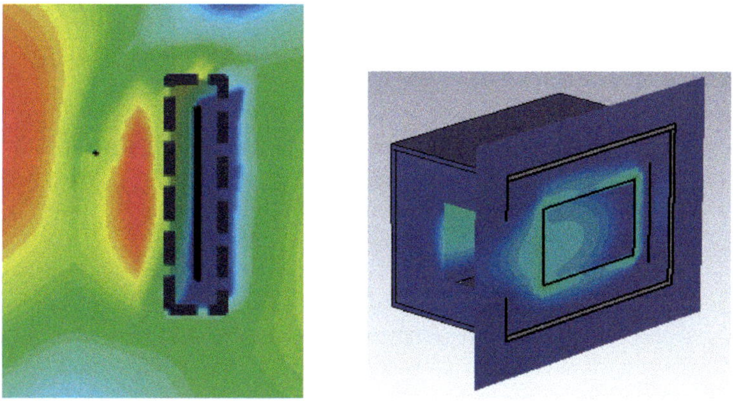

(a) (b)

Figure 3.54 (a) PCB E-near field emission: box profile, NF source, wire, PCB ground plane and dielectric substrate and (b) PCB model inside an example of enclosure

the deviation between using the PCB as measured model in the numerical simulation and the full simulation is 0.37 dB. The PCB has also been studied inside an example of an enclosure, as shown in Figure 3.54(b). In this latter case, the deviation between the PCB as considered in the numerical simulation and the full simulation is 0.20 dB.

3.5 Conclusions and discussion on advantages and limitations of the EQC technique

Equivalent sources can be calculated on an arbitrary-shaped surface fully enclosing the antenna by solving the inverse problem directly from the measured field data. As a consequence, the integral equation formulation is geometry independent, in principle being able to accommodate efficiently for atypical geometries (not uncommon in real applications) both for the radiator and the measurement range (including truncated versions of canonical ranges). The integral equation approach constitutes a complement to more established tools such as plane-wave expansion or SWE as discussed in Chapter 5. It affords a greater generality and flexibility since it allows reconstructing sources on arbitrary 3D surfaces enclosing the AUT and can be applied with measured data from different scan range. The advantages of the EQC technique are offset by a higher computational cost which for electrically large problems can be very demanding due to the number of involved unknown sources. Implementation of multilevel fast methods is one way to offset the computational burden.

EQC reconstruction has attracted considerable attention in recent years due to a host of promising applications for antenna improvement techniques, including antenna and measurement diagnostic, echo suppression, spatial filtering, NF/NF and NF/FF transformation. The wide range of application in different fields has been illustrated with a few practical examples in this book.

As pointed out through the examples in the previous section, the integral reconstruction allows a wide range of analysis and diagnostics applications:

- *Antenna diagnostics*
  - Investigate details of the antenna radiation.
  - Array element failure and/or excitation errors, loose connections.
  - Discover antenna electrical and/or mechanical errors including surface shape and material properties.
  - Sources of asymmetry/unwanted coupling/cross-polarization.
  - Antenna electrical and/or mechanical errors.
- *Measurement diagnostics/Filtering*
  - Detection and spatial filtering removal of interaction between antenna and the surroundings (positioner, mounting structure, fixture, cables, etc.).
  - Detect antenna feeding leakage in measurement systems and interaction with cables.
- *Echo suppression/Spatial filtering*
  - Remove unwanted stray signals from the surrounding structures that could be represented physically by the techniques and successively removed.
- *EMC problems*
  - Troubleshoot EMC shielding, EMS investigation—Pinpoint sources (bright spots) on the shielding structure allowing for the identification of unwanted radiation sources.
- *NF/FF transformation/Back transformation/Power density computation*
  - Investigation of field and power levels close to the antenna from NF or FF measurements;
  - NF to FF, NF to NF and FF to NF transformations;
  - Evaluate power densities very close to the radiating antenna surface with NF to NF transformation.
- *Data interpolation/Extrapolation/Truncation*
  - Sparse and missing acquisition data can be compensated with the EQC approach.
  - Mitigate the effect of the truncation error in measurement systems.
- *Source for numerical computation*
  - The EQC is a highly accurate source for numerical computations of the antenna in a larger EM problem.

References

[1] J. L. A. Quijano, G. Vecchi, L. Li, *et al.* "3D Spatial Filtering Applications in Spherical Near Field Antenna Measurements", AMTA 2010 Symposium, October 2010, Atlanta, GA, USA.

[2] L. J. Foged, L. Scialacqua, F. Mioc, *et al.* “Echo Suppression by Spatial Filtering Techniques in Advanced Planar and Spherical NF Antenna Measurements”, AMTA, October 2012, Seattle, WA, USA.

[3] S. Burgos, M. Boumans, P. O. Iversen, C. Veiglhuber, U. Wagner, and P. Miller, “Hybrid Test Range in the ESTEC Compact Payload Test Range”, 35th ESA Antenna Workshop on Antenna and Free Space RF Measurements ESA/ESTEC, September 2013, The Netherlands.

[4] A. Montesano, F. Monjas, L. E. Cuesta, and A. Olea, “GALILEO System Navigation Antenna for Global Positioning”, 28th ESA Antenna Workshop, Vol. 1, p. 185–191, June 2005, Noordwijk.

[5] L. Salghetti Drioli, L. J. Foged, F. Saccardi, and L. Scialacqua, “Verification of Complex Excitation Coefficients from Measured Space Array Antenna by the Equivalent Current Technique”, AMTA, 12–17 October 2014, Tucson, AZ, USA.

[6] L. Salghetti Drioli, L. J. Foged, F. Saccardi, and L. Scialacqua, “Analysis of Spacecraft Antenna Farm Interaction with Equivalent Current Technique”, EuCAP, 12–17 April 2015, Lisbon, Portugal.

[7] L. Salghetti Drioli, L. J. Foged, F. Saccardi, and L. Scialacqua, “Analysis of Coupling Phenomena between Spacecraft Antennas Based on Equivalent Current Technique” AMTA, 11–16 October 2015, Long Beach, CA, USA.

[8] J. E. Hansen (ed.), Spherical Near-Field Antenna Measurements, Peter Peregrinus Ltd., on behalf of IEE, London, United Kingdom, 1988.

[9] F. Saccardi, A. Giacomini, and L. J. Foged, “Probe Correction Technique of Arbitrary Order for High Accuracy Spherical Near Field Antenna Measurements”, AMTA, 30 October–4 November 2016, Austin, TX, USA.

[10] F. Saccardi, A. Giacomini, and L. J. Foged, “Comparative Investigation of Spherical NF Measurements with Full and First Order Probe Correction Using Calibrated or Simulated Probe”, EuCAP, 19–24 March 2017, Paris, France.

[11] T. Laitinen, S. Pivnenko, J.M. Nielsen, and O. Breinbjerg, “Theory and Practice of the FFT/Matrix Inversion Technique for Probe-Corrected Spherical Near-Field Antenna Measurements with High-Order Probes”, IEEE TAP, Vol. 58, No. 8, p. 2623–2631, 2010.

[12] L. J. Foged, A. Giacomini, and S. Pivnenko, “Wide Band Dual Polarized Probes for Near and Farfield Measurement Systems”, AMTA 2007, 4–9 November 2007, St. Louis, MO, USA.

[13] L. J. Foged, A. Giacomini, and R. Morbidini, “Probe Performance Limitation Due to Excitation Errors in External Beam Forming Network”, AMTA, 2011, Colorado, USA.

[14] A. Newell, S. Gregson, D. Gentle, and P. Miller, “The Effect of the Absorber Collar on Open Ended Waveguide Probes”, AMTA, 2009.

[15] C. A. Balanis, “Antenna Theory Analysis and Design”, John Wiley & Sons, Inc., Publication, 3rd edition, 2005. Hoboken, New Jersey.

[16] J. L. Araque Quijano, G. Vecchi, and L. J. Foged, “Efficient Filtering of Cable Interaction in Small Antenna Measurements”, Antenna Technology

and Applied Electromagnetics & the American Electromagnetics Conference ANTEM-AMEREM, July 2010, Ottawa, Ontario, Canada.

[17] L. Liu, Y. F. Weng, S. W. Cheung, T. I. Yuk, and L. J. Foged, "Modeling of Cable for Measurements of Small Monopole Antennas", Loughborough Antennas & Propagation Conference, November 2011, Loughborough, UK.

[18] L. J. Foged, L. Scialacqua, A. Scannavini, F. Saccardi, J. L. Araque Quijano, and G. Vecchi, "Experimental Investigation of Radiating Current Distribution and Measurement Cable Interaction on Wireless Devices", 6th European Conference on Antennas and Propagation, EuCAP 2012, March 2012, Prague, Czech Republic.

[19] E. Martini, S. Maci, and L. J. Foged, "Spherical Near Field Measurements with Truncated Scan Area", 5th European Conference on Antennas and Propagation, EuCAP 2011, 11–15 April 2011, Rome, Italy.

[20] E. Martini, S. Maci, and L. J. Foged, "Reduction of Truncation Errors in Spherical Near Field Measurements", AMTA, 10–15 October 2010, Atlanta, GA, USA.

[21] L. J. Foged, L. Scialacqua, F. Saccardi, *et al.*, "Comparative Investigation of Methods to Reduce Truncation Errors in Partial Spherical Near-Field Antenna Measurements" EuCAP, 26–30 March 2012, Prague, Czech Republic.

[22] F. Saccardi, F. Rossi, L. Scialacqua, and L. J. Foged, "Truncation Error Mitigation in Free-Space Automotive Partial Spherical Near Field Measurements", AMTA, 15–20 October 2017, Atlanta, GA, USA.

[23] L. J. Foged, A. Giacomini, F. Saccardi, and L. Scialacqua, "Investigation of Spherical Higher Order Modes Sources in Antenna Measurement Probe Design" EuCAP, 12–17 April 2015, Lisbon, Portugal.

Chapter 4

Holographic techniques to improve antenna measurements

Francisco José Cano Fácila[1] and Manuel Sierra Castañer[2]

Electromagnetic measurement results can be improved using source reconstruction techniques since the contribution from spurious sources can be immediately filtered out, once the origin of the field is known. In the previous chapters, equivalent currents have been calculated using integral equation methods, and in the following chapters, cylindrical or spherical modes processing is used to separate the contributions from the antenna under test (AUT) with respect to other non-desired signals. Holographic techniques can be considered an intermediate point between equivalent currents and radiated field modes. The sources on different planes are reconstructed; therefore, if the field is reconstructed on the antenna aperture plane, this is a source reconstruction method. On the other hand, it is used in the decomposition on the radiated field in plane waves (the plane wave spectrum (PWS)): therefore, this method can be considered similar to the ones that exploit the relation between spherical or cylindrical modes of the radiated field and the dimensions of the antenna.

The main advantages of this method are the mathematical simplicity and the reduced calculation time, since Fourier transforms are used to calculate the PWS of the radiated field. Also, the field filtering is performed, back-propagating the field to the antenna aperture and cancelling all the radiation originated out of that aperture. The main problem is that, if only the radiated fields are used (visible spectrum), the minimum resolution is half wavelength. In any case, this resolution uses to be good enough to recover the radiation pattern, as it is shown in this chapter. Also, iterative methods can be employed to enhance this resolution and to extract sources with a separation lower than half wavelength. Obviously, the processing time becomes larger, but the resolution can be drastically improved. Another limitation is that the method presents good results for aperture antennas, since the sources are reconstructed on those apertures. The method can be used for general antennas, but the resolution cannot be as good as for aperture antennas, since it is not as clear the relation between antenna surface and electromagnetic

[1]Airbus Defence and Space, Madrid, Spain
[2]Information Processing and Telecommunications Center, ETSI Telecomunicación, Universidad Politécnica de Madrid, Madrid, Spain

currents. Finally, since PWS is used, only half sphere is used; therefore, the radiation pattern can be reconstructed only in the angular range from 0 to 90 degrees. These two limitations make this method useful for directive antennas, and not for general cases of small antennas, where the back radiation can be important and desirable.

This technique has been used for different applications: first, source reconstruction itself of the sources on the antenna aperture. In this case, direct methods of calculation of the PWS and back-propagation give a resolution of half wavelength if the complete visible range is used. As it was shown, using iterative methods, the resolution can be increased, if information in the non-visible zone is recovered. This is the basic aspect of this technique and will be explained in Section 4.2. But, being able to understand the origin of the desired signal and other spurious signals, this method is employed for noise reduction, echo suppression, detection and cancellation of leakage and extrapolation of the radiation pattern from truncated measurements. In the case of noise reduction, it exploits the fact that the noise is spread out the complete plane, while the AUT contribution is focused on the antenna aperture. Therefore, if the radiated field is back propagated to a large plane, the signal-to-noise improvement will be, in first approximation, the ratio between the field-reconstructed plane and AUT aperture plane.

In the case of echo-reduction, it is easy to understand that, if the field is back propagated and focused on AUT aperture, all the contributions coming from other areas of the plane are due to spurious signals, and particularly reflections. This is easy to understand for classical planar measurement systems on a ground plane (without absorbers). For each position of the probe, the received fields are the addition of two contributions: the one from the AUT and the one from the AUT image, under the ground plane. Since the probe is moving on different positions on the scanner plane, the source of the AUT will be focused, while the echo will be unfocused but, in many cases, easy to filter out. As it will be explained in the corresponding section, this is not as clear for the ranges where the AUT is rotated during the measurements, since the echo is dispersed. The case of leakage detection and cancellation is similar to the previous ones. Different methods depending on the origin of the source of leakage will be shown in the section.

Finally, the radiation pattern can be extrapolated up to an angular range from theta 0 to 90 degrees using iterative techniques. Using only information of general geometry of AUT (dimensions), the results for aperture antenna are really good. This is due to the fact that for the radiation pattern, only the information of the antenna with a resolution of half wavelength it is required. For other applications, as fine calculation of the antenna sources, a higher resolution is necessary, but this can be obtained exploiting other dimensional aspects of the geometry of the AUT.

The chapter is divided into the following sections: Section 4.1 summarizes the state of the art, Section 4.2 explains the basic theory of holographic techniques, Section 4.3 to Section 4.6 detail the different applications of this technique for noise, echo, truncation errors and leakage reductions respectively and Section 4.7 concludes the main ideas of this chapter.

4.1 Use of holographic techniques for reducing errors in antenna measurements

The main purpose of antenna measurements is to obtain a good representation of the radiation properties of the AUT. The most important parameters are gain, directivity, polarization, radiation pattern and the parameters related to this last one as side lobe level or beamwidth. Those parameters can be obtained, directly in far-field or from near-field measurement using modal expansion methods [1,2] or calculating the electric or equivalent magnetic currents using a source reconstruction method [3,4].

Near-field measurements are usually indoor, minimizing the effect of reflections and other interferences from the environment. However, near-field measurements are affected by other errors that do not appear in far-field ranges. All these sources of errors make the real value different from the measured value, and an uncertainty analysis [5] is required to assess the effect of measurement errors on the accuracy of the far-field results. The main objective of this chapter is to propose some methods using holographic techniques to reduce the undesired effect of some of the errors in antenna measurement: noise, reflections, measurement area truncation and leakage. The main aspects for each application are detailed in following subsections.

4.1.1 Potential of source reconstruction techniques

A source reconstruction technique obtains the extreme near-field of the AUT from the knowledge of its radiated field in near or far field. These currents can be employed to identify errors, and they have been widely used in the identification of electrical errors in arrays or mechanical errors in reflectors. This technique is also known as diagnostic technique for that reason. Even if the presence of these errors can be detected directly in the measured far-field, their identification is really difficult in that domain, making necessary the inspection of the fields in the extreme near field.

Therefore, diagnostic technique gives a complete electromagnetic characterization of the antenna and its environment, being the basis of other applications, like the characterization of the effect of the radomes, the near-to-far-field transformations and the calculation of radioelectric coverage by an electromagnetic source.

In this chapter, the information provided by the source reconstruction is used to reduce the effect of noise, reflection, leakage or truncations errors using the information of the PWS.

Modal expansion methods require a less computational complexity, but, because the measured field has to be expressed as a superposition of orthogonal functions (vector wave solutions), they only can be used in particular situations, when the measurement is performed over canonical surfaces. There are six coordinate systems that provide the orthogonality between the solutions of the Maxwell's equations. However, just three of them (planar, cylindrical and spherical) are usually used due to their mechanical simplicity. Depending on the coordinate system, one particular kind of expansion (plane wave expansion, cylindrical wave expansion or spherical wave expansion) is carried out. Then, the PWS is calculated from the corresponding expansion and, unlike integral equation methods, the extreme near field can be efficiently computed by means of the fast Fourier transform (FFT) of this PWS. In the following sections, the different effects to be improved through the use of diagnosis techniques are explained.

4.1.2 Noise

White noise limits the accuracy of far-field results, particularly, when measuring in low dynamic range applications: millimetre or submillimetre frequencies, low side lobes or cross-polar. Some previous studies for the characterization of this noise in near-field measurements have already been presented for particular measurement system (spherical, planar and cylindrical) due to the different algorithms of the near-field-to-far-field transformation processes. For the planar case, two independent analyses with similar results have been proposed [5,6], introduce random errors in the planar near-field and get the expressions to represent the signal-to-noise ratio (SNR) in far field as a function of the near-field noise power. A similar study for cylindrical near-field (CNF) measurements was carried out in [7,8], where the authors derived the expression to transfer the noise power from the near field to the far field. The cylindrical case was also discussed in [9], where the authors investigated the improvement of the SNR with the application of the CNF to far-field transformation.

The work presented in [10] focuses on the planar case. In this case, the statistical analysis is completed with an algorithm based on noise filtering, without affecting the desired signal, to reduce the far-field noise power. The acquired data is assumed to be corrupted by white Gaussian and space-stationary noise. This information is employed to obtain a PWS referenced to the scanning plane. The PWS is back-propagated from the scan plane to the AUT plane, and the information out of AUT aperture is filtered out. Finally, the reconstructed field is calculated by means of the Fourier transform of the filtered spectrum. This method is based on the fact that the desired field theoretically does not extend beyond the AUT dimensions while the noise is spread throughout the reconstructed surface. The filtered field is then used to obtain a new PWS with less noise power. A similar method to increase the SNR by reducing the noise power was also proposed in [11,12]. That technique can be applied to spherical near-field (SNF) measurements, requiring an oversampling to obtain a higher number of spherical modes [13], and applying a modal filtering.

4.1.3 Reflections

The second effect to be studied is the reduction of the effects of reflections. During the last years, the number of algorithms to analyse and cancel the effects of echoes has increased.

We can classify the methods in two different kinds: the ones to characterize the chambers by a figure of merit, usually called the reflectivity level, using frequency or time domain. In frequency domain [14,15], the reflectivity is obtained from the interference pattern, moving one of the antennas over the quiet zone. The main drawback of this technique is the impossibility to separate different reflections among them, being very difficult to identify test site imperfection. However, with time-domain characterization, it is possible to isolate the different reflected signals from the direct signal [16]. In this case, a short pulse is transmitted separating the signals using the delay in the receiver. Thus, this approach not only gives the reflectivity level of the measurement system but also finds the worst places, being easy to improve the facility performance solving those hot spots. The main drawback is the requirement of expensive equipment. Reference [17] presents another method based on measurements at several distances. In this case, the averaging of the different patterns compensates the different reflections effects.

Time-gating techniques can also be employed as an option to remove reflected contributions. In this case, the information is not used to estimate the reflectivity of an anechoic chamber as in the previous case but to gate the direct contribution and cancel the echo effects. In [18,19], the measurement is directly performed in the time domain by utilizing pulse generators and sampling oscilloscopes. Once the direct contribution is gated using a time filtering, a time-to-frequency transformation is done to calculate the frequency response. Reference [20] proposes a similar technique, reducing the complexity of the equipment using frequency domain measurements and an inverse Fourier transform. The FFT algorithms require a constant frequency step. Therefore, when there are frequency regions where the field varies very quick, a small step has to be employed over the entire bandwidth to obtain a good characterization in those regions, increasing the acquisition time considerably. There are other techniques, using finite impulse response filters that avoid the use of uniform spacing [21] to obtain the time response from frequency-domain measurements. This technique allows an optimum characterization of the frequency range with a minimum number of samples.

Other approaches include frequency decomposition techniques, where the signals in the frequency-domain are divided into a set of propagating components, including the direct and all the reflected contributions. Filtering out the undesired components, a result equivalent to the one obtained in a fully anechoic chamber may be obtained. First, decomposition in a sum of complex exponentials was proposed. Polynomial methods, including Prony's algorithm, were the first methods used to carry out such decomposition [22,23]. The main limitation was that accurate results were only achieved with noiseless data. A variation of that approach was the pencil-of-function method [24]. However, even though a better performance to the noise is achieved, this method is not computationally very efficient since, a two-step process is required: solving a matrix equation and finding the roots of a polynomial. Evolutions of these methods solve those problems by employing the generalized pencil-of-function method [25] or its improved version, the matrix pencil method [26,27]. Reference [28] presents a higher resolution analysis, where the reflected components are removed in the time-frequency plane once an oversampled Gabor transform decomposition.

Other techniques used the cylindrical or spherical modes decomposition. The technique called mathematical absorber reflection suppression was introduced by the company NSI to cancel undesired reflected signals. This technique is based on the filtering of unwanted cylindrical modes [29] or spherical modes [30], depending on whether the AUT has been measured in a CNF or SNF range, respectively. A similar procedure (IsoFilter) is used by the company MI-Technology [31] to suppress reflected waves in antenna measurements. Both techniques require an oversampling to later apply a modal filtering, and they can be used to mitigate other effects as noise and diffraction.

4.1.4 Measurement scan truncation

Equivalent principle used for the calculation of the exact far-field patterns from near-field acquisition requires the acquisition of the tangential components of the electric or magnetic field on an arbitrary surface that encloses the AUT. In that case, the field can be obtained anywhere outside the measurement surface, including the far-field region, solving an integral over the surface on which the fields are known [2]. The solution of this transformation is impractical to determine the far field unless the scan surface were a plane, a cylinder of a sphere; the three classical near-field antenna measurement systems. There are some differences among the three coordinate systems: the planar and cylindrical scanning geometries are mechanically simpler than in the case of the spherical, since in the first case, the AUT is fixed, and in the second case, it only rotates on an azimuth axis. The SNF-to-far-field transformation is more complex, requiring complicated formulation to obtain the far-field pattern from the acquired data. But the main difference is that SNF system is the only measurement set-up where the AUT can be fully enclosed by the acquisition surface, getting the highest accuracy for standard antennas. Therefore, there are no truncation errors in the calculated far-field pattern (if complete acquisitions are performed). In the planar near-field (PNF) and CNF measurements, because of the finite size of the scan surface, the closed surface condition is never fulfilled, and, consequently, the true far-field pattern is never known in the whole angular region. Another effect comes from the Gibbs Effect in the near-to-far-field transformation: the discontinuity of the measured field at the edge of the scan surface creates a ripple within the reliable angular region.

Since this truncation effect is an unavoidable error in PNF and CNF measurements, most of the approaches to reduce this kind of error have been specifically proposed for the two first configurations. However, there are cases where the SNF acquisition is not complete, in order to save measurement time. In those cases, the same procedures can also be applied to this acquisition.

There are different approaches to this problem: the first one reduces the ripple within the reliable region through the application of a proper window function to the near-field data. In [32–35], different methods to reduce that ripple are presented. The approaches of the second group attempt to estimate the true pattern outside the reliable region. One of these approaches is called the equivalent magnetic current approach [3]. This method computes the equivalent magnetic currents in the AUT plane solving the system of integral equations that relate these currents and the measured field. Once these currents are known, the correct far fields from AUT in the whole region is directly calculated. The main limitation of that method is the computational complexity that is required to solve the system of equations. This complexity is reduced using a magnetic dipole array approximation instead of the equivalent magnetic current approach. There are other strategies to reduce truncation errors like the one presented in [36], where the AUT is rotated about one or more axes, measuring in different planes and combining them to increase the maximum validity angle. In [37], a priori information of the AUT is used to extrapolate the radiation pattern, estimating the near-field data outside of the scanning area before calculating the far-field pattern. The *a priori* information is employed to obtain a non-redundant and non-uniform representation of the samples that are taken over the measurement surface [38]. Thanks to this distribution of the samples, a large number of samples outside of the scan plane can be recovered with good accuracy. This strategy was experimentally validated in [39], obtaining good far-field results for the case of cylindrical scanning. The same idea of non-redundant and non-uniform representation of electromagnetic fields was also used in [40] extrapolating the data outside of a plane-polar scanning and demonstrating that these new expansions guarantee a better reconstruction of the samples outside the measurement surface for a given number of measurement points. Reference [41] proposes another method based on the same principle (the use of non-redundant sampling in the acquisition). In this case, the truncation error is reduced addressing points in surfaces external to the actual scanning area. Therefore, this method can be applied whenever the set-up allows the variation of the distance between the AUT and the probe. Also, [42] presents another alternative to increase the reliable region by extrapolating the planar near field. In this case, the extrapolation is achieved by back-propagating the measured field to the AUT plane. Then, the samples within the AUT aperture are retained and are used to restore the external samples in terms of a diffraction integral along the aperture rim. Finally, the field is transformed back to the measurement plane to obtain new-field samples outside of the measurement region. The main drawback of this last approach is the impossibility of recovering samples over an infinite surface with good accuracy, and, therefore, the truncation error is not completely removed for large elevation angles. Reference [43] uses *a priori* information of the AUT in the planar near-field scan and performs the iterative Gerchberg [44]–Papoulis [45] algorithm to extrapolate the reliable portion of the calculated far-field pattern.

As it was mentioned before, these algorithms can also be applied to incomplete SNF acquisitions, because of the prohibitive measurement time or because the impossibility to get reliable data over the whole sphere, due, for instance, to support-structure blockage when measuring small devices. In [46], the truncated SNF data are used to calculate an equivalent currents distribution of a set of dipoles that reproduce the radiation of the AUT, and, once that distribution is known, the far-field pattern is easily computed in the entire sphere calculating the field radiated by this set of dipoles. When measuring over a partial sphere, the spherical modal functions are not orthogonal, and the calculated modal coefficients are incorrect.

The use of a new basis function set orthogonal over the truncated angular domain can solve this problem. The solution was presented for two-dimensional CNF/SNF scanning in and for three-dimensional acoustic SNF scanning in [47]. Another method based on a least squares technique that significantly reduces truncation errors for not very large antennas is employed to calculate the spherical wave coefficients using forward hemisphere data only [48]. The same problem is also addressed in [49] where the band-limited property of the spherical wave coefficients is used in an iterative algorithm that replaces the unreliable portion of the measurement sphere with new samples at each iteration.

4.1.5 Leakage

The last effect to be studied in this chapter is the leakage. There are several sources of origin of the leakage: crosstalk between the reference and the measurement channels in the receiver, faulty cables or connectors and the bias error from the receiver's quadrature detectors. The first source of leakage is usually suppressed using high-quality receivers with a good isolation between channels. In the second, the faulty components with poor isolation act as new transmitters interfering in the far-field pattern. In the last one, the bias error introduces an imbalance between the two channels, adding a complex constant to every near-field data sample very difficult to remove from the near-field data because the level of that constant is usually 60–70 dB below the maximum near-field level. However, this error is augmented after the planar near-to-far-field transformation since all the contributions are sum in the broadside direction.

During the last years, several methods have been proposed to detect and cancel leakage signals. Leakage due to faulty components can be detected terminating the lines connected to the AUT or the probe and measuring the signal picked up by the receiver [50]. Then, one easy way to cancel leakage sources is to replace those elements or to wrap coaxial connections with electromagnetic material until the leakage level was acceptable. However, this technique requires several measurements to detect and cancel the effect of those faulty components. Other alternatives try to reduce the undesired effects without any extra measurement using analytical compensation techniques [51]. These techniques mathematically suppress the effect of extraneous fields created by reflections or leakage sources, increasing the accuracy of the radiation pattern. For the case of the leakage from the receiver bias error, it is not possible to reduce changing the instrumentation (usually it is in the receiver itself). The different methods to cancel this kind of leakage predict the constant added to the measured data, and that estimated constant is subtracted from the measured data. If that constant has been estimated correctly, its effect on the far field should disappear. There are options to determine that constant averaging all the measured data below a threshold [52] or located outside a certain region [53] in the near-field data.

The methods presented in this chapter are the ones used to cancel leakage from cables. These methods were initially proposed in [54] and use diagnostic technique to determine the field distribution over the AUT plane.

4.2 Theory of holographic techniques in electromagnetics

This technique is called in the literature backward-transformation technique, and it is the most employed in antenna diagnostic technique due to its mathematical simplicity and low processing time. Holographic technique is based on the direct relationship between the field acquired in the forward hemisphere of the far-field region and the visible portion of the PWS [55]. Then, the field distribution over the AUT plane is calculated performing a Fourier transform to the PWS.

This technique is applicable to a large range of antennas [56,57] for diagnostics purposes, because it provides satisfactory results and the simplicity and computationally efficiency of the FFT algorithms. The main limitation is the resolution that can be achieved: this spatial resolution in the extreme near field is limited to half wavelength. In the last years, several methods to extrapolate the information in the invisible region from the information contained in the visible region have been proposed. The most successful is based on the Gerchberg–Papoulis algorithm, initially extended to the electromagnetic problems by Rahmat-Samii [58]. This algorithm was validated using simulated data by Narasimhan and Preetham Kumar [59], using measured data by Johansson and Svensson [60]. Also, as is presented in [61], for radial line slots antennas, the use of information related the position, length and tilt of the slots allows enough quality to reconstruct antenna sources separating a quarter wavelength.

In the case of planar near-field measurements, the PWS is obtained performing an inverse Fourier transform of the measured samples. Since this PWS is referred to the measurement plane, a back-propagation to the antenna aperture plane is used to determine the appropriate PWS [2,43]. Next, the sources or equivalent currents are obtained performing the Fourier transform of the PWS. In practical approaches, the PNF measurement is performed at a few wavelengths from the AUT; this makes that the contributions coming from the evanescent modes negligible. Therefore, there is no information about the PWS in the invisible region, as in the far-field case. Therefore, the spatial resolution cannot be lower than half a wavelength. Finally, since the measured field is acquired on a finite plane, the PWS is only reliable in an angular range that depends on the dimensions of the scan and AUT and the distance between AUT and plane. Therefore, a truncation error appears in the extreme near-field and the resolution is worse than in a general case: the separation between reliable samples is higher than half wavelength.

In SNF acquisitions, the classical method is first the calculation of the far field, second the calculation of the PWS from the far field and then the application of the FFT to calculate the equivalent sources from that PWS. In this case, the information contained in the evanescent modes is lost in the near-to-far-field transformation and, as a consequence, the spatial resolution obtained in the diagnostic process is limited to half a wavelength, as stated before. However, Devaney and Wolf [62] demonstrated that a rigorous transformation from the spherical wave expansion to the plane wave expansion was possible, being able to retrieve the information of the

PWS both in the visible and invisible region from the knowledge of the coefficients of the spherical wave expansion. In 2007, Capellin [63,64] proposed the use of this transformation as the basis of a diagnostic technique for SNF measurements without moving to far field, and, therefore, without losing the spatial resolution thanks to the information that can be obtained from the spherical wave coefficients (SWC) in the invisible region of the PWS.

The relation between field on the aperture and PWS is shown in [2]. Once the PWS is obtained, the calculation of the field over the antenna aperture is directly obtained through a Fourier transform (4.1):

$$\vec{E}_{ap}(x,y) = \frac{1}{2\pi}\int_{-\infty}^{\infty}\int_{-\infty}^{\infty}\vec{P}(k_x,k_y)\cdot e^{-j(k_x x+k_y y)}dk_x dk_y \tag{4.1}$$

where $\vec{P}(k_x,k_y)$ stands for the electric field PWS in the (k_x,k_y) direction.

In planar near-field acquisitions, the PWS is directly obtained from the acquired samples through a Fourier transform. Of course, this PWS is referenced to the measurement plane, $\vec{P}(k_x,k_y,z=d)$, and a back-propagation to the AUT plane has to be used during the process to determine the appropriate PWS:

$$\vec{P}(k_x,k_y) = \vec{P}(k_x,k_y,z=d)e^{jk_z d} \tag{4.2}$$

where d is the distance between the planar scanner and the AUT.

For the SNF case, as it was mentioned before, a transformation from a spherical wave expansion to a plane wave expansion (SWE-to-PWE) is required. This transformation was developed by [63] and relates spherical and plane coefficients:

$$\vec{P}(k_x,k_y) = \sum_{n=1}^{\infty}\sum_{m=-n}^{n}\left\{Q_{1mn}^{(4)}\vec{T}_{1mn}(k_x,k_y) + Q_{2mn}^{(4)}\vec{T}_{2mn}(k_x,k_y)\right\} \tag{4.3}$$

where $Q_{1mn}^{(4)}$ and $Q_{2mn}^{(4)}$ are the spherical coefficients calculated from the measured data and $\vec{T}_{1mn}(k_x,k_y)$ and $\vec{T}_{2mn}(k_x,k_y)$ are the functions detailed in [63].

When the measurement is performed in a cylindrical scanner, there is not any direct relation between the cylindrical samples and the PWS. In this case, it is necessary to calculate the far field, and from that far field to calculate the PWS solving (4.4):

$$\begin{cases} E_\theta(r,\theta,\phi) = j\dfrac{k_0 e^{-jk_0 r}}{r}\left\{\cos\phi\cdot P_x(k_x,k_y) + \sin\phi\cdot P_y(k_x,k_y)\right\} \\ E_\phi(r,\theta,\phi) = j\dfrac{k_0 e^{-jk_0 r}}{r}\cos\theta\left\{\cos\phi\cdot P_y(k_x,k_y) - \sin\phi\cdot P_x(k_x,k_y)\right\} \end{cases} \tag{4.4}$$

where $E_\theta(r,\theta,\phi)$ and $E_\phi(r,\theta,\phi)$ are the far-field components, and $P_x(k_x,k_y)$ and $P_y(k_x,k_y)$ are the PWS components.

Therefore, microwave holographic technique can be used in far-field region or with any kind of classical near-field acquisition, after applying the corresponding near-to-far-field transformation [65].

The FFT algorithm is used to calculate the field distribution over the AUT plane from the PWS. FFT is computationally very efficient, but it requires samples distributed on a regular grid in both domains: the samples of the extreme near-field are obtained on a regular x–y grid, and the PWS samples must be known on a regular k_x–k_y grid (or u–v grid). In the PNF case, the PWS is directly obtained in that required grid from the acquired fields after applying the inverse FFT on a regular x–y grid in the scan plane. In the general case, when the algorithm uses the far-field information, the samples are given on a regular θ–ϕ grid. In those cases, the usual technique is to perform a two dimensions interpolation to calculate the samples in the required grid. This solution introduces an interpolation error that has to be taken into account, but it does not require a great computational cost. There are other studies that improve these interpolation methods, reducing the corresponding error.

4.3 Noise reduction using holographic techniques

Noise always appears in electromagnetic measurements. In particular, in antenna measurements, the noise acquired in near field is propagated to far field when the near-to-far-field transformation is applied. The near-field-acquired noise comes from data digitalization or the receiver additive noise. In most of the applications, this random noise is not the dominant factor due to the use of modern receivers and a well-designed radiofrequency system. However, in some cases, the noise may limit the accuracy of the far-field pattern, as it happens in measurements of very low side lobe level (SLL), low cross-polar or the measurement at millimetre or submillimetre wavelengths and beyond. In those cases, the noise is one of the errors to take into account the uncertainty budget.

The method proposed in this section to reduce noise is based on filtering it in the AUT domain, after obtaining the equivalent currents using holographic techniques. Since, in the initial domain, the noise and signal are overlapped, the noise cannot be filtered out. In this new domain, the noise can be reduced without cancelling the contribution of the AUT (see Figure 4.1). In the AUT plane, the desired contribution coming from the AUT is concentrated in the region where the antenna is located. In contrast, the noise contribution is spread over the whole plane, including the AUT surface, as it is explained in [10].

In the case of the planar near field, due to the linearity of the operators involved in the holographic transformation, the noise and the AUT signal can be analysed independently. Once the planar near-field noise has been back-propagated, the noise is complex, white Gaussian, with zero mean and constant variance [10]. In that case, the noise is stationary since its power is equally distributed on the reconstructed surface. On the other hand, the desired contribution from the AUT is concentrated in the region where the AUT is placed. Therefore, the noise is reduced if a spatial filtering is applied in this new domain, keeping the information of AUT equivalent currents. The increase in the SNR depends on the ratio between the scan surface (back propagated to AUT plane) and the AUT surface. In a practical

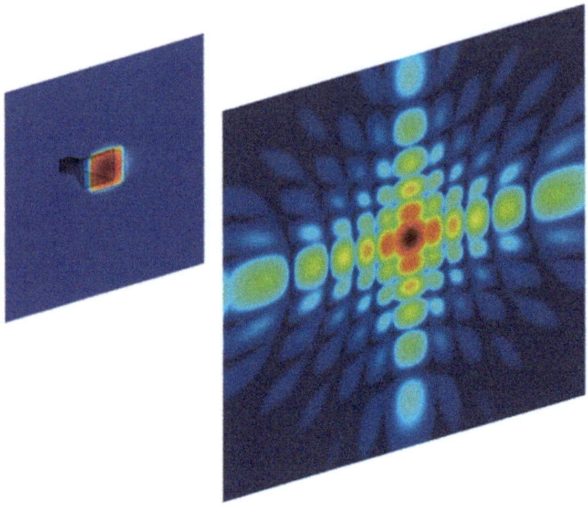

Figure 4.1 Measured planar field and field propagated to the AUT plane

situation, the scan plane is much larger than the AUT surface [43]. The definition of filtering is explained in (4.5), where the required *a priori* information is the geometry and the position of the AUT:

$$F(x,y) = \begin{cases} 1(x,y) \in \Omega_{AUT} \\ 0(x,y) \in \Omega_R - \Omega_{AUT} \end{cases} \tag{4.5}$$

where Ω_R and Ω_{AUT} represent the reconstructed region and the AUT region depicted in Figure 4.2:

$$\Delta k_x = \frac{2\pi}{\Delta x M_x} \tag{4.6}$$

$$\Delta k_y = \frac{2\pi}{\Delta y M_y} \tag{4.7}$$

where M_x and M_y represent the number of planar near-field samples in the x- and y-directions and Δx and Δy are the sampling steps in the mentioned directions.

The signal level in the PWS due to this filtering does not change because the desired contribution is not suppressed, since AUT currents are concentrated on the AUT plane. As a result, the SNR improvement is identical to the noise reduction. The original noise power in the PWS is given by

$$N_{wo\_filter} = M\left(\frac{\Delta x \Delta y}{2\pi}\right)^2 \sigma^2 \tag{4.8}$$

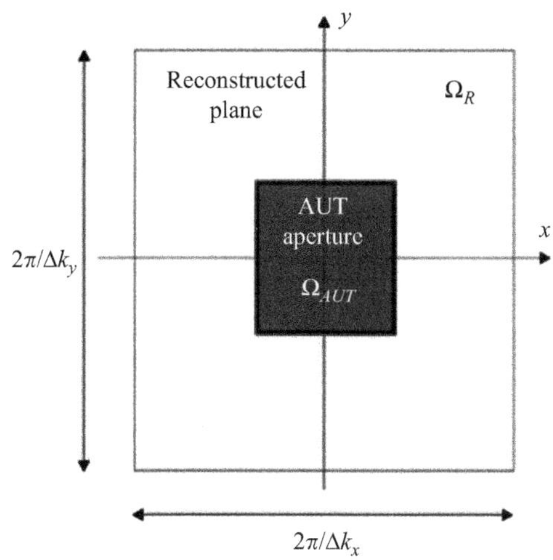

Figure 4.2 Regions used to define the spatial filtering on the AUT plane in PNF

where M is the total number of samples in the planar acquisition and σ^2 represents the variance of the near-field noise. After the spatial filtering, the noise power of the new PWS is reduced to this expression:

$$N_{w\_filter} = M\left(\frac{\Delta x \Delta y}{2\pi}\right)^2 \frac{\Omega_{AUT}}{\Omega_R}\sigma^2 \tag{4.9}$$

Therefore, the SNR improvement achieved with the proposed spatial filtering is

$$\Delta SNR = \frac{S/N_{w\_filter}}{S/N_{wo\_filter}} = \frac{\Omega_R}{\Omega_{AUT}} \tag{4.10}$$

In fact, (4.10) calculates the improvement for both PWS components. Since far-field components are obtained as a linear combination of the PWS components, the improvement obtained in the far field is also given by the same expression (4.10).

Even if theoretically, it would be possible to obtain an infinite reconstructed surface on the AUT plane; in a practical way, there are some limits. By one side, increasing the reconstructed surface is equivalent to reducing the step in the spectral domain. If the reconstructed surface has to be larger than the scanned surface, the spectral step required is smaller than that one obtained when inverse Fourier transforming the measured data and an interpolation process are needed. That interpolation introduces a correlation among noise samples in the PWS and the result or (4.10) is not valid.

On the other hand, the assumption that the desired field is concentrated on the AUT region is not completely correct. Always, a small-field contribution exists outside the AUT. In fact, since the measurement is performed in planar near-field, there is a truncation error that expands the field over the AUT plane when the measured samples are back-propagated. In the cases where a filtering as defined in (4.5) is applied, a portion of the desired signal is cancelled, introducing an error, mainly in the side lobes. This effect is avoided applying the spatial filtering over a larger area than AUT, integrating more noise power. In conclusion, a compromise between the noise reduction and an accurate far-field representation is required.

Figure 4.3 shows the steps used for this noise reduction using holographic techniques.

The steps in the algorithm are as follows:

- The PNF data is acquired in a planar scanner.
- The PWS (referenced to the scan plane) is calculated through an inverse Fourier transform.
- The PWS is back-propagated to the AUT plane.
- The sources are reconstructed on the AUT plane performing the Fourier transform of the previous PWS.
- The field distribution is spatially filtered cancelling all the samples outside the AUT aperture.
- The AUT field is inverse Fourier transformed to the spectral domain, obtaining a PWS where the noise power is reduced.
- The far-field components are obtained from the new PWS.

To validate the proposed method, an example that uses information of an antenna measurement is performed in the PNF range of Universidad Politécnica de Madrid (UPM). In this experiment, the probe is a corrugated conical-horn antenna and the AUT is a 5 by 7 cm pyramidal-horn antenna. The antennas were separated 1.57 m and, a measurement over a planar scanner of 240 $\times$ 240 cm^2 (equal to 104λ by 104λ at 13 GHz) with a spatial sampling in x and y equal to 1 cm (0.43λ at 13 GHz) was performed. Gaussian noise with a power of 30 dB lower than the maximum of the acquired data was mathematically added to the measured field. This high noise power was introduced to be much higher than the actual measurement noise. The far field obtained from the measured data without additive noise can thus be used as a reference to compare results before and after noise filtering.

Figure 4.4(a) and (b) shows the reconstructed field for the acquired data without and with added noise. The noise, in the second case, is equally distributed

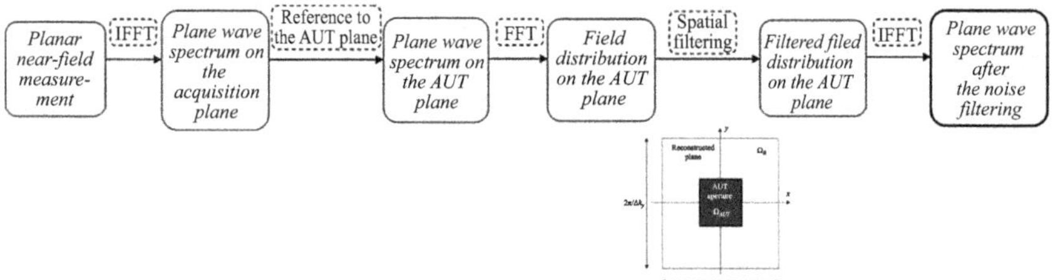

Figure 4.3 Diagram of noise reduction based on an AUT spatial filtering from PNF measurements

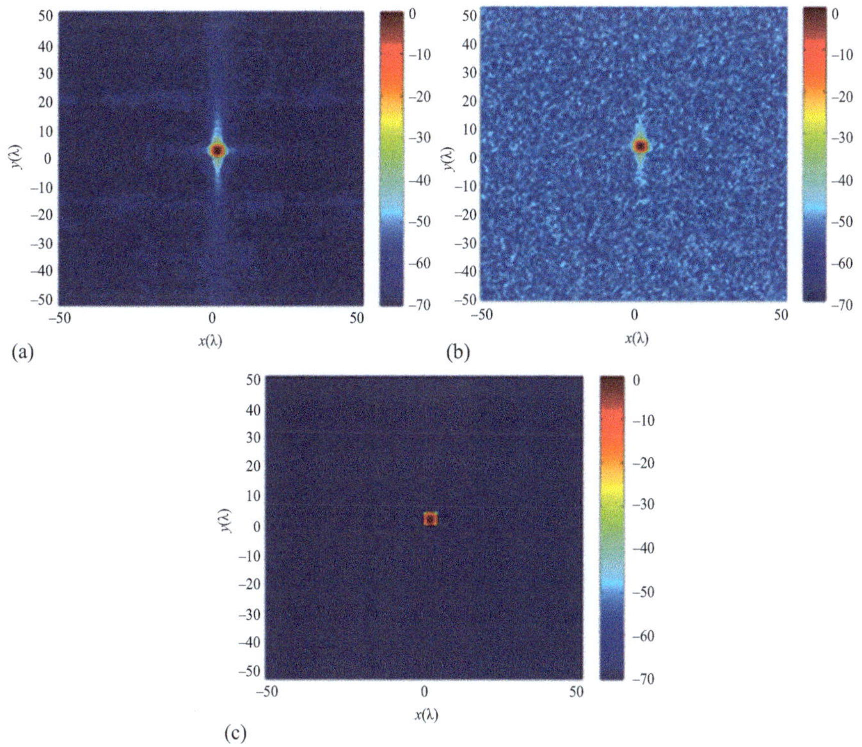

Figure 4.4 Extreme near-field of the pyramidal-horn obtained from the planar scanner without additive noise (a), with additive noise without spatial filtering (b) and with spatial filtering (c) over a 240 cm × 240 cm reconstructed surface

over the whole reconstructed surface, as predicted in the theoretical analysis [10]. Also, it is noticed that the AUT contribution extends beyond the AUT aperture, as it was explained before. Therefore, the applied window is larger than the antenna aperture. The extreme near-field filtered using this spatial filtering is depicted in Figure 4.4(c). In these cases, the size of the reconstructed surface is equal to the size of the scan surface, 240 by 240 cm. Therefore, the SNR improvement will be close to the maximum improvement that can be obtained.

Figure 4.5 illustrates the far-field patterns obtained in the three previous cases (without additive noise, with additive noise and after spatial filtering). Due to the finite size of the scanner, the pattern is not known in the whole visible region. Therefore, the filtering is able to get the reference pattern with good accuracy only within the reliable region (with similar quality of the radiation pattern calculated with the reference data). Iterative techniques would be used to retrieve the pattern in the whole visible region.

Figure 4.6 shows the horizontal cut of the pattern of the measured antenna, where it is observed a great improvement in both the co-polar and the cross-polar

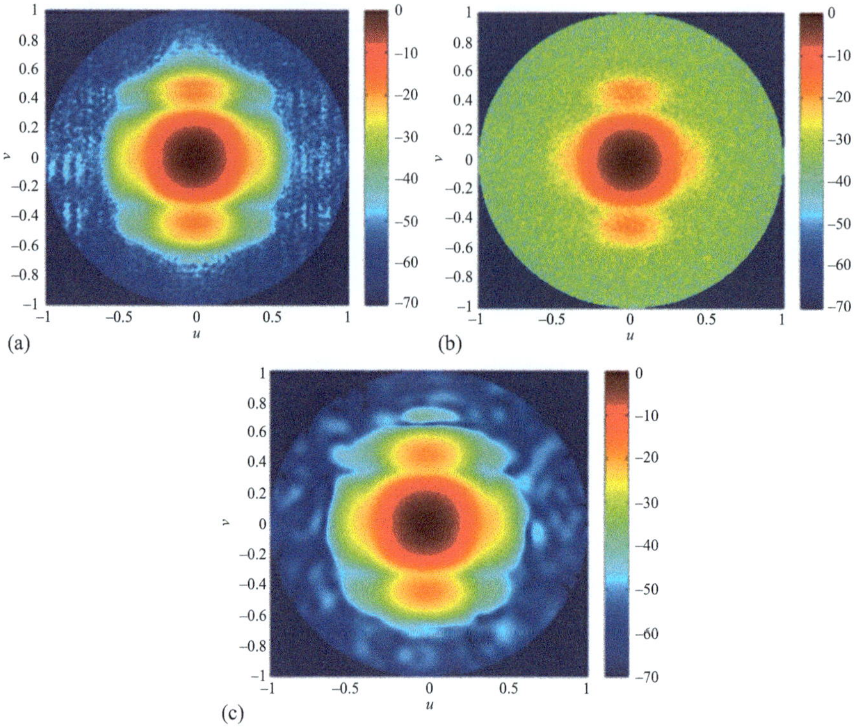

Figure 4.5 Reference far-field pattern (a), far-field pattern with additive noise (b) and far-field pattern with noise filtering on a 240 by 240 cm reconstructed surface (c)

components. For this particular case, the SNR improvement defined by (4.10) is 32.2 dB.

The weighted logarithmic difference [66] is used to compare the similarity of the previous patterns with the reference one, and the difference is plotted in Figure 4.7 for the horizontal plane. Also, the error is reduced from 0.35% to the very low value of 0.012% in the valid region.

Figure 4.8 shows the effect of the SNR improvement with the reconstructed surface. It is observed that the theoretical improvement defined in (4.10) is limited to the maximum size of the scan surface, as it was mentioned before. Therefore, two conclusions can be extracted from this figure. First, the theoretical expression deduced from the statistical analysis gives a very good estimation of the improvement achieved with the method when the size of the reconstructed surface is small. And second, when the reconstructed surface is larger than the scan area, the improvement remains constant and the theoretical expression is no longer valid. In fact, the size from which the SNR improvement does not increase any more is slightly larger than the size of the scan plane, and this is due to the interpolation errors, since this interpolation introduces a certain correlation among noise samples, and consequently, the

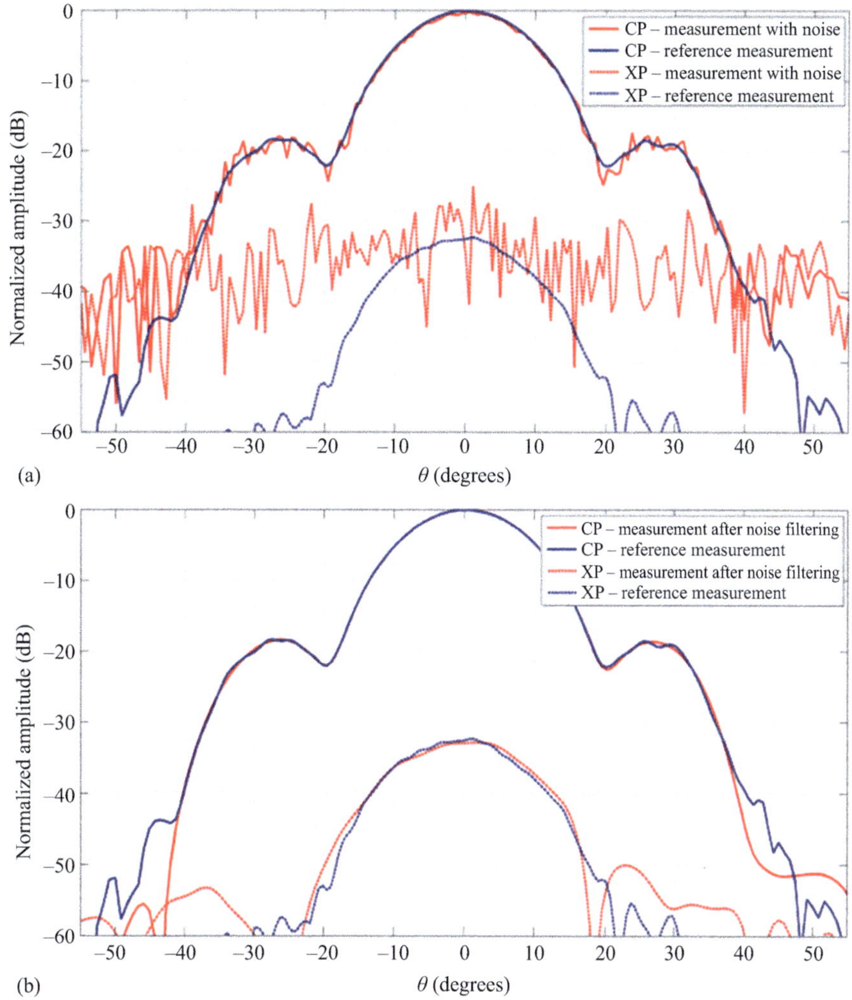

Figure 4.6 Far-field comparison between the reference pattern and the noisy pattern (a) and the reference pattern and the noise filtered pattern (b) for the $\phi = 0°$ cut

noise in the extreme near-field is not stationary (the noise power is not constant in all the samples) reducing the effectiveness of the spatial filtering.

A similar procedure can be applied in CNF measurements, applying the field transformation that allows filtering out a portion of noise without modifying the desired signal. In this example, the far field is calculated from the CNF. Then, the PWS can be easily calculated from the far field and the extreme near-field is obtained through the Fourier transform of that spectrum.

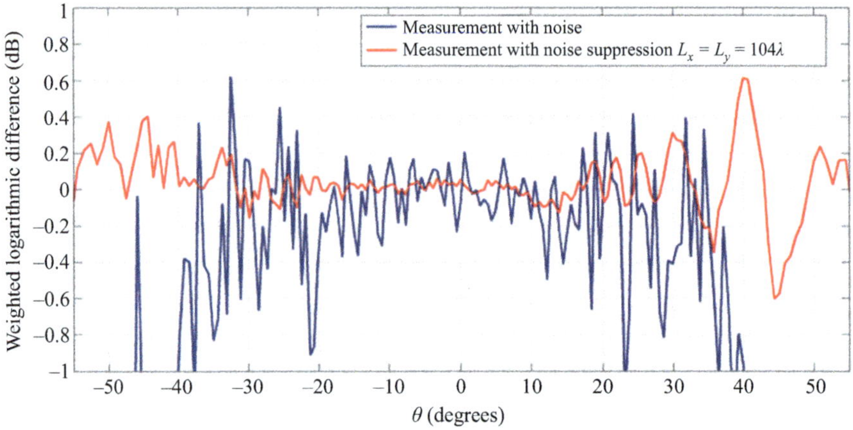

Figure 4.7 Weighted logarithmic difference between the far-field patterns before and after applying noise filtering for the horizontal cut

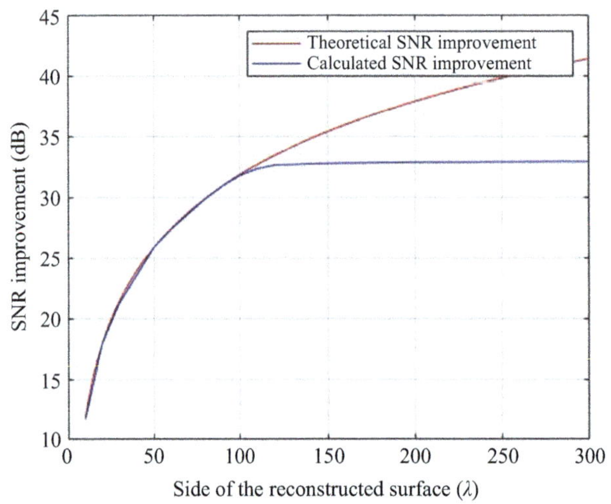

Figure 4.8 Theoretical (Equation (4.10)) and calculated SNR improvement

An exhaustive statistical study of the extreme near-field noise obtained through the cylindrical transformation of the measured near-field noise over a cylinder was presented in [67]. From this study, it was concluded that the noise in the extreme near-field is correlated in both x and y coordinates. However, all the samples have the same noise power, and the noise is stationary and identically distributed over the whole reconstructed surface. Thanks to this latter characteristic, a filtering similar to the one presented in (4.5) can be applied and similar expression for the SNR improvement will be achieved (4.10).

The schematic diagram for the CNF case is depicted in Figure 4.9.

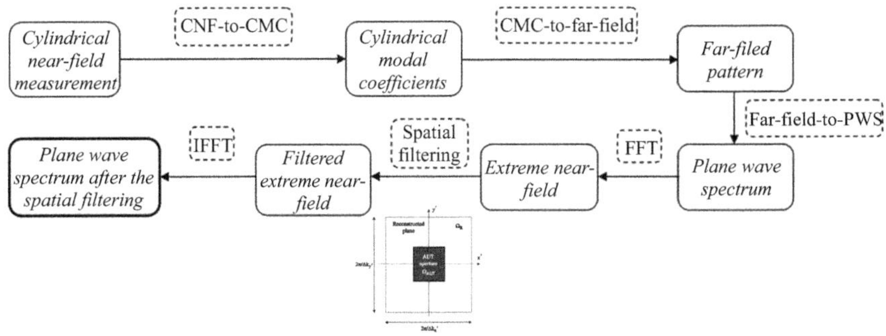

Figure 4.9 Diagram of the noise reduction method for CNF measurements based on a spatial filtering on the AUT plane

The steps of this method are as follows:

- The CNF samples are acquired.
- The cylindrical mode coefficients are calculated from the measured CNF.
- The far-field components are calculated using the cylindrical mode coefficients. The components obtained in this way are known on a regular $k_z - \phi$ grid.
- The PWS components are calculated from the far-field components of the previous point on a regular $k_{x'} - k_{y'}$ grid. For this transformation, it is necessary a change of coordinates and an interpolation to the desired size of the reconstructed surface.
- The field distribution on the AUT plane is obtained using a Fourier transform of the PWS.
- The field distribution is spatially filtered cancelling all the samples located outside the AUT aperture.
- The PWS is calculated through an inverse Fourier transform of the field distribution.
- The far-field components are calculated from the PWS components.

The main drawback of this method is the impossibility of improving the SNR in all the region, since the use of the PWS limits the information of the far field to the forward hemisphere.

This noise-reduction method is validated through the measurements of a Ku-band reflector antenna working at 14 GHz. The process is equivalent to the previous case: the original pattern is used as reference, and this information is corrupted with additive white noise. Then the spatial filtering is applied to improve the SNR. The equivalent currents of this antenna are obtained from the three cases, and they are depicted in Figure 4.10.

In Figure 4.10(a), it is possible to see the field diffracted at the edges of the reflector. This field contribution is filtered out in Figure 4.10(c) and is not taking into account to calculate the new far-field pattern. Therefore, the exact far-field pattern cannot be retrieved. In this case, it would have been preferable to use one of

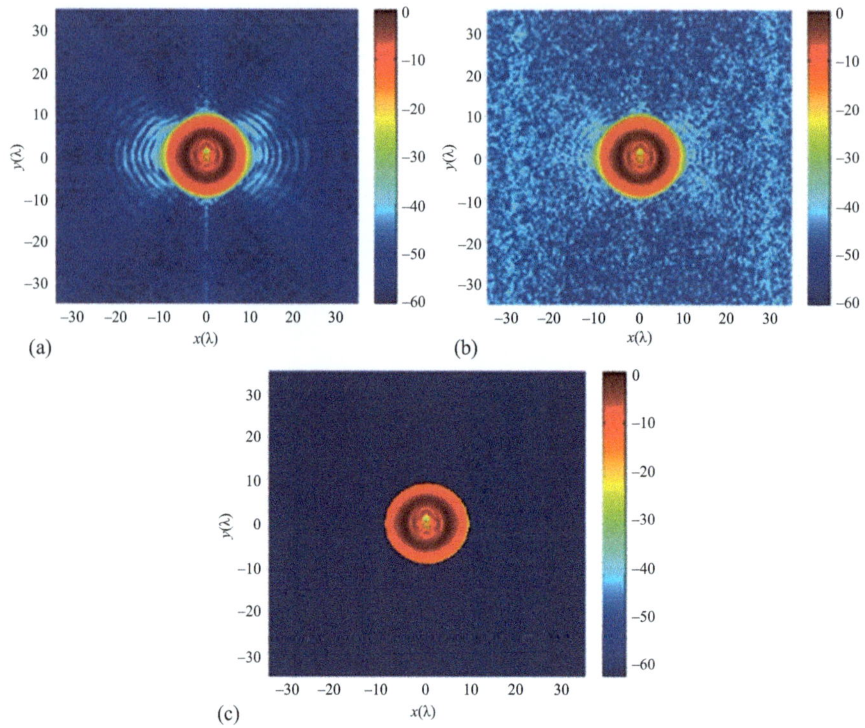

Figure 4.10 Extreme near field of a Ku-band reflector antenna obtained from the CNF data reference case (a), with additive noise (b) and with spatial filtering (c) over a 150 cm × 150 cm (70λ by 70λ at 14 GHz) reconstructed surface

the adaptive filtering defined in [61]. Figure 4.11 shows the co-polar and cross-polar far-field patterns for the three cases.

Despite the diffracted field in the edge is not well reconstructed, it is evident that there is an important noise reduction and the retrieved far-field has a very good accuracy. Figure 4.12 compares the horizontal patterns for the co-polar far-field pattern before and after applying the noise filtering. The weighted logarithmic difference is represented in Figure 4.13 in order to quantify the noise reduction of in the far-field pattern. The per cent error between the noisy and filtered patterns with respect to the reference pattern is reduced from 0.25% to 0.03%. Finally, the average SNR improvement is 12.5 dB.

A similar noise-reduction procedure can be applied when the measurement has been performed on an SNF system.

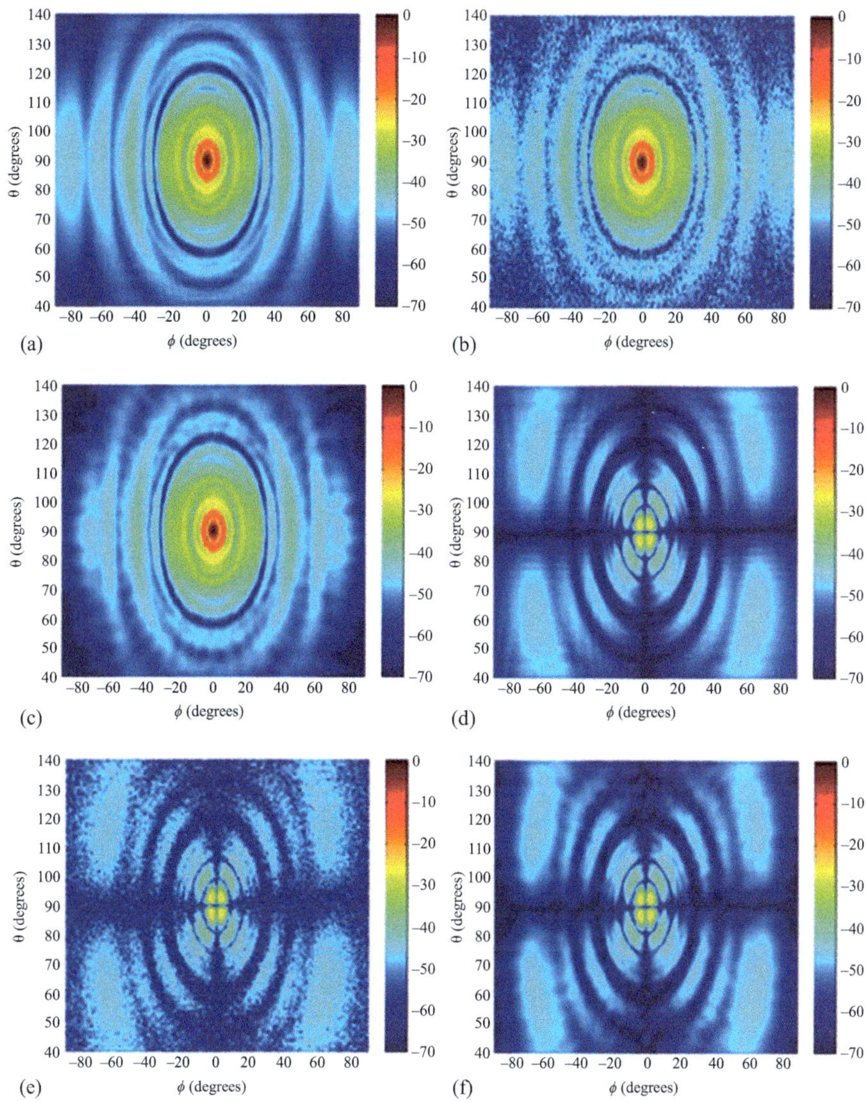

Figure 4.11 *Reference co-polar far-field pattern (a), co-polar far-field pattern with noise (b), co-polar far-field pattern with noise filtering (c), reference cross-polar far-field pattern (d), cross-polar far-field pattern with noise (e) and cross-polar far-field pattern with noise filtering (f) of the Ku-band reflector antenna at 14 GHz*

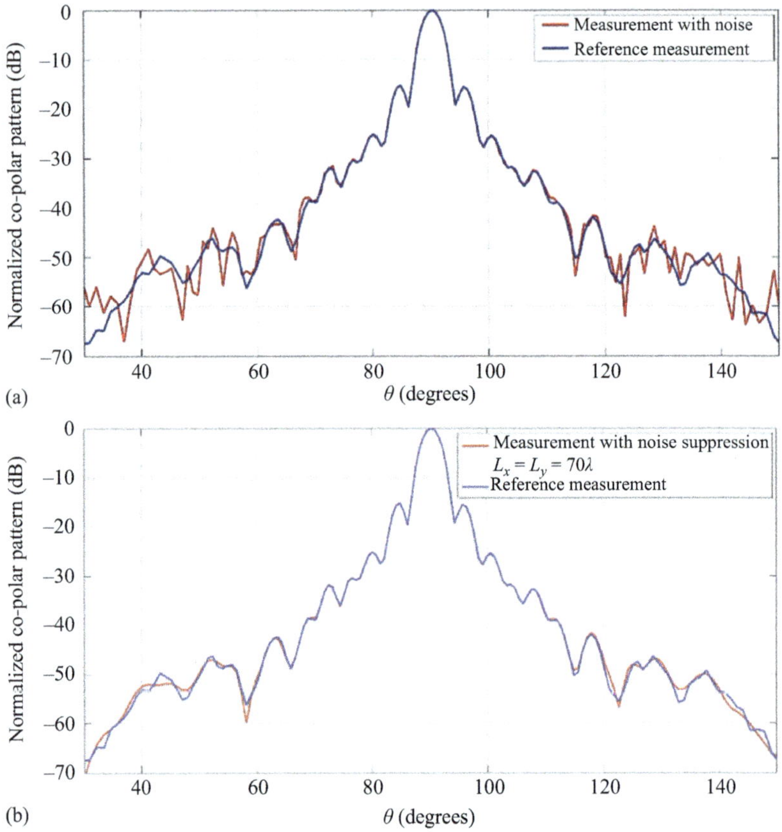

Figure 4.12 Far-field comparison between the reference pattern and the noisy pattern (a) and the reference pattern and the filtered pattern (b) for the horizontal cut

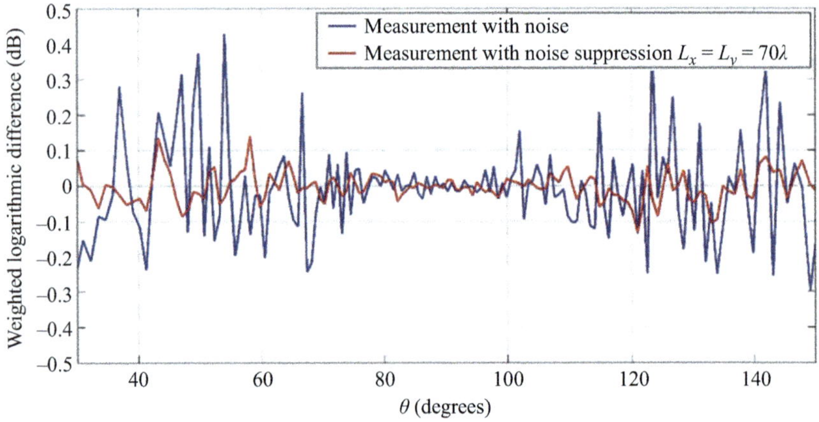

Figure 4.13 Weighted logarithmic difference between the noisy and filtered far-field pattern and the reference far-field pattern for the horizontal cut

4.4 Echo reduction using holographic techniques

Antenna measurements are usually carried out in anechoic chambers to reduce as much as possible unwanted contributions, such as reflections or diffractions from the environment. However, there are special cases, as the case of measurement of large antennas, in which the use of those measurement set-ups is not possible. In those cases, a semi-anechoic chamber or an outdoor measurement system is usually employed. In those situations, the final results are affected by the presence of undesired components coming mainly from the ground, fences or other objects in the environment. Echoes may also appear in anechoic chambers due to imperfections in the facility or simply because the measurement frequency is below the minimum working frequency of the absorbing material. In any case, when the measurement is not performed in a fully anechoic environment, the unwanted contributions could significantly modify the actual antenna properties, producing some ripples in the radiation pattern.

During the last years, the number of approaches to analyse and to cancel the effects of reflections has increased. There are also methods that are employed to characterize the chambers by a figure of merit, usually called the reflectivity level [14–16].

In this chapter, the method for cancelling the echoes in antenna measurements is based on diagnosis technique on a huge surface area and filtering out the contributions from image sources [68]. In these methods, the data are taken over a surface in frequency domain, and, in contrast to the time-gating techniques or frequency decomposition techniques, it is not necessary to measure more than one frequency, reducing considerably the measurement time and employing standard vector network analysers.

As in all the applications presented in this work, this echo suppression method is based on microwave holographic techniques, obtaining the reconstructed field over a surface larger than the AUT size. Thus, it is possible to identify virtual sources out of the antenna aperture, which appear due to the presence of reflections. This can be explained using the image theory (Figure 4.14). Cancelling the virtual or unwanted sources, the echoes can be suppressed, and, finally, employing this filtered reconstructed field, the radiation pattern without echoes is obtained.

It is remarkable that the identification of the virtual sources is not possible using a conventional holographic diagnostic technique, where the field is only reconstructed over the antenna aperture. However, if the field reconstruction is performed over a surface larger than the antenna dimensions, the aforementioned fictitious sources can be found and cancelled with a spatial filtering. Thus, the proposed method has to calculate the reconstructed field in a region whose size in each direction depends on the distance between the AUT and the potential reflective surfaces, being at least twice as large as that distance, to ensure the correct image sources detection. With that previous concept in mind, the procedure to obtain the reconstructed field over a larger surface is the reduction of the spectral steps, using an interpolation in the PWS domain.

Applying the appropriate interpolation to the PWS samples, the extended reconstructed field can be found. In this reconstructed domain, real sources and

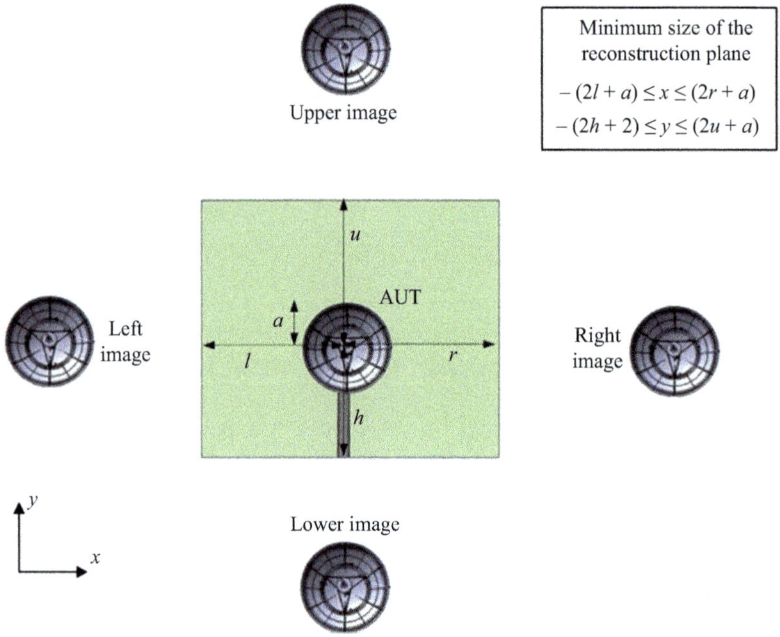

Figure 4.14 Image theory in a general case with reflections on the floor, the ceiling and both side walls

virtual sources are not spatially coincident, so each of them belongs to a different subspace, B_{real} or $B_{virtual}$, respectively:

$$B_{real} = \left\{ \vec{s}(x,y) \in L_R : \vec{s}(x,y) = \vec{0}, \forall (x,y) \in Z_V \right\} \tag{4.11}$$

$$B_{virtual} = \left\{ \vec{s}(x,y) \in L_R : \vec{s}(x,y) = \vec{0}, \forall (x,y) \notin Z_V \right\} \tag{4.12}$$

where $\vec{s}(x,y)$ is a generic reconstructed field within the space of the reconstructed signals represented by L_R, and Z_V represents the zone where virtual sources are placed. Therefore, any extended reconstructed field can be decomposed as follows:

$$\begin{aligned} &\vec{E}_{ap}(x,y) = \vec{E}_{real}(x,y) + \vec{E}_{virtual}(x,y) \\ &with\ \vec{E}_{real}(x,y) \in B_{real}, \vec{E}_{virtual}(x,y) \in B_{virtual} \end{aligned} \tag{4.13}$$

where B_{real} and $B_{virtual}$ are orthogonal subspaces, as demonstrated in (4.14):

$$\begin{aligned} \langle \vec{s}_1(x,y), \vec{s}_2(x,y) \rangle &= \iint_{Z_R} \vec{s}_1^{\,*}(x,y) \cdot \vec{s}_2(x,y) dxdy \\ &= \iint_{Z_R - Z_V} \vec{s}_1^{\,*}(x,y) \cdot \underbrace{\vec{s}_2(x,y)}_{=\vec{0} \text{ in } Z_R - Z_V} dxdy \\ &\quad + \iint_{Z_V} \underbrace{\vec{s}_1^{\,*}(x,y)}_{=\vec{0} \text{ in } Z_V} \cdot \vec{s}_2(x,y) dxdy = 0 \\ &\quad \forall\, \vec{s}_1(x,y) \in B_{real}, \forall\, \vec{s}_2(x,y) \in B_{virtual} \end{aligned} \tag{4.14}$$

where Z_R is the complete region where the field is reconstructed.

Therefore, $\vec{E}_{real}(x,y)$ and $\vec{E}_{virtual}(x,y)$ are orthogonal, and thus an orthogonal projection operator can be applied to $\vec{E}_{ap}(x,y)$ to obtain each of those components separately. In this case, it is desirable to get $\vec{E}_{real}(x,y)$ from $\vec{E}_{ap}(x,y)$, which is achieved by applying $P_{B_{real}}$ to $\vec{E}_{ap}(x,y)$:

$$P_{B_{real}}\left(\vec{E}_{ap}(x,y)\right) = \begin{cases} \vec{E}_{ap}(x,y) & \forall (x,y) \notin Z_V \\ 0 & \forall (x,y) \in Z_V \end{cases} \tag{4.15}$$

This process of suppressing virtual sources can also be viewed as a spatial filtering. The filter, $F^{spatial-AUT}(x,y)$, which is applied to the reconstructed field, is easily deduced from the orthogonal projection operator:

$$\begin{aligned} P_{B_{real}}\left(\vec{E}_{ap}(x,y)\right) &= F^{spatial-AUT}(x,y) \cdot \vec{E}_{ap}(x,y) \\ F^{spatial-AUT}(x,y) &= \begin{cases} 1 & \forall (x,y) \notin Z_V \\ 0 & \forall (x,y) \in Z_V \end{cases} \end{aligned} \tag{4.16}$$

Since the echoes are directly related to those virtual sources, once the sources are cancelled out, the reflections are suppressed, and the radiation pattern can be improved. In a practical way, all the sources not coming from AUT surface can be filtered out, having also the effect of noise cancellation seen in the previous section. Then, the corrected PWS is obtained by taking the inverse Fourier transform of this filtered field distribution. And, finally, the far-field pattern is calculated from the last PWS through the asymptotic evaluation of the plane-wave expansion.

In summary, this method is based on a modification of the existing diagnostic techniques to determine the field distribution on a surface larger than the AUT dimensions. In this surface, the desired contribution coming from the AUT and the contribution of the reflections are spatially non-coincident, and, therefore, the reflection suppression is just the application of a spatial window to remove the contribution related to the potential echoes.

To verify the accuracy of the proposed method, several experimental tests were performed at the antenna measurement facilities of the UPM. The measurements were obtained using the UPM planar scanner. For the experiment, the probe and the AUT were the two horns used in the experiment of the noise section: a corrugated conical-horn antenna and a pyramidal-horn antenna, respectively, and they were separated from each other by 1.57 m. Now the pyramidal-horn is placed vertically to better appreciate the effects. The reference measurement over a 2.4 m × 2.4 m acquisition plane was performed in order to have a pattern with which to compare the results with echoes and with echoes suppression. For the echoes, a rectangular metallic plate was placed in the planar system (see Figure 4.15) to introduce large reflections in the measurement. The AUT is placed at a height of 1.3 m above the plate. Figure 4.16 shows the measured planar near-field in both cases, without (reference) and with the reflective surface (corrupted).

Figure 4.15 Experimental set-up with a metallic plate in the planar scanner of UPM

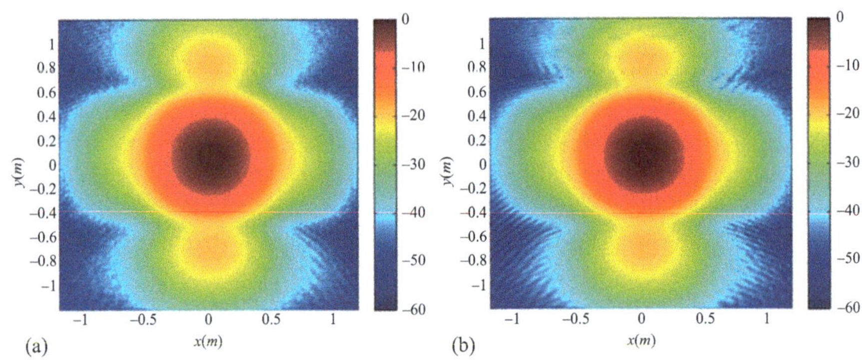

Figure 4.16 Measured planar near-field data of the pyramidal-horn antenna without the metallic plate – reference (a) and with the metallic plate – corrupted (b)

Figure 4.17(a) and (b) shows the extreme near field and the far-field patterns for the three cases: reference measurement, pattern with the large reflector and filtered pattern. In all the cases, the AUT contribution is located in the centre, and all the contributions that extend beyond the AUT dimensions are due to the truncation of the measurement surface and the echoes (when exist). These echoes are observed in the lower part of the reconstructed plane, since the metallic plate is in floor. The filtered reconstructed field (Figure 4.17(c)) cancels out those virtual sources.

Figure 4.17(d)–(f) shows the radiation patterns in the three cases. However, it is difficult to appreciate the improvement achieved with the method, even, clearly, some echoes are cancelled. A better comparison is presented in Figure 4.18; first, the far-field pattern without reflection suppression is compared with the reference far-field pattern

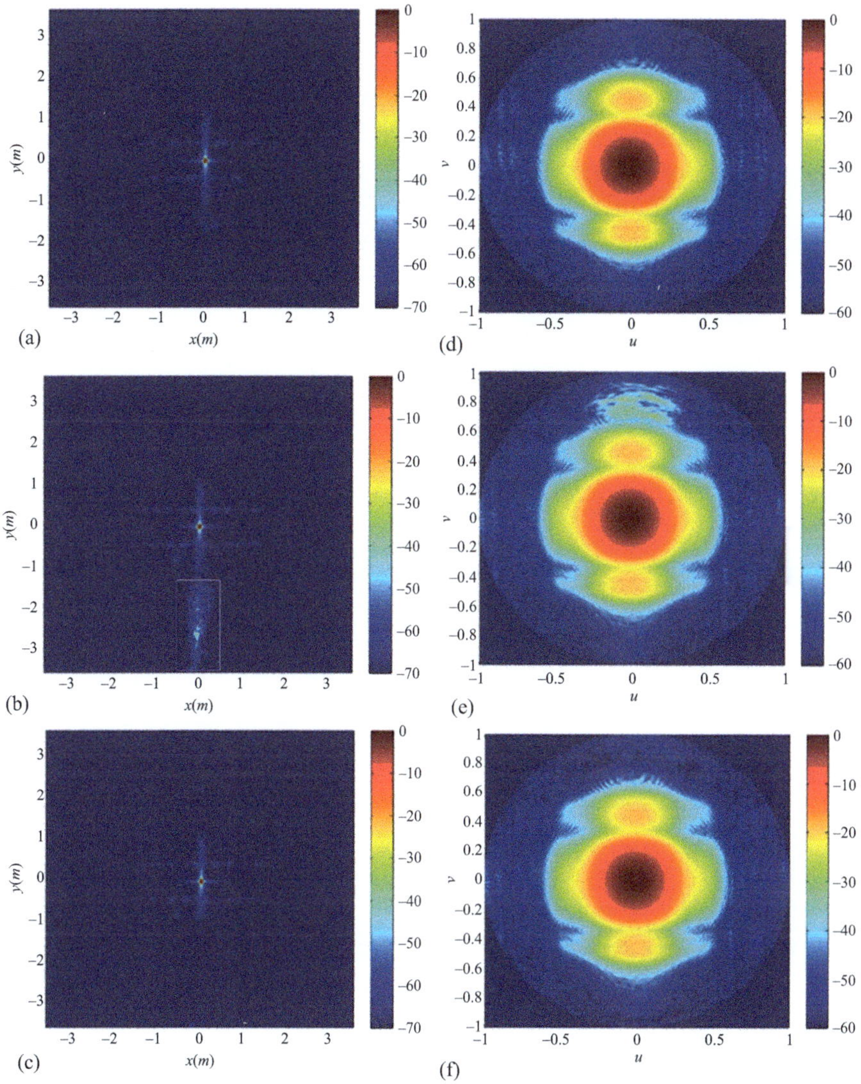

Figure 4.17 Extended reconstructed field of the pyramidal-horn antenna in the reference situation (a), with reflections on the metallic plate (b), and after the spatial filtering (c), far-field pattern of the pyramidal-horn antenna in the reference situation (d), with reflection on the metallic plate (e) and after the spatial filtering (f)

for the vertical cut, seeing that a large ripple appears in the upper lateral lobe as a consequence of the disturbance generated by the reflective surface. The same comparison is performed for the far-field pattern calculated from the filtered reconstructed field, which shows a much better agreement with the reference pattern.

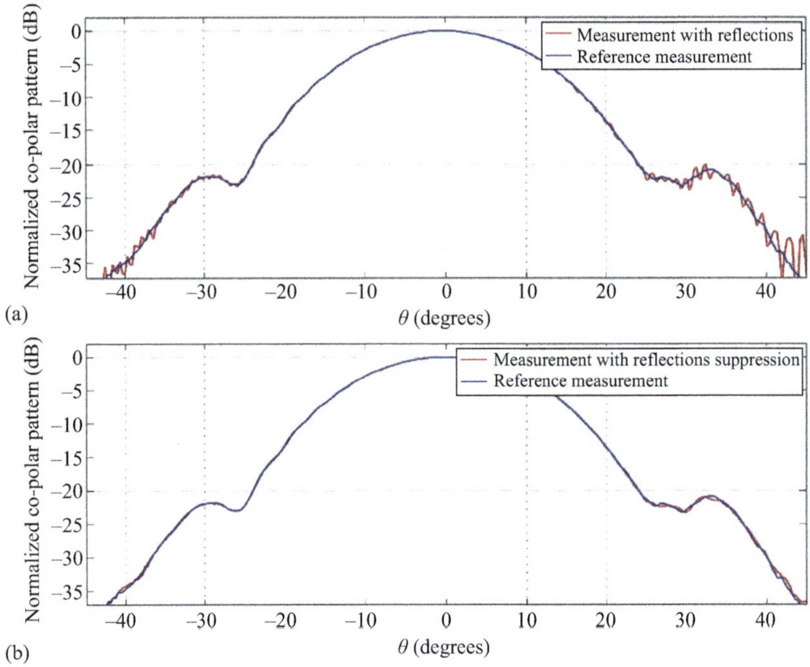

Figure 4.18 Far-field comparison for the pyramidal-horn antenna between the reference pattern and the corrupted pattern (a) and between the reference pattern and the filtered pattern for the vertical cut (b)

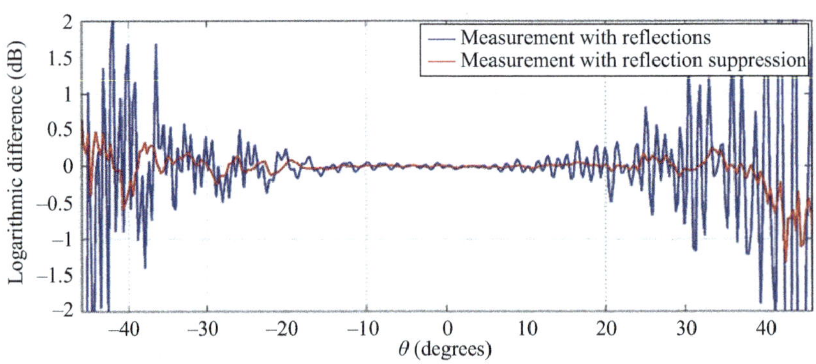

Figure 4.19 Logarithmic difference between the reference far-field pattern and the corrupted and filtered far-field pattern for the vertical cut

Figure 4.19 presents the differences between the reference pattern and the corrupted and filtered patterns. In this case, it is observed that the improvement introduced by this echo suppression method is shown as a reduction of the maximum error from approximately 2 to 0.3 dB within the valid region (angular range of

37 degrees). Also, the error, defined as difference between both patterns, is reduced from 0.012% to 0.005%.

Even if the previous example is applied to an anechoic environment, obviously, it is not the set-up where reflections are important. This algorithm is designed to be applied for semi-anechoic chambers or outdoor systems, in the situations where indoor systems are not available, for instance, in the case of large antennas. This algorithm is applied to the measurement of large L-band radar antennas, measured in an outdoor cylindrical measurement system. In this case, the size of the AUT aperture is 5.3 m × 2.1 m (Figure 4.20). The measurement of this antenna was performed in the outdoor CNF measurement system of the research institute ITM-CEAR located in Guadalajara (Spain). The near-field data is acquired on a cylinder of 15.7 m height, and the distance from the probe tower to the rotation axis of the radar is 5.69 m.

As it can be observed in Figure 4.20, the reflections in the ground can create important distortions in the radiation pattern. In fact, during the measurement campaign, these effects were detected as ripples in the far-field pattern, due to the lobes steering to the ground.

The applied process to this situation equivalent to the previous case. In this case, the CNF is transformed to far field, and the extreme near field on the projection of the AUT aperture on a plane parallel to the measurement tower is calculated through a microwave holographic technique. Figure 4.21 shows this near field, where it is possible to detect the virtual sources related to the reflections in the ground. These sources are located at a distance equal to twice the height of the AUT over the ground, as deduced from the image theory.

The sources from these images of the AUT were cancelled, and the far field is reconstructed from the filtered sources, through an inverse Fourier transform. Figure 4.22 shows the co-polar component of the original and reconstructed patterns.

(a) (b)

Figure 4.20 (a) L-band radar antenna composed and (b) ITM-CEAR outdoor cylindrical near-field measurement system

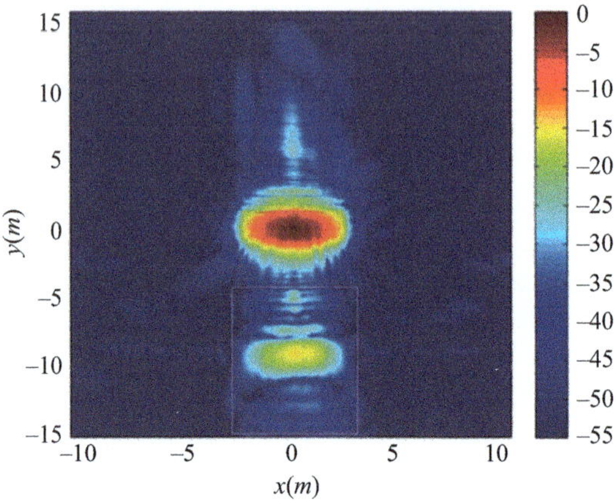

Figure 4.21 Reconstructed near field of the L-band radar antenna with reflections

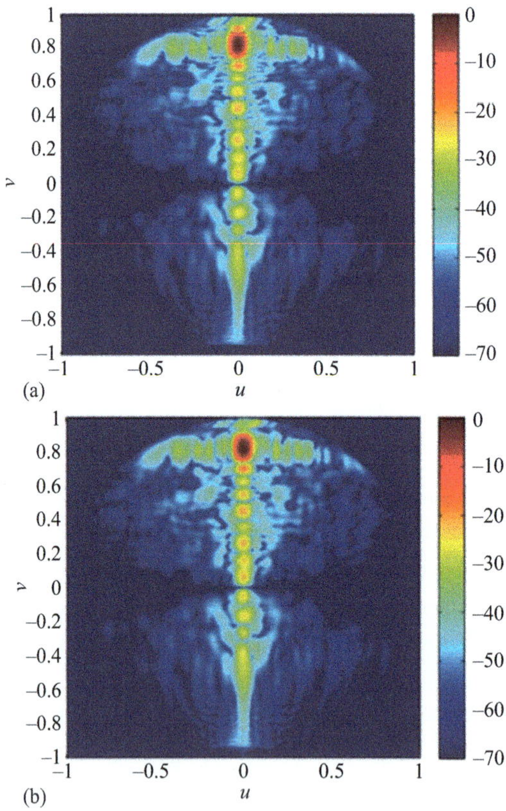

Figure 4.22 L-band radar antenna. Co-polar component of the far-field pattern with reflections in the ground (a) and co-polar component of the far-field pattern after reflection filtering (b)

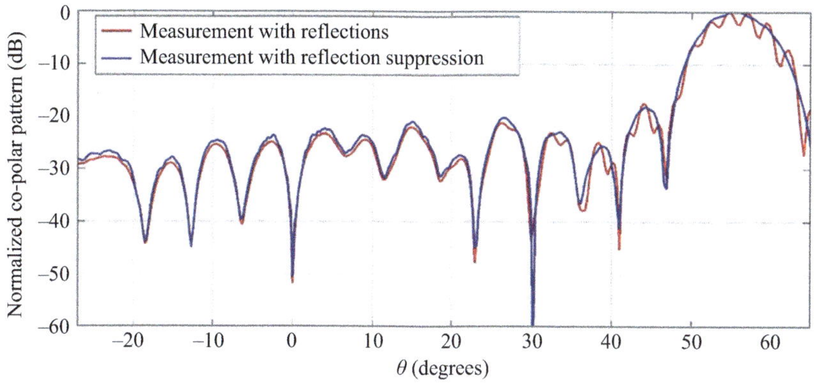

Figure 4.23 Far-field comparison for the L-band radar antenna for the vertical pattern

Figure 4.23 shows the vertical cut of the pattern, where the ripple is better observed. Although in this example there is no reference measurement, it is clear that the result obtained after the echo filtering is much better, with a clear reduction of the ripple created by the echoes.

4.5 Extrapolation of the radiation pattern and reduction of truncation errors

Near-field theory requests the measurement of the two tangential components of the electric or fields on a surface that fully encloses the AUT. In this situation, the field can be obtained for any point outside the measurement surface, being particularly interesting the far field. In planar or cylindrical acquisitions, due to the finite size of the scan surface, this condition is never fulfilled, and, therefore, the true far-field pattern cannot be characterized in the whole sphere. In those cases, the far-field pattern is only valid within the called reliable region [43]. In the case of SNF measurements, if the acquisition is complete, there is no truncation error. However, for the measurement of small antennas where the shadow of the tower distorts the field, or for the measurement of very large antennas, when the measurement time can be very long, the sphere is truncated during the measurements, measurement only on a partial sphere. In this case, a truncation error appears in the far-field pattern, and the far-field cannot be reconstructed on the whole angular range [46].

In this chapter, a method to extrapolate the radiation pattern and to reduce truncation errors when the measurements are taken on truncated surfaces is presented. The method is based on the Gerchberg–Papoulis iterative algorithm proposed in [44,45], and it was already applied to the planar case in [43]. However, this iterative method can be applied to any scanning geometry, and this chapter presents a summary of this technique for all the possible truncated surfaces. In this method, it is not necessary to take samples out of the actual scanning area, in comparison with other approaches to reduce truncation errors. This advantage makes this algorithm very convenient, since the measurements are performed in the classical way.

For the application of this method, two operators are required. The first operator is applied in the spectral domain and defines the reliable portion of the PWS (4.17):

$$A(k_x,k_y) = \begin{cases} 1 & \forall (k_x,k_y) \in \Omega_R \\ 0 & \forall (k_x,k_y) \notin \Omega_R \end{cases} \tag{4.17}$$

where $A(k_x, k_y)$ is the spectral-truncation operator and Ω_R represents the reliable region.

The second one is called the band-limited operator, and it is applied to the near field on the AUT plane:

$$h(x,y) = \begin{cases} 1 & \forall (x,y) \in \omega_{AUT} \\ 0 & \forall (x,y) \notin \omega_{AUT} \end{cases} \tag{4.18}$$

where $h(x,y)$ stands for the band-limited operator and ω_{AUT} is the AUT region. In the spectral domain, this second operator is expressed in (4.19):

$$BP(k_x,k_y) = P(k_x,k_y) \times H(k_x,k_y) \tag{4.19}$$

being B the band-limited operator defined in the spectral domain, $P(k_x, k_y)$ represents the PWS and $H(k_x, k_y)$ is the inverse Fourier transform of the operator $h(x,y)$.

Figure 4.24 shows the diagram of the method. The different steps are as follows:

- Two tangential components of near-field data on a truncated surface (plane, cylinder or sphere) are collected.
- Far field is calculated using a classical near-to-far-field transformation. Then, the PWS is calculated from that far field. Since an FFT is used in the iterative algorithm, the samples of the PWS should be defined on a regular $k_x - k_y$ grid.
- The unreliable portion of the PWS is filtered out by applying the operator defined in (4.17).
- The field distribution over the AUT plane is obtained through the Fourier transform of the filtered PWS.
- The near-field distribution is spatially filtered using the band-limited operator defined in (4.18).
- This filtered near-field is transformed to the spectral domain to obtain an auxiliary PWS.
- A new reconstructed PWS is calculated substituting the unreliable portion of the initial PWS for the portion of the PWS calculated in the previous step.
- When the new PWS fulfils the termination condition, the algorithm stops. If not, a new iteration starting from the step 3 is performed, calculating a new-field distribution on the AUT plane.

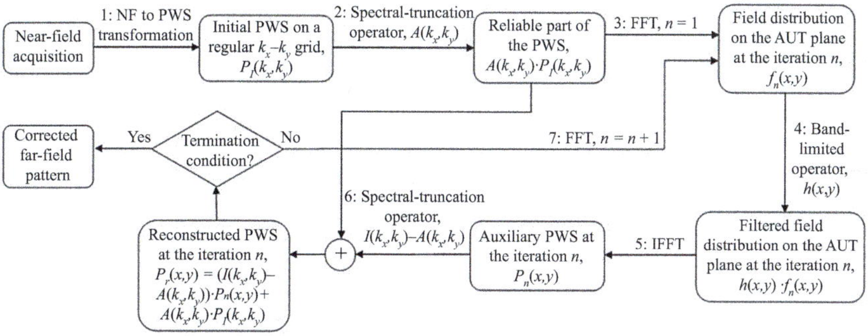

Figure 4.24 Diagram of the method to extrapolate the radiation pattern

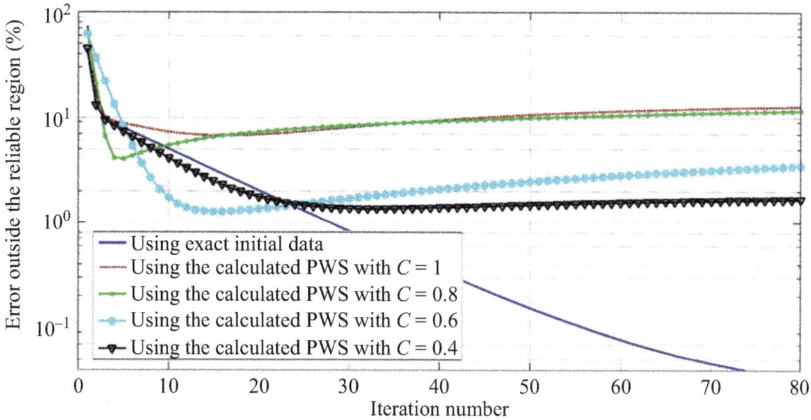

Figure 4.25 Convergence of the error outside the reliable region for different sizes of that region

An exhaustive convergence analysis of the proposed algorithm was carried out in [69]. From this analysis, it was concluded that the global error decreases in the initial iterations, but after a certain large number of iterations, the error smoothly increases due to errors in the initial PWS within the reliable region, since the truncation also introduces ripples within the reliable region. Therefore, when starting from error-free data within the reliable spectral region, the proposed method converges to the correct solution, but in a general case, the error decreases with the iteration number, but after several iterations, it starts to increase. Therefore, the establishment of an accurate termination condition is necessary.

A key point is the definition of the reliable spectral region: usually, it is based on geometrical optics, although the error in the far-field pattern is never zero due to the truncation effect. Figure 4.25 shows the convergence of the algorithm for an ideal case and for real cases, taking a percentage (C) of the reliable spectral region calculated with the classical geometrical optics.

Therefore, an important point in this algorithm is to calculate the proper termination point of the iterative method. An algorithm to find that point was proposed in [43] and further developed in [69], and it is based on the following principle: when different initial truncated surfaces are used, the initial error within the reliable region will be also different, and the error in the iterative procedure will vary in a different way. Therefore, comparing the iterative results obtained in both cases, a minimum will be present when both the results have minimum error. This algorithm does not require additional measurements since two different subsets of the same measured near-field information can be used for the iterative method.

Figure 4.26 shows the results of the application of this algorithm for a rectangular horn antenna at 11 GHz, measured in the planar near field of UPM in a scan surface of 2.7 m × 2.7 m. The probe, a corrugated conical-horn antenna and the AUT were separated from each other by a distance of 1.3 m. The reference pattern is calculated through the measurement of the same AUT in the UPM SNF range. Figure 4.26 shows the comparison between the truncated and reconstructed far-field patterns and the difference between both patterns before and after applying this iterative method.

Figure 4.27 shows the cut comparison of the vertical pattern for the reference case (spherical), original pattern (planar) and reconstructed one using the iterative

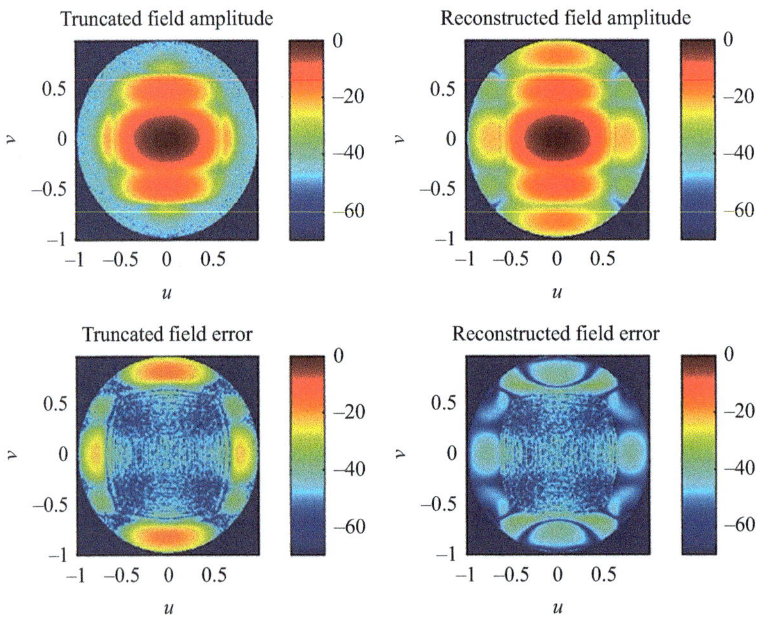

Figure 4.26 Far-field pattern (dB) and truncation error (dB) for the regular measurement in the planar scanner and the reconstructed one of the pyramidal-horn antenna

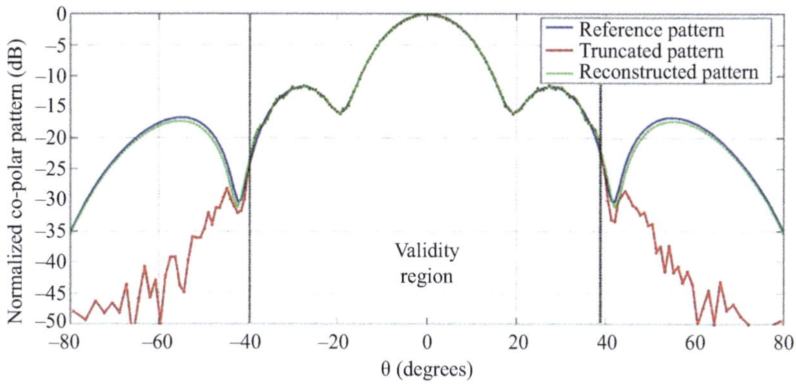

Figure 4.27 Comparison among the reference, truncated and reconstructed vertical cuts of the far-field patterns for the rectangular horn

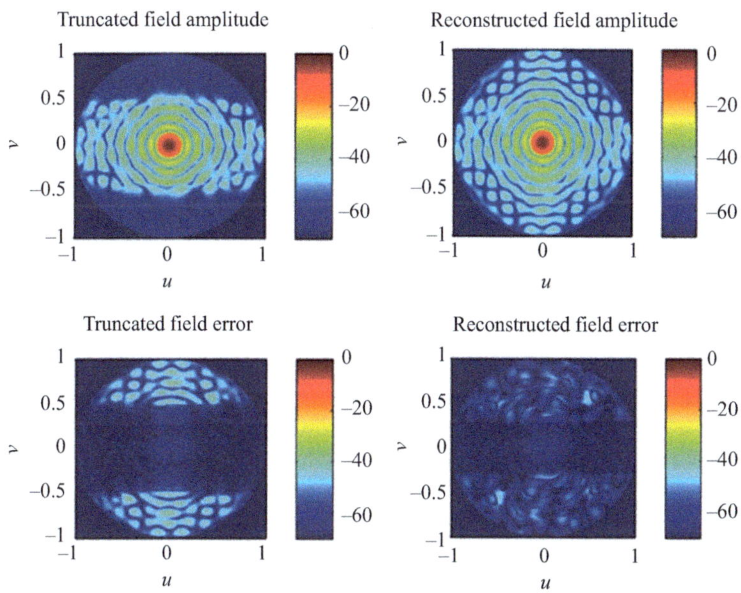

Figure 4.28 Far-field pattern (dB) and error (dB) before and after applying the iterative method for the Ku-band reflector measured in CNF

algorithm. A great improvement of the accuracy outside of the reliable region is achieved, reducing the per cent error from 79.3% to 7.1%.

The method was also validated with cylindrical and partial spherical acquisitions. This example shows the case of the measurement of a 40 cm diameter Ku-band reflector in the UPM CNF scanner. The results obtained in this case can be observed in Figures 4.28 and 4.29. Again, the far field can be perfectly extrapolated

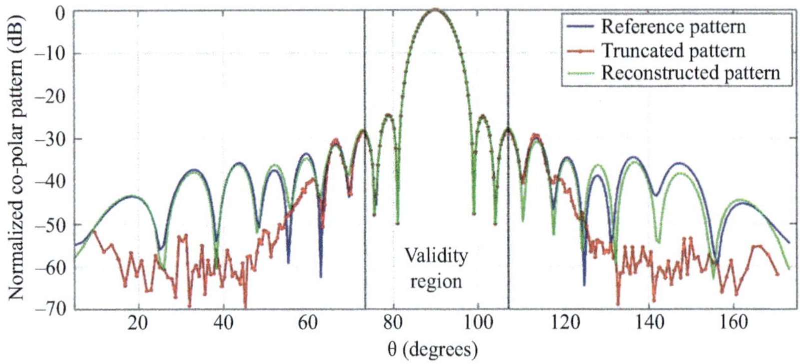

Figure 4.29 Comparison among the reference, truncated and reconstructed for the vertical far-field patterns for the Ku-band reflector in the LEHA-UPM CNF scanner

for the forward hemisphere, and the ripple created by the truncation effect in the reliable region is reduced. Similar examples of SNF are analysed in [69].

4.6 Detection and suppression of leakage

One of the most common leakage sources appeared due to some imperfections in the measurement system, as faulty connectors or rotary joints, coaxial cables or any other damaged or bad isolated component. These elements can transmit or receive undesired radiation, interfering with the measurements. Because of this perturbation, the measured field is distorted and the accuracy of the far-field pattern obtained from the measurement is degraded.

One of the usual techniques to detect these leakages sources is based on measuring by terminating either the transmission lines connected to the AUT or the probe into a load [53]. Then, the result of this measurement is near-field-to-far-field transformed and compared to the far-field pattern obtained in a normal test configuration, making possible to identify which the regions are more affected by the leakage. If a large leakage is detected in the far-field pattern, the faulty components can be substitutes. However, this technique may be a long and a difficult task, requiring, first, an additional measurement and then, a search process, tightening connectors or shielding other components.

In this section, a method to detect the exact position of the sources of leakage, and therefore to suppress its undesired effect on the far-field pattern, is presented. This method does not required additional measurements and it is based on the use of the information on the AUT plane obtained from microwave holography.

As commented, the undesired sources that create the leakage may be transmission lines on either the transmitting or receiving side that radiate or receive radio frequency (RF) energy: faulty connectors, rotary joints or any other bad isolated RF component. Therefore, assuming that the AUT signal is interfered by

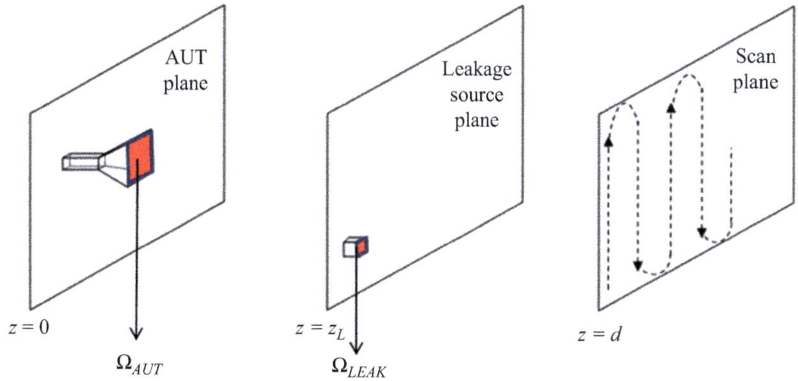

Figure 4.30 Planar near-field measurement set-up including a leakage source

this undesired contribution, the following relationships can be established for both the x- and y-components of the field:

$$E_{AUT}(x, y, z = 0) = 0, \quad \forall (x, y, z = 0) \notin \Omega_{AUT} \tag{4.20}$$

$$E_{LEAK}(x, y, z = z_L) = 0, \quad \forall (x, y, z = z_L) \notin \Omega_{LEAK} \tag{4.21}$$

where E_{AUT} and E_{LEAK} represent the field radiated by the AUT and the source of leakage, respectively; Ω_{AUT} is the region within the AUT plane, which generates the desired contribution; and Ω_{LEAK} is the region within the plane where the leakage is located and creates the radiation of the unwanted sources. These regions are depicted in Figure 4.30.

In any other plane, there will be contributions from both the leakage source and the AUT (4.22)–(4.23):

$$E_{AUT}(x, y, z \neq 0) \neq 0, \quad \forall (x, y, z \neq 0) \tag{4.22}$$

$$E_{LEAK}(x, y, z \neq z_L) \neq 0, \quad \forall (x, y, z \neq z_L) \tag{4.23}$$

In all those planes, the measured field can be expressed as a superposition of both sources (4.24), and therefore, all the measured samples are corrupted by the field radiated from the leakage source:

$$E_{meas}(x, y, z = d) = E_{AUT}(x, y, z = d) + E_{LEAK}(x, y, z = d) \tag{4.24}$$

The diagram of the method proposed for leakage detection and suppression is depicted in Figure 4.31. Depending if the method is employed to detect or cancel the effect of the leakage, some differences appear. The different steps are as follows:

- The measured near-field data are employed to calculate the PWS.
- The PWS is back-propagated to the AUT plane, multiplying by an exponential factor.
- The extreme near-field is obtained through a Fourier transform of the PWS.

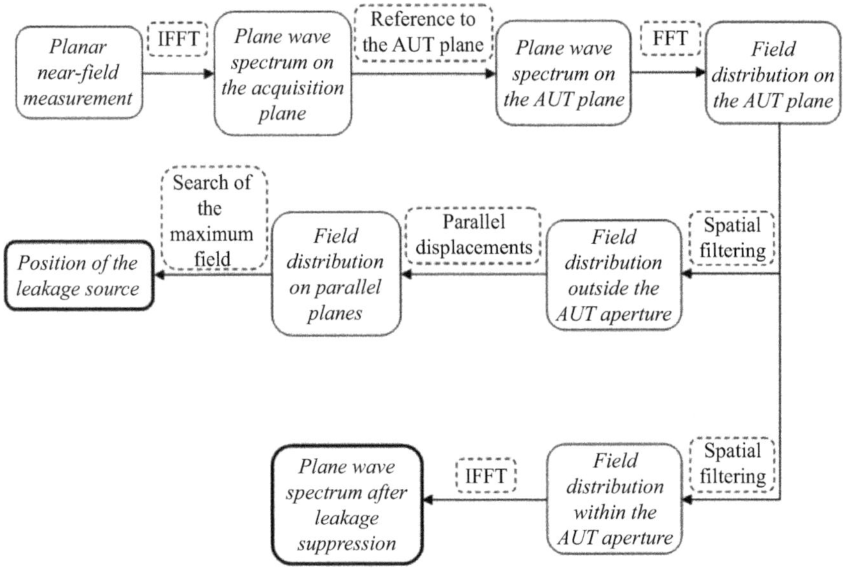

Figure 4.31 Diagram of the method to detect and cancel leakage errors from faulty components in planar measurements

If the objective is to detect the position of the leakage source, the following steps are followed:

- A spatial filtering is used to suppress the AUT contribution.
- The filtered field distribution is employed to calculate the field on different parallel planes. This is done propagating the PWS to different z-coordinates and applying a Fourier transform for each plane.
- The maximum of the field in all those planes will give the position of the leakage source.

If the effect of the leakage source on the far-field pattern wants to be cancelled, the steps are as follows:

- A spatial filtering sets all the samples out of the AUT to zero.
- From those sources, the PWS and far field are calculated.

This method was validated through an experiment in the planar-range measurement system at UPM. Again, the 5 cm $\times$ 7 cm pyramidal-horn antenna was measured, separated 1.57 m from the probe antenna. A second pyramidal-horn antenna, with an additional attenuator of 10 dB, was placed at the position $x = 0.82$ m, $y = -1.6$ m, $z = 0.55$ m (Figure 4.32) in order to simulate the leakage source.

The reference was the measurement without the leakage source over a 2.4 m $\times$ 2.4 m plane. The second measurement, on the same plane, was performed

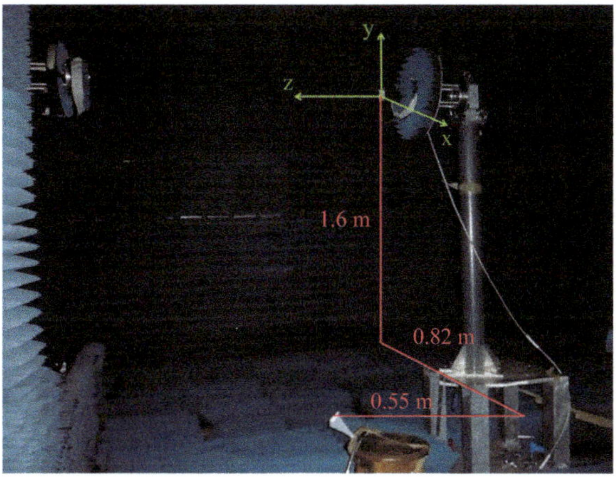

Figure 4.32 Planar near-field measurement set-up with including the leakage source

with the presence of the leakage source. The spatial sampling in both measurements was equal to 1 cm (0.43λ at 13 GHz). Figure 4.33 shows the planar near-field acquisitions obtained without (reference) and with the additional leakage source (corrupted signal).

Figure 4.34 shows the results of the first steps, when the extreme near field on the AUT aperture is calculated.

It is possible to observe that the information about the AUT is mostly concentrated within the AUT aperture (Figure 4.34(a)), except a small portion of the field that is located outside that region due to the truncation error. When introducing the leakage source, the reconstructed field does not include only the AUT contribution but also the leakage contribution. Although this last contribution is not focussed because the field is back-propagated to a plane behind the leakage source, both contributions are not spatially coincident (in this situation), and, therefore, they can be properly separated using the spatial filters.

When the algorithm is used to detect the position of the leakage source, it starts cancelling the AUT contribution. After this spatial filter, the field is calculated on a set of parallel planes, as shown Figure 4.35. The field originated by the leakage source is focused exactly on the plane where that source is located. Looking for the maximum of the signal is possible to determine the exact position of the faulty element. Figure 4.36 represents the maximum of that spurious signal for the different z-planes: the position of that maximum corresponds to the z-position of the leakage source.

This method is used to locate the possible leakage sources and, therefore, to improve the quality of the measurement system replacing the faulty components or shielding the cables.

This unwanted effect of the leakage can be also suppressed without any modification in the measurement set-up, just filtering out the leakage contribution

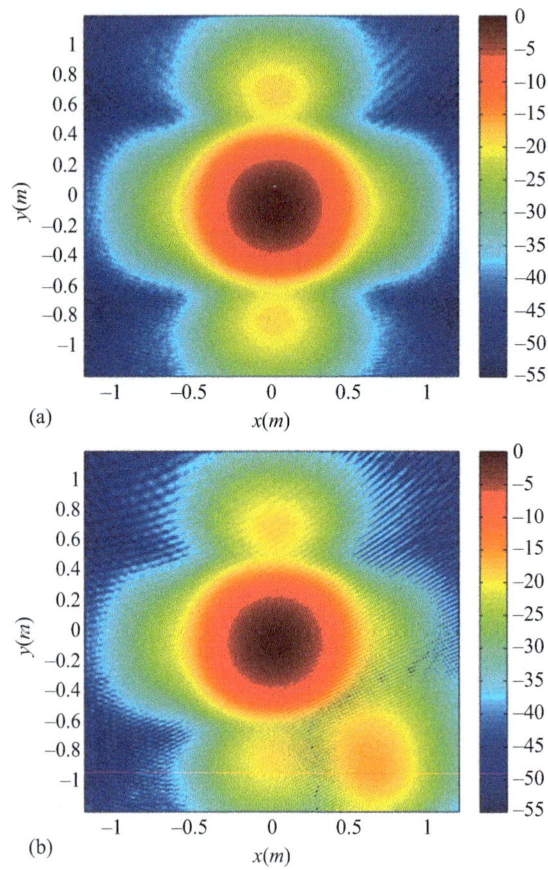

Figure 4.33 Planar near-field acquisition: reference (a) and with leakage (b) of the pyramidal-horn antenna

located outside the antenna aperture on the AUT plane, as it was done in previous sections. The new PWS is obtained through an inverse Fourier transform of this filtered field. Figure 4.37 shows the results for this experiment. Clearly, the effect of the leakage has been cancelled. Also, the reference pattern is noisy, due to a poor dynamic range during the measurement, and the corrected one also reduces the noise.

Similar technique can also be used to cancel the effect of the leakage originated by the detectors. In [70], a method based on holographic techniques estimates the constant associated to the bias error averaging the field outside the AUT aperture, after a diagnosis on a large plane. Using the information in this new domain, the constant obtained is more accurate than the determined with previous methods, as the ones presented in [52,53], since the samples used in this new estimation only contain noise and leakage contribution, and not the information from AUT.

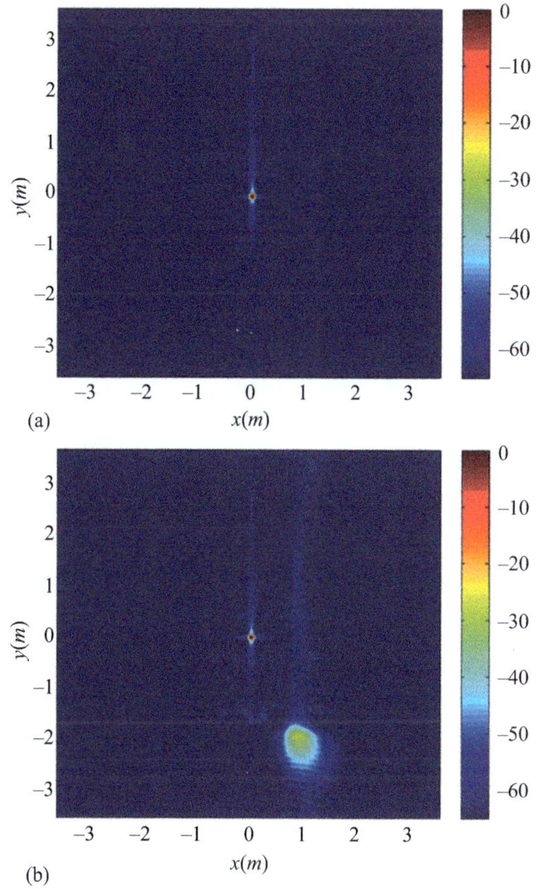

Figure 4.34 Extreme near field for the reference (a) and with leakage (b) in the measurement of the pyramidal-horn antenna

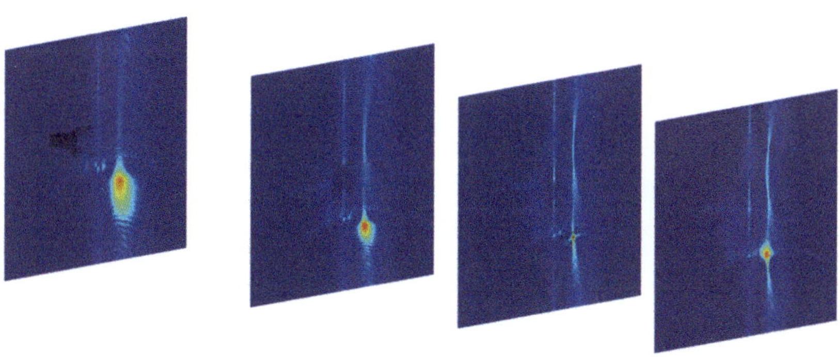

Figure 4.35 Leakage contribution on the different parallel planes between AUT and planar scanner

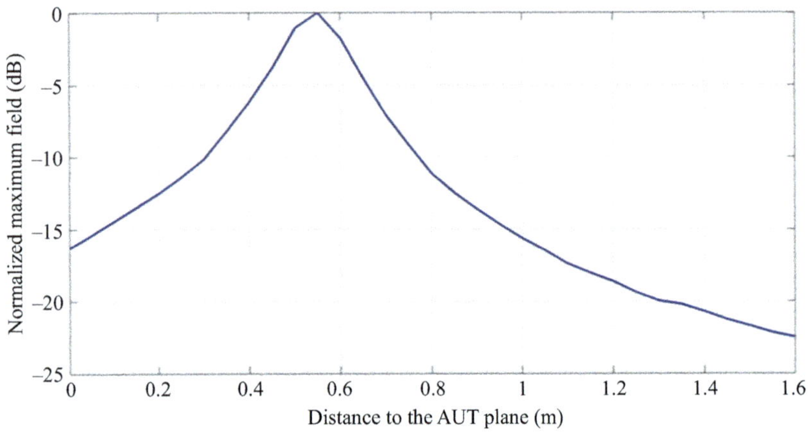

Figure 4.36 Maximum of the field for the different planes

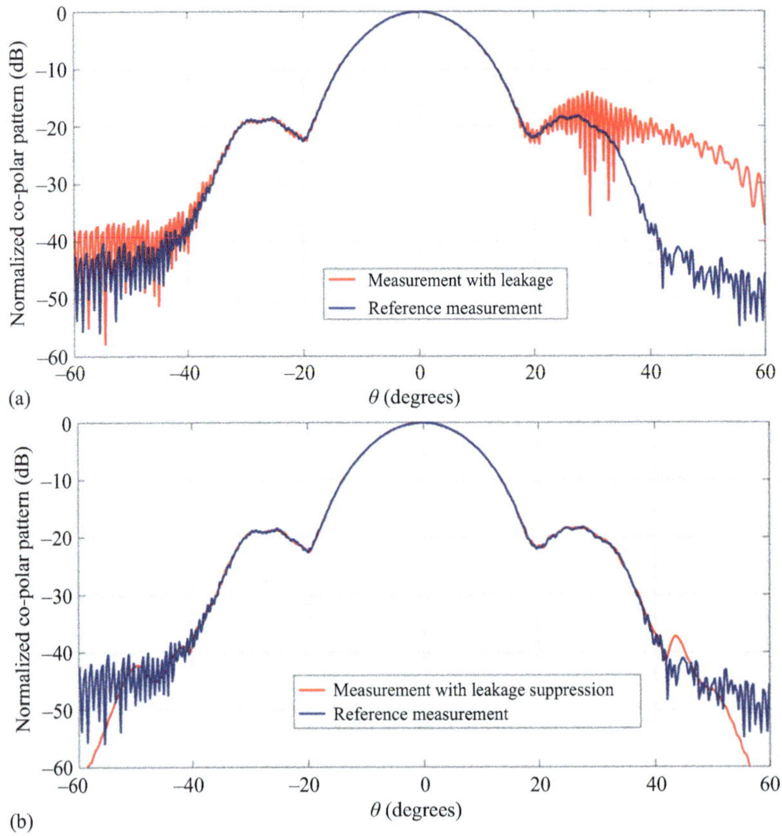

Figure 4.37 Far-field comparison between the reference pattern and the pattern corrupted by leakage (a) and between the reference pattern and the pattern with leakage suppression (b)

4.7 Conclusions

This chapter has analysed the use of holographic techniques for the improvement of the quality of antenna measurements through post-processing techniques. Holographic techniques and the use of plane wave expansion of the radiated field allow an easy way of expression of the radiated field of the antenna. This plane wave expansion is directly related with the sources of the antenna in a plane through an FFT. The use of the FFT allows a very fast calculation, and the results can be practically obtained in real time.

The chapter has explored the use of these techniques for echo reduction, increase of SNR, reduction of truncation errors and leakage detection and cancellation. The main advantage of this technique is the required time and reduced computer requirements. These advantages make this technique very useful for some applications. The main drawbacks are the use of planar surfaces, which makes this technique very appropriate for aperture or planar antennas, but not so good for other kind of antennas. However, the sources can be reconstructed in a plane in front of the antenna. Also, the use of holographic techniques makes these techniques appropriate for antennas radiating in one hemisphere. Again, aperture or planar antennas are the most pertinent for the use of these algorithms. In the previous chapter, a more general technique calculating electromagnetic currents on a surface enclosing the antenna has the advantage of the application for a general antenna, but the required time is much higher.

Finally, it is important to mention the high number of experiments and simulations including in this chapter. The authors have wanted to show the different cases of applications of these techniques emphasizing both advantages and limitations.

More examples of the application of this holographic technique for the improvement of these effects are presented in [70].

References

[1] R. C. Jonhson, H. A. Ecker, and J. S. Hollis, "Determination of far-field antenna patterns from near-field measurements," *Proc. IEEE*, vol. 61, no. 12, pp. 1668–1694, 1973.

[2] A. D. Yaghjian, "An overview of near-field antenna measurements," *IEEE Trans. Antennas Propag.*, vol. AP-34, no. 1, pp. 30–44, 1986.

[3] P. Petre and T. K. Sarkar, "Planar near-field to far-field transformation using an equivalent magnetic current approach," *IEEE Trans. Antennas Propag.*, vol. 40, no. 11, pp. 1348–1356, 1992.

[4] T. K. Sarkar and A. Taaghol, "Near-field to near/far-field transformation for arbitrary near-field geometry utilizing an equivalent electric current and MoM," *IEEE Trans. Antennas Propag.*, vol. 47, no. 3, pp. 566–573, 1999.

[5] A. C. Newell and C. F. Stubenrauch, "Effect of random errors in planar near-field measurements," *IEEE Trans. Antennas Propag.*, vol. 36, no. 6, pp. 769–773, 1988.

[6] J. B. Hoffman and K. R. Grimm, "Far-field uncertainty due to random near-field measurement error," *IEEE Trans. Antennas Propag.*, vol. 36, no. 6, pp. 774–780, 1988.
[7] J. Romeu, L. Jofre, and A. Cardama, "Far-field errors due to random noise in cylindrical near-field measurements," *IEEE Trans. Antennas Propag.*, vol. 40, no. 1, pp. 79–84, 1992.
[8] J. Romeu and L. Jofre, "Effect of random errors in cylindrical near-field measurements," in *Proc. 1991 IEEE/AP-S Int. Symp.*, London, ON, 1991, pp. 1450–1453.
[9] J. Romeu, L. Jofre, and A. Cardama, "An approximate expression to estimate the signal-to-noise ratio improvement in cylindrical near-field measurements," *IEEE Trans. Antennas Propag.*, vol. 42, no. 7, pp. 1007–1010, 1994.
[10] F. J. Cano-Fácila, S. Burgos, and M. Sierra-Castañer, "Novel method to improve the signal-to-noise ratio in far-field results obtained from planar near-field measurements," *IEEE Antennas Propag. Mag.*, vol. 53, no. 2, pp. 215–220, 2011.
[11] P. Koivisto, "Reduction of errors in antenna radiation patterns using optimally truncated spherical wave expansion," *Prog. Electromagn. Res.*, vol. pier-47, pp. 313–333, 2004.
[12] L. J. Foged and M. Faliero, "Random noise in spherical near-field systems," in *Proc. 2009 Antenna Measurement Techniques Assoc., AMTA*, Salt Lake City, UT, 2009, pp. 135–138.
[13] F. Jensen and A. Frandsen, "On the number of modes in spherical wave expansion," in *Antenna Measurement Techniques Association, AMTA 2004*, Stone Mountain, GA, Oct., 2004, pp. 489–494.
[14] J. Appel-Hansen, "Reflectivity level of radio anechoic chambers," *IEEE Trans. Antennas Propag.*, vol. 21, no. 4, pp. 490–498, 1973.
[15] A. Lehto, J. Touvinen, and A. Räisänen, "Reflectivity level of anechoic chambers at 183 GHz," in *Antennas Propag. Soc. Int. Symp.*, Dallas, TX, 1990, pp. 1310–1313.
[16] E. N. Clouston, P. A. Langsford, and S. Evans, "Measurement of anechoic chamber reflections by time-domain techniques," *IEE Proc. H, Microwaves Antennas Propag.*, vol. 135, no. 2, pp. 93–97, 1988.
[17] M. Nagatoshi, M. Hirose, H. Tanaka, S. Korokawa, and H. Morishita, "A method of pattern measurement to cancel reflection waves in anechoic chamber," in *Antennas Propag. Soc. Int. Symp.*, San Diego, CA, 2008, pp. 1–4.
[18] J. D. Young, D. E. Svoboda, and W. D. Burnside, "A comparison of time- and frequency-domain measurement techniques in antenna theory," *IEEE Trans. Antennas Propag.*, vol. 21, no. 4, pp. 581–583, 1973.
[19] G. A. Burrell and A. R. Jamieson, "Antenna radiation pattern measurement using time-to-frequency transformation (TFT) techniques," *IEEE Trans. Antennas Propag.*, vol. 21, no. 5, pp. 702–704, 1973.
[20] S. Loredo, M. R. Pino, F. Las-Heras, and T. K. Sarkar, "Echo identification and cancellation techniques for antenna measurement," *IEEE Antennas Propag. Mag.*, vol. 46, no. 1, pp. 100–107, 2004.

[21] M. Dadic and R. Zentner, "A technique for elimination of reflected rays from antenna measurements performed in echoic environment," *AEÜ – Int. J. Electron. Commun.*, vol. 61, no. 2, pp. 90–94, 2007.

[22] M. L. Van Blaricum and R. Mittra, "A technique for extracting the poles and residues of a system directly from its transient response," *IEEE Trans. Antennas Propag.*, vol. 23, no. 6, pp. 777–781, 1975.

[23] T. K. Sarkar and O. Pereira, "Using the matrix pencil method to estimate the parameters of a sum of complex exponentials," *IEEE Antennas Progap. Mag.*, vol. 37, no. 1, pp. 48–55, 1995.

[24] T. K. Sarkar, J. Nebat, D. D. Weiner, and V. K. Jain, "Suboptimal approximation/identification of transient waveforms from electromagnetic systems by pencil-of-function method," *IEEE Trans. Antennas Propag.*, vol. 28, no. 6, pp. 928–933, 1980.

[25] Y. Hua and T. K. Sarkar, "Generalized pencil-of-function method for extracting poles of an EM system from its transient response," *IEEE Trans. Antennas Propag.*, vol. 37, no. 2, pp. 229–234, 1989.

[26] R. S. Adve, T. K. Sarkar, O. M. C. Pereira-Filho, and S. M. Rao, "Extrapolation of time-domain responses from three-dimensional conducting objects utilizing the matrix pencil technique," *IEEE Trans. Antennas Propag.*, vol. 45, no. 1, pp. 147–156, 1997.

[27] B. Fourestié, Z. Altman, J. Wiart, and A. Azoulay, "On the use of the matrix-pencil method to correlate measurements at different test sites," *IEEE Trans. Antennas Propag.*, vol. 47, no. 10, pp. 1569–1573, 1999.

[28] B. Fourestié and Z. Altman, "Gabor schemes for analyzing antenna measurements," *IEEE Trans. Antennas Propag.*, vol. 49, no. 9, pp. 1245–1253, 2001.

[29] S. Gregson, A. Newell, and G. Hindman, "Reflection suppression in cylindrical near-field antenna measurement systems-cylindrical MARS," in *Antenna Measurement Techniques Association, AMTA 2009*, Salt Lake City, UT, 2009, pp. 119–125.

[30] G. Hindman and A. Newell, "Reflection suppression in large spherical near-field range," in *Antenna Measurement Techniques Association, AMTA 2005*, Newport, RI, 2005, pp. 270–275.

[31] D. W. Hess, "The IsoFilter technique: a method of isolating the pattern of an individual radiator from data measured in a contaminated environment," *IEEE Antennas Propag. Mag.*, vol. 52, no. 1, pp. 174–181, 2010.

[32] P. R. Rousseau, "The planar near-field measurement of a broad beam antenna using a synthetic subarray approach," in *IEEE Antennas and Propag. Symp.*, 1997, pp. 160–163.

[33] T. Al-Mahdawi, "Planar near-field measurement of low directivity embedded antennas using a synthetic array probe approach," in *IEEE Antennas Propag. Symp.*, 1998, pp. 1311–1314.

[34] E. B. Joy, "Windows'96 for planar near-field measurements," in *Antenna Measurement Techniques Association, AMTA 1996*, 1996, pp. 80–85.

[35] E. B. Joy, C. A. Rose, and A. H. Tonning, "Test-zone field quality in planar near-field measurements," in *Antenna Measurement Techniques Association, AMTA 1997*, Williamsburg, VIR, 1997, pp. 206–210.

[36] S. F. Gregson, C. G. Parini, and J. McCormick, "Development of wide-angle pattern measurements using a probe-corrected polyplanar near-field measurement technique," *IEE Proc. Microwaves Antennas Propag.*, vol. 152, no. 6, pp. 563–572, 2005.
[37] O. M. Bucci, G. D. D'Elia, and M. D. Migliore, "A new strategy to reduce the truncation error in near-field far-field transformation," *Radio Sci.*, vol. 35, no. 1, pp. 3–17, 2000.
[38] O. M. Bucci, C. Gennarelli, and C. Savarese, "Representation of electromagnetic fields over arbitrary surfaces by a finite and nonredundant number of samples," *IEEE Trans. Antennas Propag.*, vol. 46, no. 3, pp. 351–359, 1998.
[39] J. C. Bolomey, O. M. Bucci, L. Casavola, G. D'Elia, M. D. Migliori, and A. Ziyyat, "Reduction of truncation error in near-field measurements of antennas of base-station mobile communications systems," *IEEE Trans. Antennas Propag.*, vol. 52, no. 2, pp. 593–602, 2004.
[40] F. D'Agostino, F. Ferrara, C. Gennarelli, R. Guerriero, and G. Riccio, "An effective technique for reducing the truncation error in the near-field-far-field transformation with plane-polar scanning," *Prog. Electromagn. Res.*, vol. 73, pp. 213–238, 2007.
[41] O. M. Bucci and M. D. Migliore, "A new method for avoiding the truncation error in near-field antennas measurements," *IEEE Trans. Antennas Propag.*, vol. 54, no. 10, pp. 2940–2952, 2006.
[42] L. Infante, E. Martini, A. Peluso, and S. Maci, "The shadow boundary integral for the reduction of the truncation error in near-field to far-field transformations," in *IEEE Antennas Propag. Soc. Int. Symp.*, Washington, DC, 2005, pp. 416–419.
[43] E. Martini, O. Breinbjerg, and S. Maci, "Reduction of truncation errors in planar near-field aperture antenna measurements," *IEEE Trans. Antennas Propag.*, vol. 56, no. 11, pp. 3485–3493, 2008.
[44] R. W. Gerchberg, "Super-resolution through error energy reduction," *Opt. Acta*, vol. 21, no. 9, pp. 709–720, 1974.
[45] A. Papoulis, "A new algorithm in spectral analysis and band-limited extrapolation," *IEEE Trans. Circuits Syst.*, vol. 22, no. 9, pp. 735–742, 1975.
[46] M. Serhir, J.-M. Geffrin, A. Litman, and P. Besnier, "Aperture antenna modeling by a finite number of elemental dipoles from spherical field measurements," *IEEE Trans. Antennas Propag.*, vol. 58, no. 4, pp. 1260–1268, 2010.
[47] K. T. Kim, "Truncation-error reduction in 2D cylindrical/spherical near-field scanning," *IEEE Trans. Antennas Propag.*, vol. 58, no. 6, pp. 2153–2158, 2010.
[48] R. C. Wittmann, C. F. Stubenrauch, and M. H. Francis, "Spherical scanning measurements using truncated data sets," in *Antenna Measurement Techniques Association, AMTA 2002*, Cleveland, OH, 2002, pp. 279–283.
[49] E. Martini, S. Maci, and L. J. Foged, "Spherical near-field measurements with truncated scan area," in *5th European Conference on Antennas and Propagation, EuCAP 2011*, Rome, 2011, pp. 3412–3414.

[50] G. Hindman and D. Slater, "Error suppression techniques for near-field antenna measurements," in *Antenna Measurement Techniques Association, AMTA 1989*, 1989.
[51] P. S. H. Leather and D. Parsons, "Equalization for antenna-pattern measurements: established technique-new application," *IEEE Antennas Propag. Mag.*, vol. 45, no. 2, pp. 154–161, 2003.
[52] P. R. Rousseau, "An algorithm to reduce bias errors in planar near-field measurement data," in *Antenna Measurement Techniques Association, AMTA 1999*, Monterey, CA, USA, 1999, pp. 269–271.
[53] A. Newell, "Methods to estimate and reduce leakage bias errors in planar near-field antenna measurements," in *Antenna Measurement Techniques Association, AMTA 2002*, Cleveland, OH, USA, 2002.
[54] F. J. Cano-Fácila, S. Burgos, and M. Sierra-Castañer, "New methods to reduce leakage errors in planar near-field measurements," in *European Conference on Antennas and Propagation, EuCAP 2011*, Rome, 2011.
[55] H. G. Booker and P. C. Clemmow, "The concept of an angular spectrum of plane waves and its relation to that of polar diagram and aperture distribution," *Proc. Inst. Electr. Eng.*, vol. 97, no. 1, pp. 11–16, 1950.
[56] O. M. Bucci, M. D. Migliore, G. Panariello, and M. Sgambato, "Accurate diagnosis of conformal arrays from near-field data using the matrix method," *IEEE Trans. Antennas Propag.*, vol. 53, no. 3, pp. 1114–1120, 2005.
[57] D. Sánchez, M. Baquero, V. M. Rodrigo, and B. Bernardo, "Currents reconstruction using a modal expansion of near-field measurements for synthesis error detection," *Microwave Opt. Technol. Lett.*, vol. 49, no. 8, pp. 2043–2047, 2007.
[58] Y. Rahmat-Samii, "Surface diagnosis of large reflector antennas using microwave holographic technique: an iterative approach," *Radio Sci.*, vol. 19, no. 5, pp. 1205–1217, 1984.
[59] M. S. Narasimhan and B. Preetham Kumar, "A technique of synthesizing the excitation currents of planar arrays or apertures," *IEEE Trans. Antennas Propag.*, vol. 38, no. 9, pp. 1326–1332, 1990.
[60] M. Johansson and B. Svensson, "Array antenna diagnosis and calibration," in *Antenna Measurement Techniques Association, AMTA 1990*, 1990, pp. 27–32.
[61] M. Sano, M. Sierra-Castañer, T. Salmerón-Ruiz, J. Hirokawa, and M. Ando. "Reconstruction of the field distribution on slot array antennas using the Gerchberg–Papoulis algorithm". *IEEE Trans. Antennas Propag.*, vol. 63, no. 8, pp. 3441–3451, 2015.
[62] A. J. Devaney and E. Wolf, "Multipole expansions and plane wave representations of the electromagnetic field," *J. Math. Phys.*, vol. 15, pp. 234–244, 1974.
[63] C. Capellin, "Antenna diagnostics for spherical near-field antenna measurements," Ph.D. Dissertation. Danmarks Tekniske Universitet, Copenhagen, 2007.

[64] C. Capellin, O. Breinbjerg, and A. Frandsen, "Properties of the transformation from the spherical wave expansion to the plane wave expansion," *Radio Sci.*, vol. 43, RS1012, 2008. doi:10.1029/2007RS003696.

[65] J. E. Hansen. Spherical Near Field Antenna Measurements. *IEE Electromagnetic Series*. London: Peter Peregrinus Ltd, 1988.

[66] S. Pivnenko, J. E. Pallesen, O. Breinbjerg, *et al.*, "Comparison on antenna measurement facilities with the DTU-ESA 12 GHz validation standard antenna within the EU Antenna Centre of Excellence," *IEEE Trans. Antennas Propag.*, vol. 57, no. 7, pp. 1863–1878, 2009.

[67] F. J. Cano Facila, and M. Sierra-Castañer, "Methods for noise reduction in far-field patterns obtained from cylindrical near-field antenna measurements," *IEEE Trans. Antennas Propag.*, vol. 60, no. 12, pp. 5635–5644, 2012.

[68] F. J. Cano-Fácila, S. Burgos, F. Martín, and M. Sierra-Castañer, "New reflection suppression method in antenna measurement systems based on diagnostic techniques," *IEEE Trans. Antennas Propag.*, vol. 59, no. 3, pp. 941–949, 2011.

[69] F. J. Cano-Fácila, S. Pivnenko, and M. Sierra-Castañer, "Reduction of truncation errors in planar, cylindrical, and partial spherical near-field antenna measurements," *Int. J. Antennas Propag.*, vol. 2012, Article ID 438727. 2012.

[70] F. J. Cano Fácila, "Contribution to the quality improvement of the antenna measurement using post-processing techniques," Ph.D. Thesis Dissertation. Universidad Politécnica de Madrid, July 2012.

Chapter 5

Antenna measurement improvement using spherical or cylindrical mode expansion

Francesco Saccardi[1], Manuel Sierra Castañer[2], and Lars J. Foged[1]

Most near field (NF) antenna measurements are performed in shielded anechoic chambers to emulate free-space conditions. High-quality absorbing materials are generally used on all surrounding surfaces that otherwise would interfere with the measurement by transmitting energy toward the test region through a combination of reflections and/or diffractions.

In most measurement cases with well-designed anechoic chambers, the measurement uncertainty contribution from these additional errors named echoes or stray signals can be considered negligible [1]. However, in special measurement conditions and/or when very high precision measurements are needed, the above consideration is no longer valid. Significant echo contributions can occur in the case of large angle of incidence even on absorber-covered surfaces. Furthermore, if the anechoic chamber is used near or below the lower frequency limit of the absorbers, the quality of the test region is compromised by strong echo due to diminishing absorption properties of the absorbing material in these frequency ranges. Another source of test zone echo pollution is when the antenna under test (AUT) is placed too close to the chamber walls or close to objects that cannot effectively be covered by absorbers. An example of measurement situation where echoes or stray signal might be not negligible is shown in Figure 5.1, where the Hybrid European RF and Antenna Test Zone range at ESA-ESTEC are shown. As can be seen, in such antenna measurement range, a dual-reflector compact range and NF system share the same measurement environment (the large reflectors may act as disturbance while performing NF measurements).

To mitigate the errors originated from the above mentioned extraneous signals, and thus improve the measurement accuracy, the spherical and cylindrical modal expansions, objective of this chapter, can be effectively exploited taking advantage of their spatial filtering properties. When the spherical (SWE) or cylindrical wave expansion (CWE) is applied, the measured filed is projected over a set of orthogonal (spherical or cylindrical) basis functions and a corresponding set of complex

[1]Microwave Vision Italy (MVI), Pomezia, Italy
[2]Information Processing and Telecommunications Center, ETSI Telecomunicación, Universidad Politécnica de Madrid, Madrid, Spain

Figure 5.1 Compact Payload Test Range (CPTR) at ESA-ESTEC (Noordvijk, The Netherlands). A Compact Range and an NF System share the same measurement environment. Echoes and stray signals may occur while performing NF measurements

coefficients, or spectrum, is computed. As we will see in detail in this chapter, the field oscillations present in the spatial (physical) domain are linked to the spectral/ modal domain by a simple and direct relation. The modal domain can thus be exploited to separate the (wanted) contributions coming from the AUT from the unwanted scattering contributions. Since the latter are more highly oscillating than the direct contributions, they will be represented by higher order modes while the former will be represented by lower order modes. A reduction of the measurement errors due to the echoes can thus be obtained by simply applying a low-pass filtering whose size purely depends on the maximum electrical extend of the AUT (i.e., AUT minimum sphere and AUT minimum cylinder).

The spatial filtering performed using spherical and cylindrical modes presents several advantages if compared to other techniques. For example, the holography spatial filtering technique described in Chapter 4 can only be applied to medium or high directive antennas, while the modal-expansion-based spatial filtering can be used with any type of radiating devices. Then, a direct consequence of the orthogonality of the basis-functions is that the expansion can be computed very efficiently by applying the fast Fourier transform (FFT). For this reason, this type of spatial filtering is often preferred to the equivalent current-based [2] spatial filtering techniques, which is much more computationally expensive.

In general, any type of spatial filtering technique does not require the measurement of many frequency points as needed instead when using time-domain filtering post-processing techniques. Spatial filtering is thus also suited for narrow-band antennas, while time-domain filter can only be applied when wide-band devices are considered.

Another advantage of the modal-expansion-based spatial filtering is that very few *a priori* knowledge of the AUT must be known, namely, the maximum physical extent of the device and its position in the reference coordinate system.

As will be described in this chapter, the oversampling (i.e., acquisition of more sampling points than the minimum required) can be exploited in order to improve the filtering capabilities of the spatial filtering technique. In the case of single probe NF systems, the oversampling generally leads to a significant increase of the measurement time unless special acquisition schemes are adopted (e.g., on-the-fly acquisitions). Conversely, the oversampling is usually not a problem in the case of multi-probe systems [3–6], where more NF sampling points are measured at the same time using an array of probes.

As will be also addressed in this chapter, the effectiveness of the spatial filtering techniques also depends on how the NF scanning is performed (stationary probe and moving AUT against stationary AUT and moving probe) but, in general, such techniques are able to mitigate the unwanted effect of almost any type scattering contributions. Some limitations are unfortunately observable in the case of multiple AUT-probe reflections (or in general higher order reflections redirected toward the AUT) because such contributions cannot be discriminated in the spatial/modal domain being spatially located on the AUT itself. To mitigate the effect of such unwanted signals, one may use time-domain techniques, which allow to separate and filter out the echoes having different time delays. Another widely adopted technique for this type of errors is the averaging of two (or more) NF acquisitions performed at a distance of $\lambda/4$ (or multiple of $\lambda/4$) from the other(s) [1] so that the echo's contributions cancel because they end up being out of phase. Of course, this technique is suited for narrow-band antennas and has the disadvantages that one, or more, acquisitions must be performed. Nevertheless, it should be noted that in many cases, such multiple reflections may consider negligible because of the large measurement distance.

In this chapter, we will focus separately on the spherical and cylindrical modal expansion techniques.

The spatial filtering based on the SWE technique is described in Section 5.1. The SWE can be applied to any type of antenna and it requires the knowledge of the full three-dimensional (3D) radiation pattern. In Section 5.1, we will see how the SWE can be exploited not only for echo reduction purposes but also for noise reduction and to mitigate the errors coming from the truncation of the scanning area.

The spatial filtering based on the CWE is instead described in Section 5.2. The main advantage of the CWE is that it can be applied also to single pattern cuts, thus it does not require the knowledge of the full 3D radiation pattern. In the same section, we will also describe how the CWE can be exploited to perform the so-called single-cut NF-to-far-field (FF) (NF/FF) transformation, which, under certain assumptions, allows to strongly reduce the measurement time while maintaining a good level of accuracy.

5.1 Improvements of antenna measurement accuracy using spherical mode expansion

In this section, we will focus on how the SWE technique can be used in order to improve the accuracy in NF antenna measurements. The primary purpose of the SWE is to perform the NF/FF transformation in spherical NF antenna measurements.

Nevertheless, as we will describe in this section, the spectrum computed by the SWE has important spatial filtering properties that can be exploited in order to mitigate possible measurement errors caused, for example, by echoes or stray signals present in the measurement environment.

To understand well how such spatial filtering can be exploited, the SWE theory will be described in Section 5.1.1 without going into too much detail (the reader may refer to [7] for an extended descriptions of the technique) but underling the main properties and trying to have a better understanding of the spherical wave spectrum. In Section 5.1.2 instead, we will show how in practice the SWE can be used to mitigate errors produced by

- random noise in spherical NF (SNF) measurements,
- echoes or stray signals present in different type of NF measurements,
- truncation of the scanning area in SNF measurement.

To do that, both simulated and real-life applicative examples will be taken into account.

5.1.1 Spherical wave expansion theory

The spherical modal/SWE is a very a well-known and established technique which is widely used to perform the NF/FF transformation in the case of SNF antenna measurements. The SWE technique is described with great details in [7].

With reference to the coordinate system drawn in Figure 5.2, let us assume $\boldsymbol{E}(r, \theta, \varphi)$ to be the electric field vector measured by an ideal probe (Hertzian dipole) over a spherical surface of radius $r = R_M$. The field samples position on the surface are denoted by the conventional (θ, φ) spherical coordinates as shown in

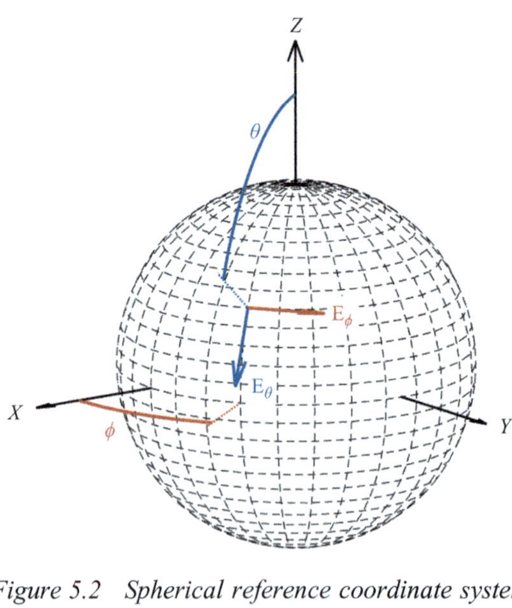

Figure 5.2 Spherical reference coordinate system

Figure 5.2. Let us also assume that two orthogonal field components tangential to the spherical surface, namely $E_\theta\hat{\theta}$ and $E_\varphi\hat{\varphi}$, are measured for each sampling point. The measured field $\boldsymbol{E}(r,\theta,\varphi)$ can be expressed as a superimposition of outward propagating spherical wave vectors, as shown in the following equation:

$$\boldsymbol{E}(r,\theta,\varphi) = \frac{k}{\sqrt{\eta}}\sum_{s=1}^{2}\sum_{n=1}^{N}\sum_{m=-M}^{M} Q_{smn}^{(4)}\boldsymbol{F}_{smn}^{(4)}(r,\theta,\varphi) \tag{5.1}$$

where k is the wavenumber ($k = 2\pi/\lambda$ with λ being the wavelength) and η is the free-space admittance. The suffix $c = 4$ indicates that, under the assumption of a time dependence of the type $e^{j\omega t}$, we are considering outward propagating waves as normally done when dealing with antenna pattern measurements [7]. $\boldsymbol{F}_{smn}^{(c)}(r,\theta,\varphi)$ are the spherical wave vector functions (SWF) defined on the same spherical surface and the same coordinate system of the measured field and $Q_{smn}^{(c)}$ are the corresponding (unknown) spherical wave coefficients (SWC) which form the spherical wave spectrum.

The SWFs form an orthogonal and complete basis of modes of different orders denoted by the index triplets *snm*. Such indexes have the following meaning:

- *s* can be 1 or 2. Spherical modes with $s = 1$ are transverse electric (TE) waves, meaning that they have no radial component. Spherical modes with $s = 2$ are instead transverse magnetic (TM) waves, meaning that they do have radial component.
- *n* index varies between 1 and N, where N, as it will be described below, is associated to the global dimension of the AUT (the AUT *minimum sphere*). Such mode indexes are in fact related to the radial extent of the antenna and they strictly depend on the θ-variation of the radiated field.
- *m* index varies between $-M$ and M, where M is associated to the dimension of the AUT, only the *xy*-plane (the AUT *minimum cylinder*). Such mode indexes are used to represent the azimuthal symmetry of the radiated field and thus, they strictly depend on the φ-variation of radiated field.

The unknown SWCs are complex and scalar numbers which are used to weight each specific SWF.

The superimposition over the three mode indices *snm* of the SWFs weighted by the corresponding SWCs gives rise to the measured field. In other words, the SWE reported in (5.1) is a projection used to decompose the measured field into well-defined basis functions. The results of this projection are the SWCs which represent the first step of the spherical NF/FF transformation. In fact, once the SWCs are known, the FF associated with the measured SNF can be computed in a second step using again (5.1) evaluating the SWFs at infinite distance ($r \to \infty$) and considering the desired output (θ,φ) sampling density.

5.1.1.1 AUT minimum sphere and minimum cylinder

As anticipated above, the dimensions of the AUT play a fundamental role in the SWE because they allow to define the N and M truncation indexes of the corresponding series which otherwise would have an infinity of modes.

In particular, let us define the AUT minimum sphere as the smallest sphere centered in the origin of the coordinate system which fully encloses the AUT. The minimum sphere allows to define the minimum number of n-order modes (N) to be computed to correctly represent the field radiated by the AUT. As demonstrated in [7,8], N is given by the following equation:

$$N = \lceil kR_{\min} + 0.045\sqrt[3]{k\,R_{\min}}(-P_{tr})\rceil \tag{5.2}$$

where $R_{\min}$ is the radius of the minimum sphere and P_{tr} is the truncated power expressed in dB which depends on the required accuracy (e.g., by setting $P_{tr} = -60$ dB, in an ideal noise-free scenario, the SWCs with $n > N$ are expected to be -60 dB below peak). The term $0.045\sqrt[3]{kR_{\min}}(-P_{tr})$ inside (5.2) should be seen as a "safety factor" needed to ensure a proper convergence of the n-series of (5.1).

Similarly, let us define the AUT minimum cylinder as the smallest cylinder aligned with the z-axis and centered in the origin of the coordinate system which fully encloses the AUT. The minimum cylinder allows to define the minimum number of m-order modes (M) to be computed to correctly represent the field radiated by the AUT which is given by the following equation:

$$M = \lceil kC_{\min} + 0.045\sqrt[3]{kC_{\min}}(-P_{tr})\rceil \tag{5.3}$$

where $C_{\min}$ is the radius of the minimum cylinder. Similar to (5.2), the term $0.045\sqrt[3]{kC_{\min}}(-P_{tr})$ inside (5.3) should be seen as a "safety factor" needed to ensure a proper convergence of the m-series of (5.1).

An example of AUT minimum sphere and minimum cylinder is shown in Figure 5.3 in a base transceiver station (BTS) antenna aligned along the z-axis. It is evident that in the case of antennas having a larger dimension along the z-axis, we will have $N > M$; otherwise, we will have $N = M$. In any case, it is highlighted that, due to the properties of the SWF [7], M cannot be greater than N.

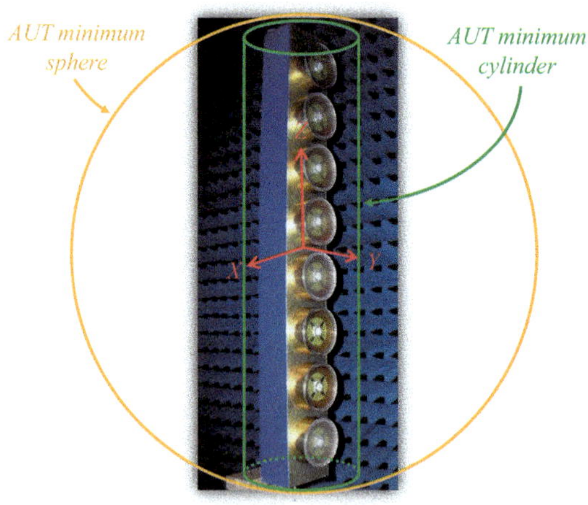

Figure 5.3 Example of AUT minimum sphere and AUT minimum cylinder in the case of a BTS antenna

5.1.1.2 Sampling criteria

When dealing with data transformations in general, a sufficient number of sample points should be acquired in the initial domain in order to avoid the well-known aliasing effect. In particular, the sampling rate, according to the Nyquist–Shannon sampling theorem for signal processing, is proportional to the inverse of twice the bandwidth of the signal to be transformed. Of course, the same criterion also applies to the spatial sampling of electromagnetic waves whose rate must be at least half wavelength ($\lambda/2$). This directly applies to planar NF (PNF) measurements [1] where the field is typically measured on a plane in front of the antenna considering the abovementioned half wavelength rate along both scanning axes.

In the case of SNF measurements, the NF/FF transformation is performed with the SWE. In the previous section, it has been shown that the minimum number of SWC to be computed (N and M) is given by (5.2) and (5.3). Intuitively, in order to correctly compute the SWCs, the minimum number of samples to be measured should be at least equal to N and M. In particular, considering a uniform sampling along the θ-axis (with $\theta = [0^\circ, 180^\circ]$) and the φ-axis (with $\varphi = [0^\circ, 360^\circ]$), the minimum angular sampling rate is given by the angular steps reported in the following equations:

$$\Delta\theta = \frac{\pi}{N} \tag{5.4}$$

$$\Delta\varphi = \frac{\pi}{M} \tag{5.5}$$

It is interesting to note that inserting (5.2) and (5.3), neglecting the two "safety factors", inside (5.4) and (5.5), respectively, gives rise to

$$\Delta\theta R_{\min} = \frac{\pi}{k} = \frac{\lambda}{2} \tag{5.6}$$

$$\Delta\varphi C_{\min} = \frac{\pi}{k} = \frac{\lambda}{2} \tag{5.7}$$

Equations (5.6) and (5.7) show that also in the case of SNF measurements, the minimum sampling rate is given by the conventional half wavelength spacing. The difference with respect to PNF measurement is that such minimum sampling should be enforced considering the AUT minimum sphere/cylinder instead of the spherical scanning surface.

Following the same argument, it is possible to define *equivalent sampling areas* based on the applied angular sampling step. In particular, the equivalent sampling sphere (ESS) and the equivalent sampling cylinder (ESC) are the "virtual" sphere/cylinder where half wavelength sampling is met and which are given by (5.8) and (5.9):

$$R_{ESS} = \frac{\lambda}{2\Delta\theta} = \frac{\pi}{k\Delta\theta} \tag{5.8}$$

$$R_{ESC} = \frac{\lambda}{2\Delta\varphi} = \frac{\pi}{k\Delta\varphi} \tag{5.9}$$

where R_{ESS} is the radius of the ESS and R_{ESC} is the radius of the ESC. Of course, if a denser sampling with respect to the minimum one is considered (oversampling), the ESS and the ESC will be larger than the AUT minimum sphere and cylinder. As we will see in the following sections, when the oversampling is applied, the spatial filtering properties of the SWE can be exploited allowing for a suppression of possible unwanted contributions such as noise and echoes.

It is finally highlighted that the ESS and the ESC cannot be larger than the radius of the measurement spherical surface (R_M). This is a direct consequence of the cutoff properties of the SWFs [7] which do not allow modes with n and m indexes higher than kR_M to propagate.

5.1.1.3 Computation of the SWC

The easiest and more intuitive way of calculating the SWC is to rewrite (5.1) as a linear system of the form $\mathbf{b} = \mathbf{Ax}$, where $\mathbf{b}$ is a column vector with i measurement *NF points*, $\mathbf{x}$ is another column vector with j unknown SWC and $\mathbf{A}$ is an $i \times j$ matrix containing the SWE basis represented by the SWF. The solution of the system by mean of $\mathbf{x}$ can be simply performed by standard inversion methods. Although very simple, this approach is computationally inefficient and in the case of large antennas, it may require a lot of memory resources and be very time-consuming.

In order to compute the SWC in an efficient way, it is convenient to express the measured signal over the spherical surface with *transmission formula* [7] shown in the following equation:

$$w(r, \chi, \theta, \varphi) = 0.5 \sum_{\substack{smn \\ \sigma\mu\nu}} Q_{smn}^{(4)} e^{jm\varphi} d_{\mu m}^{n}(\theta) e^{j\mu\chi} C_{\sigma\mu\nu}^{sn(4)}(kr) R_{\sigma\mu\nu}^{p} \tag{5.10}$$

With reference to the same coordinate system shown in Figure 5.2, such formula expresses the complex signal (w) radiated by and AUT of unknown SWC ($Q_{smn}^{(4)}$) and measured by a generic probe with known SWC ($R_{\sigma\mu\nu}^{p}$, where the $\sigma\mu\nu$ probe indexes are used to distinguish them from the ones of the AUT). The probe is moved on a spherical surface of radius $r = R_M$, in the (θ, φ) sampling points with the orientation given by χ ($\chi = 0$ for $w_\theta \hat{\theta}$ and $\chi = \pi/2$ for $w_\varphi \hat{\varphi}$). The symbols $d_{\mu m}^{n}(\theta)$ and $C_{\sigma\mu\nu}^{sn(3)}(kr)$ are, respectively, rotation and translation operators [7] that, together with the two complex exponentials ($e^{jm\varphi}$ and $e^{j\mu\chi}$), are used to describe the probe position/orientation in each measurement point. It should be noted that the knowledge of the probe SWC allows to compensate for the effect of the probe pattern (probe correction [7,9–11]). For simplicity, let us now assume a Hertzian dipole as probe. Such ideal probe has nonzero SWC only for $\sigma = 2$, $\mu = \pm 1$ and $\nu = 1$. Equation (5.10) can thus be significantly simplified by removing the σ and ν-series and considering only two terms in the μ-series as shown in (5.11). It should be noted that the latter equation is equivalent to (5.1):

$$E(r, \chi, \theta, \varphi) = \frac{k}{\sqrt{\eta}} \sum_{s=1}^{2} \sum_{n=1}^{N} \sum_{m=-M}^{M} Q_{smn}^{(4)} e^{jm\varphi} \sum_{\mu=-1}^{1} d_{\mu m}^{n}(\theta) e^{j\mu\chi} C_{2\mu 1}^{sn(4)}(kr) R_{2\mu 1}^{p} \tag{5.11}$$

In [7], it has also been shown that the rotation coefficient can be expressed in the form of a Fourier series using the following equation:

$$d^n_{\mu m}(\theta) = (-j)^{m-\mu} \sum_{m'=-n}^{n} \Delta^n_{m'\mu} \Delta^n_{m'm} e^{jm'\theta} \tag{5.12}$$

where $\Delta^n_{m'\mu}$ and $\Delta^n_{m'm}$ are the so-called delta pyramids that can be calculated very efficiently using recurrence relations also reported in [1,7].

By substituting (5.12) into (5.11), we obtain (5.13):

$$E(r,\chi,\theta,\varphi) = \frac{k}{\sqrt{\eta}} \sum_{m=-M}^{M} e^{jm\varphi} \sum_{m'=N}^{N} e^{jm'\theta} \sum_{\mu=-1}^{1} e^{j\mu\chi}(-j)^{m-\mu} \times \sum_{n=1}^{N} \Delta^n_{m'\mu} \Delta^n_{m'm} \sum_{s=1}^{2} Q^{(4)}_{smn} C^{sn(4)}_{2\mu 1}(kr) R^p_{2\mu 1} \tag{5.13}$$

As can be seen, the two exponential functions, $e^{jm\varphi}$ and $e^{jm'\theta}$, define two Fourier series for the φ and θ dependencies, respectively. Using the transmission formula, the computational complexity of the SWE can thus significantly reduce applying a 2D FFT to the measured field as suggested by (5.14), where $IFFT_m$ and $IFFT_{m'}$ indicate the inverse FFT for the mode indexes m and m', respectively:

$$E(r,\chi,\theta,\varphi) = \frac{k}{\sqrt{\eta}} IFFT_m \left(IFFT_{m'} \left(\sum_{\mu=-1}^{1} e^{j\mu\chi}(-j)^{m-\mu} \times \sum_{n=1}^{N} \Delta^n_{m'\mu} \Delta^n_{m'm} \sum_{s=1}^{2} Q^{(4)}_{smn} C^{sn(4)}_{2\mu 1}(kr) R^p_{2\mu 1} \right) \right) \tag{5.14}$$

Once the 2D FFT is applied to the measured field, the Fourier coefficients with m and m' indexes are found. From them, another transformation should be applied to get the seeked SWC with indexes s, m and n. For the sake of brevity, the details of such transformation are not reported here. For more details, the reader may refer to Section 4.3 of [7], where it is explained how the SWCs are efficiently computed from the Fourier coefficients using the recurrence relation of the delta pyramids.

5.1.1.4 Understanding the SWC

In order to have a better understanding of the SWE, the SWC associated to a full-wave-simulated horn antenna will be illustrated and analyzed in this paragraph.

The considered antenna is the MVG SH600 dual-ridge linearly polarized horn* shown in Figure 5.4. Such device works in the 0.6–9 GHz-wide frequency range and is typically used for gain calibration purposes. The directivity co-polar and

*www.mvg-world.com/en/products/field_product_family/antenna-1/dual-ridge-horns.

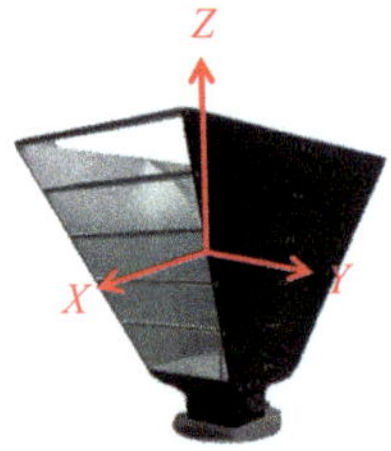

Figure 5.4 MVG SH600 dual-ridge horn

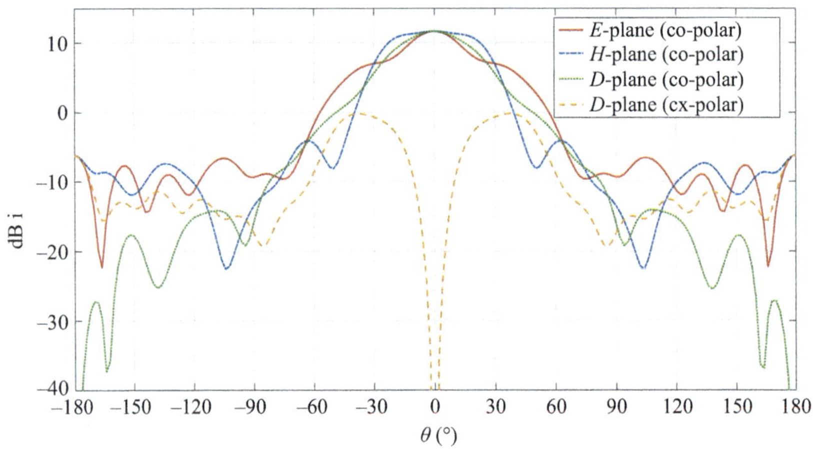

Figure 5.5 Directivity radiation pattern cuts of the MVG SH600 dual-ridge horn at 3 GHz. E-plane co-polar (solid-orange); H-plane co-polar (dash/dotted-blue); diagonal-plane co-polar (dotted-green); diagonal-plane cx-polar (dashed-yellow)

cx-polar radiation patterns, according to Ludwig's third definition [12], of the SH600 at 3 GHz along the two main cuts (*E*-plane and *H*-plane) and along one intercardinal (diagonal) plane are shown in Figure 5.5. Such patterns have been obtained with full-wave simulation of the device.[†] As can be seen, the antenna has no cx-polar along the two main cuts while it has delta cx-polar pattern in the diagonal plane. The dimension of the antenna aperture is 199 × 356 mm (*E*-plane × *H*-plane) while its height is 295 mm. Considering the origin coordinate system in the geometrical center of the antenna, the radius of the minimum sphere enclosing the antenna is $R_{min} = 252$ mm corresponding to approximately 2.5λ at 3 GHz. The radius of the minimum cylinder enclosing the antenna is $C_{min} = 204$ mm corresponding to approximately 2λ at 3 GHz.

[†]www.cst.com/.

According to (5.2) and (5.3), the minimum number of modes needed to represent such antenna at 3 GHz with an accuracy of $P_{tr} = -60$ dB is given by $N = 23$ and $M = 20$. Consequently, according to (5.4) and (5.5), the minimum sampling required to measure such antenna is approximately $\Delta\theta = 7.8°$ and $\Delta\varphi = 9°$.

The computed SWCs of the simulated horn at 3 GHz are shown in Figure 5.6 as 2D maps where the n-mode and m-mode variations are reported, respectively, along the vertical and horizontal axes. TM-modes ($Q^{(4)}_{1mn}$ or simply Q_1) are shown in

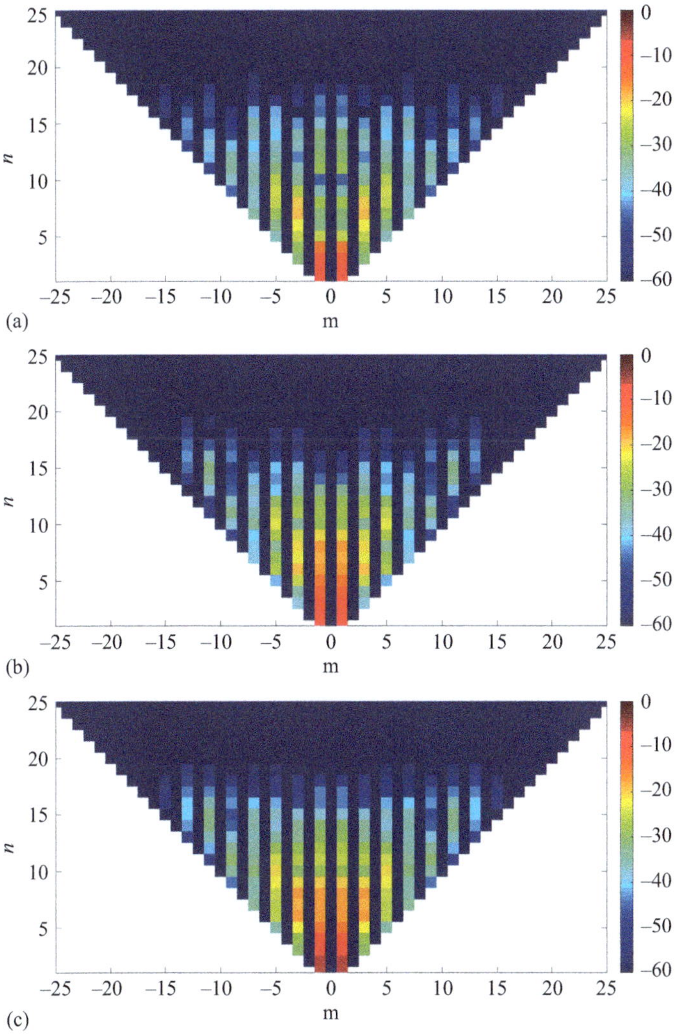

Figure 5.6 2D representation of the SWC of the full-wave simulated SH600 dual-ridge antenna at 3 GHz. Q_1 coefficients (a), Q_2 coefficients (b) and Q_{tot} coefficients (c)

Figure 5.6(a) while TE-modes ($Q_{2mn}^{(4)}$ or simply Q_2) are shown in Figure 5.6(b). Both mode maps are normalized with respect to the total radiated power given by (5.15):

$$P_{tot} = \frac{1}{2}\sum_{s}\sum_{m}\sum_{n}|Q_{smn}^{(4)}|^2 \tag{5.15}$$

and expressed in dB ($10\log_{10}\left(|Q_{smn}^{(4)}|^2/P_{tot}\right)$). The TM-modes dominate in the case of magnetic dipoles while the TE-modes dominate in the case of electric dipoles. As can be seen in this case, the level and distribution of the TM- and TE-modes are very similar. Since this is a very common situation, it is often convenient to consider a more compact mode map obtained by the squared sum of the TM- and TE-modes ($Q_{tot,\mathrm{mn}}^{(4)}$ or simply Q_{tot}). Such map is shown in Figure 5.6(c) and is also normalized with respect to P_{tot} and expressed in dB. As can be seen, modes with indexes $N > 23$ and $|M| > 20$ are as expected well below the considered dynamic.

An even mode compact and widely used way of representing the spherical modal spectrum of an antenna can be obtained using the P_n and $P_{|m|}$ (or simply P_m) power spectrum curves defined by (5.16) and (5.17), respectively:

$$P_n = \frac{1}{P_{tot}}\sum_{s=1}^{2}\sum_{m=-n}^{n}|Q_{smn}^{(4)}|^2 \tag{5.16}$$

$$P_{|m|} = \frac{1}{P_{tot}}\sum_{s=1}^{2}\sum_{n=|m|}^{N}\left(\left|Q_{s,m,n}^{(4)}\right|^2 + \left|Q_{s,-m,n}^{(4)}\right|^2\right) \tag{5.17}$$

The P_n power spectrum is obtained by the squared summation of all the m- and s-mode indexes and is thus useful for a compact description of the radial extended of the antenna. Similarly, the P_m power spectrum is obtained by the squared summation of all the n- and s-mode indexes (and of modes with positive m-index with the corresponding modes with negative m-index and) and is thus useful for a compact description of the azimuthal behavior of the antenna. Both P_n and P_m are normalized with respect to the total power.

The P_n and P_m power spectra of the simulated dual-ridge horn at 3 GHz are shown in Figure 5.7. Looking at P_n, it can be observed that modes with index $n > 23$ are, as expected, more than 60 dB below peak, meaning that a proper convergence of the n-series of the SWE is reached. From the P_m power spectrum, it can be observed that a proper convergence of m-series is reached even before the predicted $M = 20$ minimum index, meaning that the edges of the considered antenna do not contribute significantly to azimuthal behavior of the radiate field. In other words, the *effective* minimum cylinder is slightly smaller than the considered geometrical one. It is highlighted that the P_m power spectrum is very useful to check the symmetries of the antenna; i.e., antennas with some rotational symmetry around the z-axis are expected to radiate only few azimuthal modes (in the case of perfect rotational symmetry only modes with $|m| = 1$ are radiated). In this example, modes with even m-index are negligible. This is a particular situation that happens

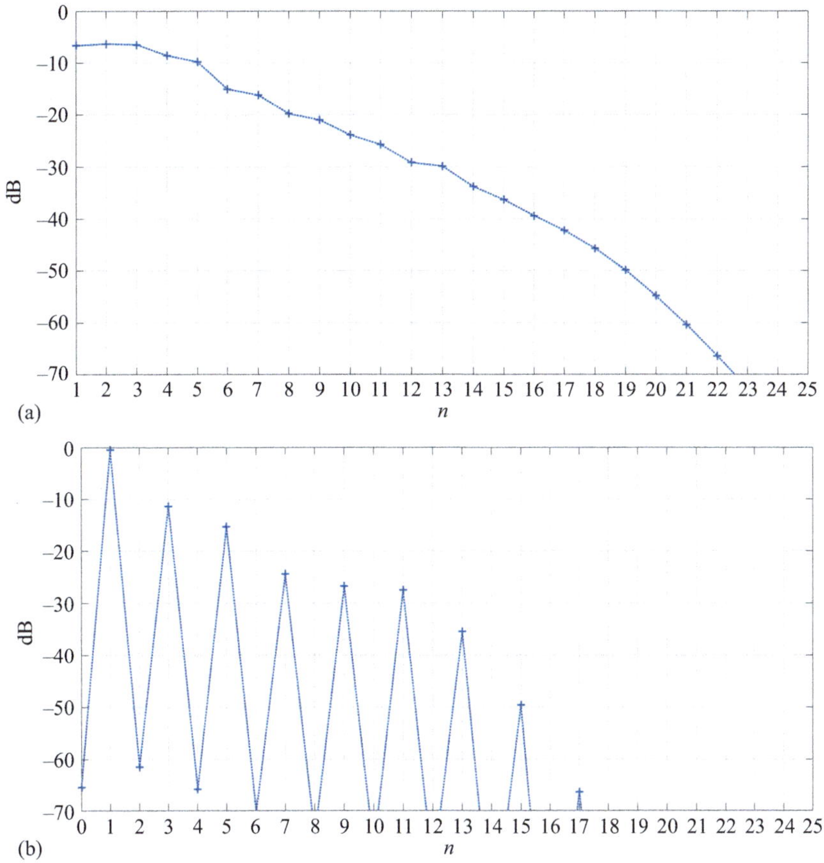

Figure 5.7 1D representation of the SWC of the full-wave simulated SH600 dual-ridge antenna at 3 GHz: P_n power spectrum (a) and P_m power spectrum (b)

when the elevation pattern cuts of the antenna are symmetric with respect to the boresight direction ($\theta = 0°$). As can be seen in Figure 5.5, the considered horn antenna at 3 GHz falls in this category.

Let us now consider the same horn antenna at 3 GHz but placed in different offset positions in the reference systems. When an antenna is translated away from the origin of the coordinate systems, its minim sphere/cylinder is increased and, consequently, the number of modes needed to describe the device and the associated number of sampling points are higher.

For example, let us consider the horn antenna with its geometrical center at $z_t = +5\lambda$, $z_t = +10\lambda$ and $z_t = +15\lambda$. The radius of the antenna minimum sphere is increased from 2.5λ to 7.5λ, 12.5λ and 17.5λ, respectively. With such translation along the z-axis, the radius of minimum cylinder is instead unchanged. The Q_{tot}

spectra associated to these three displacements of the antenna along the *z*-axis are shown in Figure 5.8. As can be seen, the increase of the antenna minimum sphere is reflected directly on the spectrum whose modes are spread along the *n*-indexes. Of course, larger translations give rise to bigger mode spreads. As expected, the

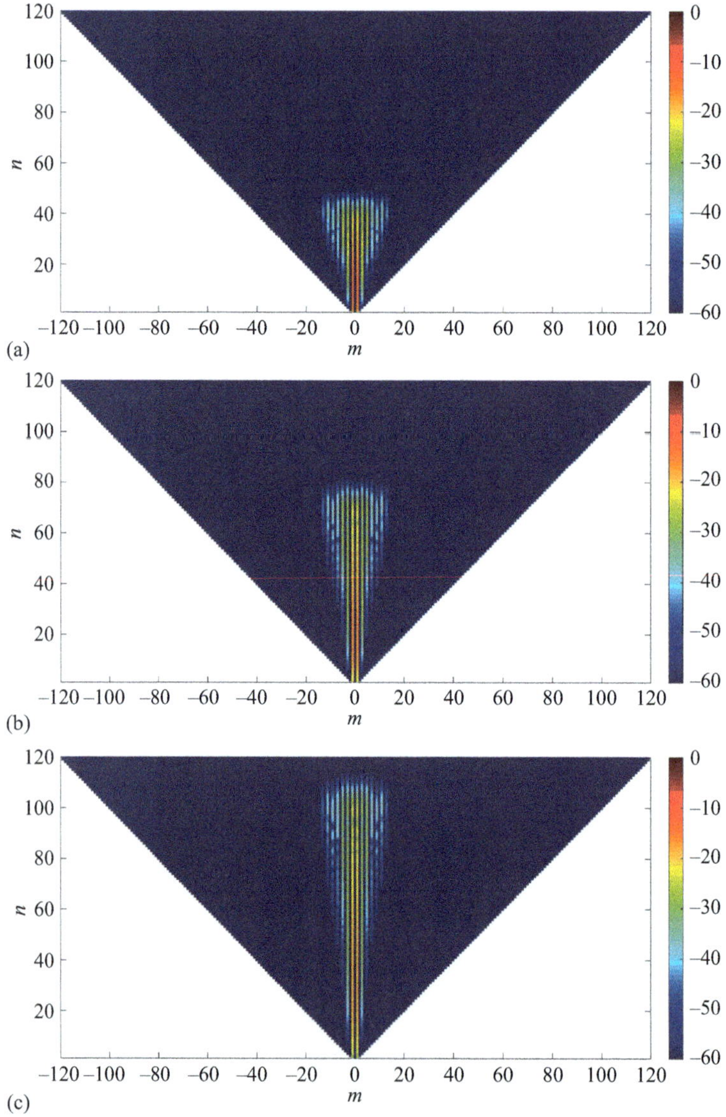

Figure 5.8 2D representation of the SWC (Q_{tot}) of the full-wave simulated SH600 dual-ridge antenna at 3 GHz, offset displaced along the z-axis at three different locations: 5λ (a), 10λ (b), and 15λ (c)

mode extension along the m-index is unchanged because the z-translation does not increase the minimum antenna cylinder.

The P_n power spectra of the horn antenna at 3 GHz at the considered different locations along the z-axis are shown in Figure 5.9 together with the spectrum of the centered antenna. It is evident how the modal bandwidth of the antenna is enlarged along the n-indexes. Moreover, it can be observed that a cutoff of the spectrum is present at approximately $N_{cutoff} = \lceil kd \rceil$ with d being the translation distance from the origin. With $d = z_t$, we in fact have $N_{cutoff} = 32$, $N_{cutoff} = 63$ and $N_{cutoff} = 94$, respectively, for the three considered z-translations. It is also observed that the decay of the spectrum after the cutoff is slower for bigger offsets. This explains the reason why the safety factor used in (5.2) also increases with the minimum sphere radius [8].

As mentioned above, a translation along the z-axis does not modify the azimuthal symmetry of the antenna (same minimum cylinder). The P_m power spectra of the considered horn antenna when the z-translations are applied are thus identical to the one shown in Figure 5.7(b) where no translation was applied.

Let us now consider the horn antenna with its geometrical center translated along the x-axis, namely, $x_t = +5\lambda$, $x_t = +10\lambda$ and $x_t = +15\lambda$. As in the previous case, the radius of the antenna minimum sphere is increased from 2.5λ to 7.5λ, 12.5λ and 17.5λ, respectively. In this case also, the radius of minimum cylinder is increased and is approximately as large as one of the minimum spheres. The Q_{tot} spectra associated to these displacements of the antenna along the x-axis are shown in Figure 5.10. The translation along the x-axis modifies the radial extent and the azimuthal symmetry of the antenna. Hence, the spherical modes are spread both along the n- and m-indexes after the translation. The same behavior is also observed

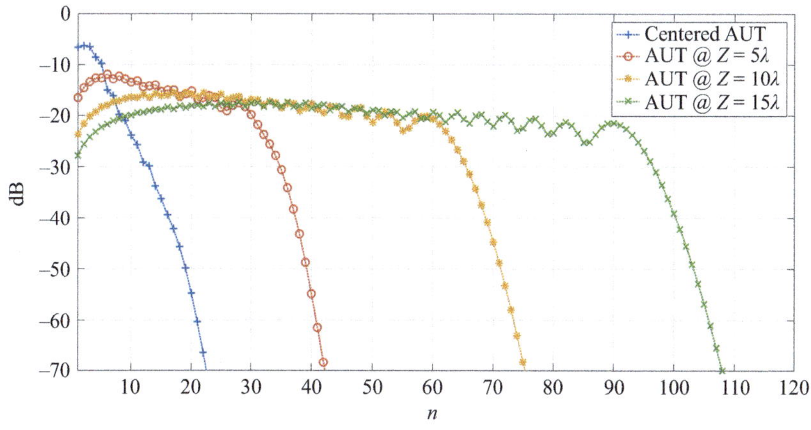

Figure 5.9 P_n power spectrum comparison of the full-wave simulated SH600 dual-ridge antenna at 3 GHz, offset displaced along the z-axis at three different locations: onset (blue "+"), 5λ (orange "o"), 10λ (yellow "") and 15λ (green "x")*

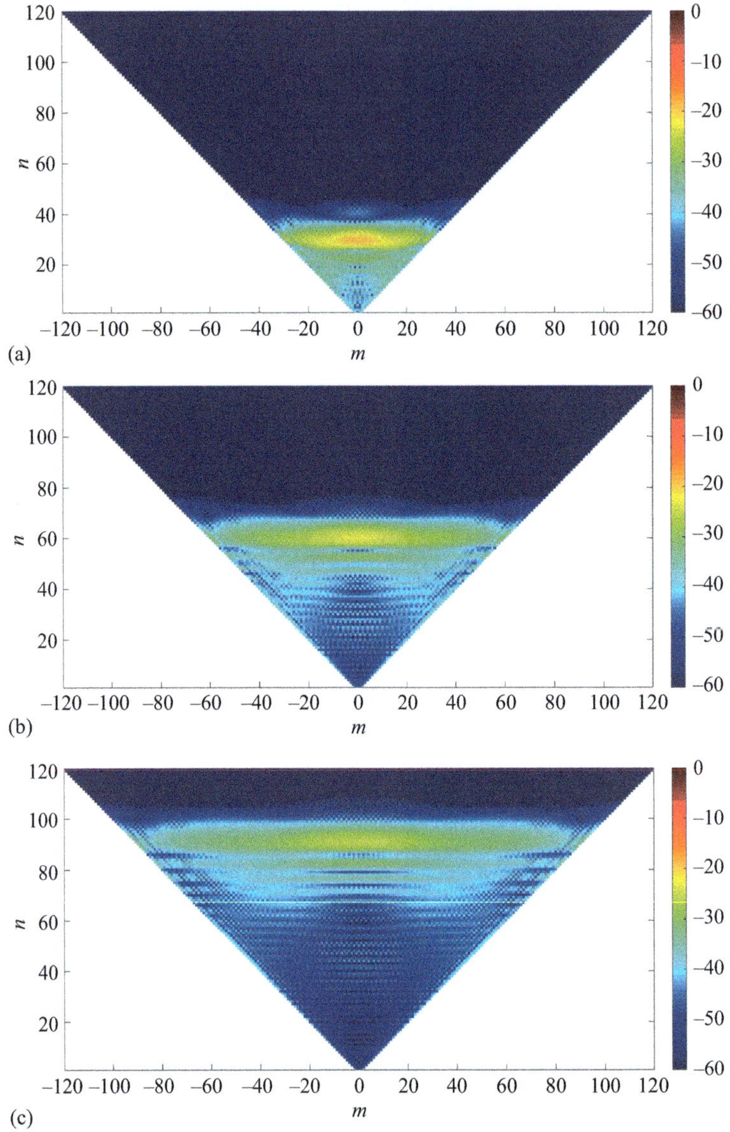

Figure 5.10 2D representation of the SWC (Q_{tot}) of the full-wave simulated SH600 dual-ridge antenna at 3 GHz, offset displaced along the x-axis at three different locations: 5λ (a), 10λ (b) and 15λ (c)

in the P_n and P_m power spectra comparison reported in Figure 5.11, where it can be clearly seen how the modal bandwidth is expanded along both indexes. A similar behavior of the spectrum would have been observed also considering a translation along the y-axis.

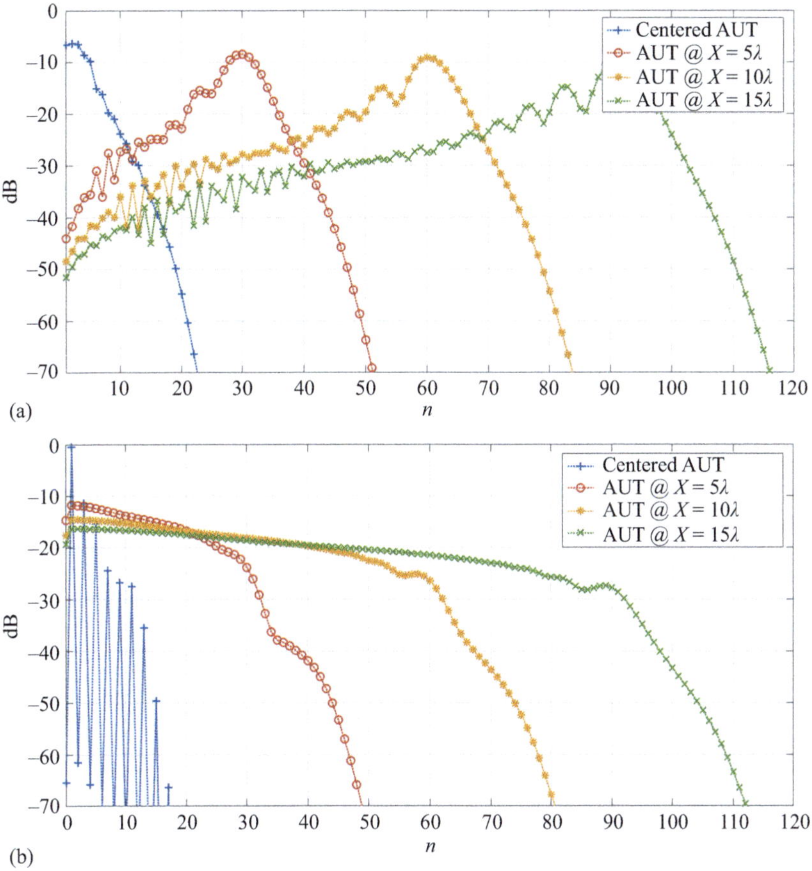

Figure 5.11 *P_n (a) and P_m (b) power spectrum comparison of the full-wave-simulated SH600 dual-ridge antenna at 3 GHz, offset displaced along the x-axis at three different locations: onset (blue "+"), 5λ (orange "o"), 10λ (yellow "*") and 15λ (green "x")*

5.1.2 SWE as spatial filter

After the introduction and examples of the SWE reported in previous paragraph, it is evident that the SWE is a very powerful tool that can be used not only to efficiently perform the spherical NF/FF transformation but also to improve the measurement accuracy, taking advantage of the properties of the spherical wave functions.

In Section 5.1.1.4, examples of spherical wave spectrum of a horn antenna in onset and different offset configurations have been presented. It has been shown that when the antenna is moved away from the origin of the reference coordinate system, the SWCs are spread over the spherical wave spectrum. In particular, it has been pointed out that the additional modes needed to describe the translated antenna are directly proportional to kd, where d is the distance from the origin of the coordinate system to the radiating source location. It is thus clear that the spherical wave spectrum is strongly related to the space domain. Unwanted contributions *spatially* located outside the volume of the antenna minimum sphere/cylinder are in fact associated to higher order spherical modes that can be easily filtered out by setting them zero. Due to the strong relation between modal and spatial domain, such filtering operation is called either *modal* or *spatial filtering*.

The spherical wave spectrum examples of the horn antenna in different positions reported in Section 5.1.1.4 also underline the importance of defining a proper reference system for the spatial filtering. Depending on the adopted reference system, the same antenna is described with a different number of SWC. Intuitively, the reference system that minimizes the number SWC of the antenna is the optimal choice since it allows for a more compact representation of the device. When this situation is met, the effectiveness of the spatial filtering is maximized applying a low-pass filter where only the modal frequencies associated to the antenna are kept.

For example, let us take into account the P_n power spectra of the horn antenna translated along the x-axis (see Figure 5.11(a)). When the onset antenna configuration is considered, modes with index $n > 23$ can be clearly filtered out. The retained modes with $n \leq 23$ are associated to a maximum equivalent radius of approximately 3.6λ, or 360 mm at 3 GHz (see (5.4) and (5.8)). The spatial filtering will be in this case very effective because the extent of the retained modes is only slightly larger than the antenna minimum sphere ($R_{\min} = 2.5\lambda$). Instead, when the antenna is translated at $x_t = 15\lambda$, only modes with index $n > 115$ can be filtered out. The retained modes with $n \leq 115$ are associated to a much larger maximum equivalent radius approximately equal to 18.3λ, or $1,830$ mm at 3 GHz. Clearly, in the latter case, much more possible unwanted contributions may fall within retained modes being associated to a wider spatial domain. To maximize the effectiveness of the modal filtering, it is thus important to consider the reference system that allows to define the smallest antenna minimum sphere. Such optimal choice is obtained when the reference system coincides with geometrical center of the antenna.

As we will see with more details in Section 5.1.2.1, when the FF condition is met, the reference system can be easily modified by applying a proper phase-shift to the phase pattern of the measured field. To exploit the spatial filtering properties of the SWE, it is thus convenient to consider the field already transformed to FF. This also allows to take advantage of the spatial filtering properties of the SWE in the case of other type of NF measurements, such as planar (PNF) and cylindrical (CNF). As a matter of fact, the FF obtained applying the proper NF/FF transformation to the NF data measured in these other types of ranges is a 3D field defined over a spherical surface with radius tending to infinity. The SWE can thus be applying considering the same transmission formula reported in (5.14) where the asymptotic approximation of the translation coefficient is considered [7].

A detailed description of the workflow that allows to apply the spatial filtering based on the SWE from a general NF measurement (either SNF, PNF or CNF) is reported in Section 5.1.2.1.

Practical applications of the SWE as spatial filter will also be addressed in this paragraph. Specifically, in Section 5.1.2.2, we will see how the SWE can be effectively used as *noise reduction* tool in spherical NF measurements. Due to the random nature of the noise present in the measurement instrumentations (thermal noise basically), the noise is in fact uniformly distributed all over the computed spectrum. If more NF samples than the ones required by the sampling criteria (Section 5.1.1.2) are acquired (*Oversampling*), the signal-to-noise ratio (SNR) can be improved after the application of the modal/spatial filtering.

In Section 5.1.2.3 instead, the application of the SWE as echo suppression technique will be discussed and different measurement examples will be presented. In such a case, we will see that the effectiveness of the spatial filtering depends a lot on how the NF acquisition is performed. For example, in PNF measurements, the AUT is stationary while the probe is moved over the scanning surface. Instead, in many single probe SNF system, the AUT is rotated around two axes while the probe is stationary. As we will see, possible echoes present in these scenarios have different effects on the measured data and they must be treated in a proper way.

Finally, in Section 5.1.2.4, we will also see how the spatial filtering properties of the SWE can be exploited to reduce the errors caused by the truncation of the scanning area in SNF measurements by using an iterative algorithm.

5.1.2.1 Spatial filtering based on SWE: workflow

The workflow of the spatial filtering based on the SWE is shown in Figure 5.12. The input is the NF dataset measurement in any of the conventional NF geometries (spherical, cylindrical or planar). Each step of the process is described below.

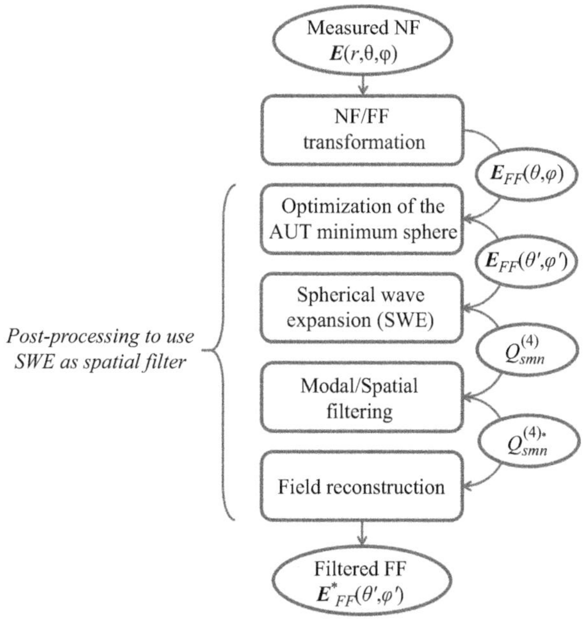

Figure 5.12 Spatial filtering based on SWE: workflow

NF/FF transformation

The FF is computed mathematically applying the NF/FF transformation. As recalled in Section 5.1.1.3 and described in [1,7] in the case of the SNF measurements, the NF/FF transformation process is computed with the SWE. In the case of CNF measurements, the transformation is typically performed with the CWE as described in [1,13]. When PNF measurements are performed, the FF is obtained by computing the plane wave spectrum as described in [1,14]. It should be noted that, in the case of equally spaced acquisitions, each of the mentioned transformation is very efficient because they all use the FFT to perform the expansion. Moreover, probe pattern compensation can be applied during the mentioned NF/FF transformation processes [1,7,9,13,14].

Optimization of AUT minimum sphere

As previously described, to maximize the effectiveness of the spatial filtering is important to consider the reference system that allows to define the smallest minimum sphere/cylinder that fully encloses the antenna. Such optimal choice is obtained when the reference system coincides with geometrical center of the antenna. In this step, the reference system is properly translated for this purpose.

It should be noted that when the FF of an antenna is available (either because the results of the NF/FF transformation or because measured at sufficiently large distance, i.e., $>2D^2/\lambda$, where D is the maximum diameter of the AUT), the reference system can be easily translated by simply applying a simple phase-shift to the phase pattern of the radiated field [15]. Naming $t = (x_0, y_0, z_0)$, the translation vector which allows to transform from an initial reference system to a second (primed) reference system, the transformation is given by the following equation:

$$\boldsymbol{E}'(r', \theta', \varphi') = \boldsymbol{E}(r, \theta, \varphi)e^{j(\sin\theta\,\cos\varphi\,x_0 + \sin\theta\,\sin\varphi\,y_0 + \cos\theta\,z_0)} \tag{5.18}$$

It is well known [15] that such formulas are valid under the assumption of FF condition which ensures that only the phase pattern varies with respect to the coordinate system, while the amplitude pattern is independent of the coordinate system.

Spherical wave expansion

The spherical wave spectrum is computed applying the SWE. In particular, since the field to be expanded is a FF, the transmission formula reported in (5.14) is inverted by mean of the $Q_{smn}^{(4)}$ coefficients considering the asymptotic approximation of the translation coefficient [7].

It should be noted that the FF obtained from a PNF measurement is defined only on the forward hemisphere. To correctly apply the SWE, the field on the backward hemisphere must be extrapolated. Typically, the AUTs measured in PNF systems have a relatively high directivity (e.g., >15 dB i); hence, in such cases, virtual points of zero intensity (zero-padding (ZP)) can be added to the missing portion of the sphere without introducing significant errors.

Modal/Spatial filtering

Once the spherical wave spectrum is available, the spatial/modal filtering is applied by setting to zero the higher order modes having indexes bigger than those defined by the AUT minimum sphere and cylinder. In other words, a low-pass filtering is applied considering as cutoff the modes index, N and M, defined, respectively, by (5.2) and (5.3).

Field reconstruction

From the filtered spherical wave spectrum, the FF is recomputed using directly the transmission formula reported in (5.14), considering again the asymptotic approximation of the translation coefficient [7].

It should be noted that in the case of SNF measurements with antenna located in the center of the measurement sphere (origin of the coordinate system), such workflow can be simplified by performing the SWE on the SNF data, applying the modal filtering on the computed spectrum and finally computing the FF from the filtered spectrum.

5.1.2.2 Noise reduction in SNF measurements using the SWE

Thermal and other random noise sources originated by the RF instrumentation give rise to an error contribution in any type of measurement systems [1]. With modern receivers and sufficient amplification in the system, this term often gives an insignificant contribution in the overall measurement uncertainty. However, in special cases, the uncertainty linked to random noise may be more significant and the proper treatment of this term is needed to evaluate the impact on overall measurement uncertainty.

As recalled in previous sections, in any NF systems, an NF/FF transformation is applied to compute the FF. Such transformations are based on the computation of a modal spectrum whose extension depends on the considered NF sampling, i.e., the higher the number acquired samples, the bigger the computed spectrum. Intuitively, if more samples are measured with respect to the minimum required by the Nyquist criteria (*oversampling*), the additional part of the computed spectrum will be only associated to the noise (and eventually other source of errors), resulting in an improvement of the *SNR* if such additional part of the spectrum is filtered out [16].

The oversampling is particularly useful in the case of low/medium directive devices under test which are typically measured in SNF systems [16]. In fact, when antennas with a higher directivity (e.g., >20 dB i and more) are measured in NF ranges, the NF/FF transformation introduces the so-called processing gain. The processing gain is a direct consequence to the fact that the beam of the antenna is much more pronounced in FF than in NF. Due to the random nature of the noise, this one will be equally spread all over the computed FF pattern resulting in an improvement of the SNR on the main beam. Instead, in the case of antennas with low/medium directivity, the FF beam is very close to the measured NF beam so the processing gain is close to zero.

In this section, we explain how the oversampling combined with the SWE modal filtering can be exploited in SNF measurements to improve the SNR of low/medium directive devices under test. To do that, the full-wave simulated SH600 dual-ridge horn shown in Figure 5.5 will be taken into account at the working frequency of 3 GHz, where its peak directivity is approximately 12 dB i (see directivity radiation pattern Figure 5.6), and different noisy measurements will be emulated.

As reported in [7], the total number of spherical wave modes, J, is given by (5.19), where N is defined given by (5.2):

$$J = 2N(N+2) \tag{5.19}$$

Due to the random nature of typical noises (e.g., thermal noise), the noise power (P_{noise}) is uniformly spread over each computed mode. The noise power of a single SWC (ΔP_{noise}) is then given by $\Delta P_{noise} = P_{noise}/J$. It is thus evident that, increasing the sampling density and consequently the number of SWC, ΔP_{noise} is reduced. The noise power that falls in the modes needed to describe the AUT is thus also reduced obtaining an improvement of the SNR after the application of the modal filtering. The theoretical improvement of the SNR, defined as the ratio between the noise power *before* and *after* the modal filtering, can be defined with (5.20):

$$SNR_{improv} = \frac{P_{noise,before}}{P_{noise,after}} = \frac{2O_sN(O_sN+2)}{2N(N+2)} \sim O_s^2 \tag{5.20}$$

where O_s is the oversampling factor. Assuming the same O_s along both scanning axis (θ and φ), the SNR improvement is in first approximation proportional to the square of the oversampling factor. For example, considering $O_s = 10$, approximately 20 dB improvement of the SNR is expected. In the case of different oversampling along the scanning axes, the improvement of the SNR will be proportional to $O_{s,\theta} \times O_{s,\varphi}$, where $O_{s,\theta}$ and $O_{s,\Phi}$ are, respectively, the oversampling factors along the θ- and φ-scanning axis.

To verify (5.20), different noisy measurements of the simulated SH600 dual-ridge horn at 3 GHz have been emulated considering different oversampling. The SWC spectrum of a measurement emulation with an SNR of 40 dB (e.g., the noise floor is at −40 dB level with respect to the peak of the NF measured pattern) and three times oversampling is shown in Figure 5.13. The random noise has been

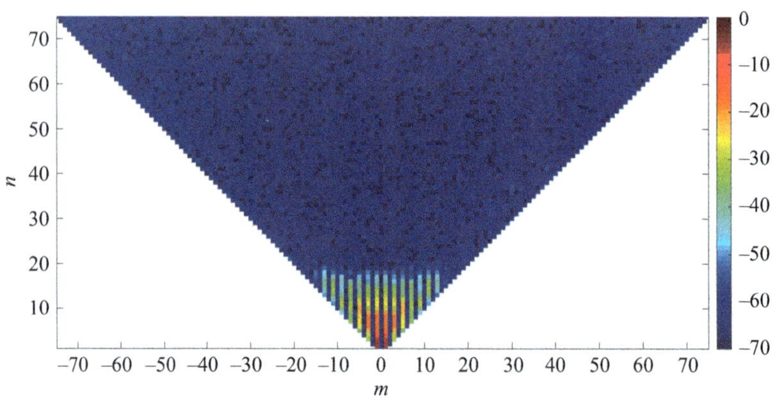

Figure 5.13 2D representation of the SWC (Q_{tot}) of an emulated noisy measurement (SNR = 40 dB) of the SH600 dual-ridge antenna at 3 GHz with three times oversampling ($O_S = 3$)

modeled considering a Gaussian distribution with zero-mean (typical assumption for noise in RF instrumentation), and the SNFs have been emulated at a measurement distance of $R_M = 1.5$ m. As can be seen, the noise is uniformly spread over the entire spherical wave spectrum.

In Figure 5.14, the FF patterns obtained from different measurement emulations with different oversampling and the same SNR of 40 dB are compared to the noise-free FF reference. The *H*-plane reference directivity pattern cut is shown by the solid black trace. The *H*-plane pattern obtained with the minimum sampling ($O_s = 1$) is shown by the dashed blue trace. As can be seen, such pattern cut shows some deviation with respect to the reference which is of course more evident at lower pattern level where the local SNR is worse. To better quantify the error, the point-wise equivalent noise level (ENL) defined by (5.21), where $E(\theta, \varphi)$ is the reference pattern and $\tilde{E}(\theta, \varphi)$ is the test pattern, is computed and reported in the same figure (in dB scale) for different oversampling:

$$ENL = 20 \log_{10}\left(\left|\frac{E(\theta, \varphi) - \tilde{E}(\theta, \varphi)}{E(\theta, \varphi)_{\text{MAX}}}\right|\right) \tag{5.21}$$

The dotted blue trace with "o" markers is the ENL in the case of $O_s = 1$. As can be seen, the error is on average around −45 dB level, thus very close to the considered SNR (40 dB). The dotted orange trace with "x" markers is the ENL in the case of $O_s = 3$ with modal filtering applied, considering the AUT minimum sphere ($R_{\min} = 2.5\lambda$) and is reduced of approximately 10 dB with respect to the previous

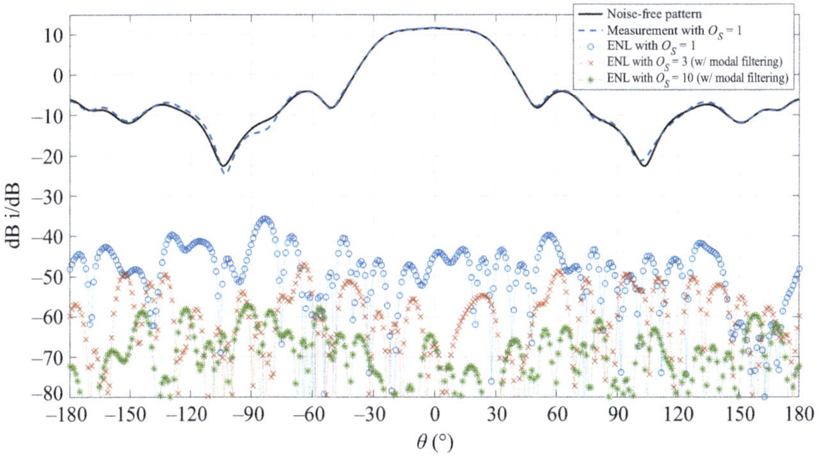

Figure 5.14 H-plane directivity pattern cut at 3 GHz of the SH600: noise-free (solid black); emulated measurement with SNR = 40 dB without O_S (dashed blue). Point-wise ENL in the case of emulated measurement with SNR = 40 dB without O_S (dotted blue "o"); with $O_S = 3$ plus modal filtering (dotted orange "x"); with $O_S = 10$ plus modal filtering (dotted green "")*

case. The dotted green trace with "*" markers is instead the ENL in the case of $O_s = 10$ with modal filtering applied. In such case, the error is in average further improved of approximately 10 dB.

A more compact illustration of the benefits obtained by the combination of oversampling and modal filtering is reported in Figure 5.15 where the global ENL, obtained with (5.22), is computed for several oversampling ranging from 1 up to 10:

$$ENL_{global} = 20\log_{10}\left(RMS\left|\frac{E(\theta,\varphi) - \tilde{E}(\theta,\varphi)}{E(\theta,\varphi)_{\mathrm{MAX}}}\right|\right) \tag{5.22}$$

In Figure 5.15, the global ENL computed on the NF data for the different oversampling is illustrated by the dashed trace with "x" markers. The ENL is constantly equal to the considered noise-floor (−40 dB), meaning that no improvement of the SNR is obtained in NF applying the oversampling. When the NF/FF is applied, but no modal filtering is considered (dotted trace with "*" marker), the SNR is improved of approximately 5 dB when up to four times oversampling is considered, meaning that the transformation itself automatically filters out some noise components. In this example, improvements for $O_s > 4$ are obtained also without modal filtering, because the considered measured radius ($R_M = 1.5$ m) generates an intrinsic modal filtering at $N = kR_M$. As also pointed out in Section 5.1.1.2, this is a direct consequence of the cutoff properties of the spherical wave functions which do not allow modes of order higher than kR_M to propagate. Finally, if the modal filtering is applied after the NF/FF considering the AUT minimum sphere, the ENL is more and more lowered increasing the oversampling.

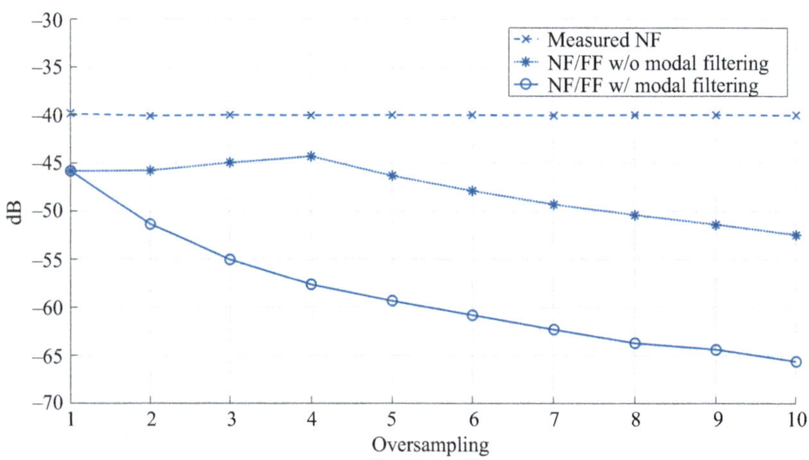

Figure 5.15 Noise reduction due to the O_S in the case of an emulated noisy measurement (SNR = 40 dB) of the SH600 at 3 GHz. ENL on NF data (dashed "x"); ENL on NF/FF transformed data without modal filtering (dotted ""); ENL on NF/FF-transformed data with modal filtering (solid "o")*

In Figure 5.16, the observed SNR improvement after the NF/FF transformation (with modal filtering) against several oversampling is shown for different NF SNR (40, 50 and 60 dB). As can be seen, the observed SNR improvements agree very well with the expected theoretical ones defined by (5.20).

A similar verification of the SNR improvement due to the oversampling considering real SNF measurements of electrically small devices can be found in [16].

5.1.2.3 Echo reduction with SWE

As introduced in the beginning of this section, the SWE can be used as spatial filter in order to mitigate the effect of the echoes (echo reduction) by considering the workflow shown in Figure 5.12. Before considering some practical examples, some considerations regarding the benefits coming from the acquisition of redundant NF data (oversampling) can be pointed out.

In Section 5.1.2.2 and in [16], it has been proven, respectively, through simulations and actual experimentation that if the SNF acquisition is performed using a redundant set of data (oversampling), the expansion and successively modal filtering will improve the system SNR with a ratio proportional to the square of the oversampling factor. This is due to the fact that random noise is completely uncorrelated, so it is spread all over the computed spectrum.

On the other hand, stray signals arising from a physical echo are highly correlated to the wanted signals radiated by the AUT (especially if their source is relatively close to the AUT minimum sphere). Therefore, it is not easy to know if oversampling can in such cases improve the effectiveness of the spatial filtering.

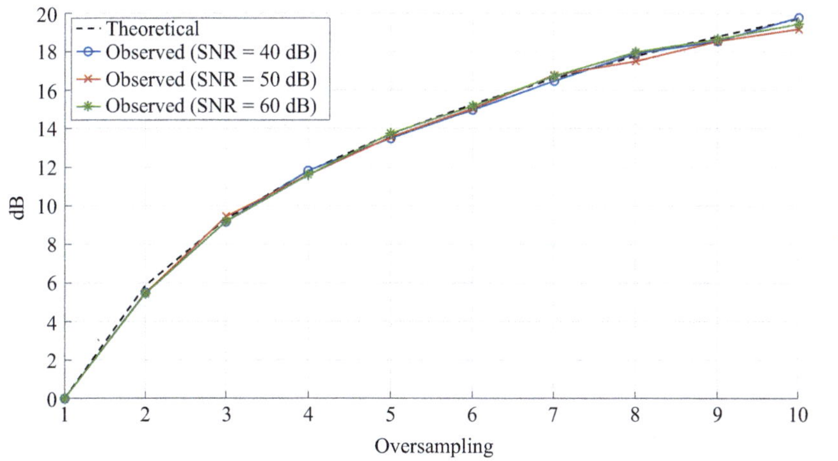

Figure 5.16 Improvement of the SNR due to the O_S in the case of emulated noisy measurements of the SH600 at 3 GHz. Theoretical (dashed black). Emulated measurements with SNR = 40 dB (solid blue "o"); SNR = 50 dB (solid orange "x"); SNR = 60 dB (solid green "")*

The benefits of the oversampling in the spatial filtering of stray signal sources have been numerically investigated considering as antenna the same dual-ridge horn shown in Figure 5.4. SNF measurements have been emulated considering echo signals modeled as electrically small sources located at different distances from the antenna. For each configuration, different oversampling factors with respect to the AUT minimum sphere have been tested and the improvement of the signal-to-interference ratio (SIR) after the application of the spatial filtering has been computed similarly to what has been done for the random noise, using the global ENL defined by (5.22).

The SIR improvements are reported in Figure 5.17 for echo sources located at 7, 10 and 13 wavelengths from the sources. For a more direct comparison with the error-source location, the considered oversampling is expressed in the abscissa of the graph as radius of ESS defined by (5.8). For comparison, the theoretical improvement of the SNR defined by (5.20) is also shown in Figure 5.17. From the results, it can be seen that the oversampling allows the improvement of the pattern estimation only to a certain limit. After this limit has been reached, no further improvement can be expected. As expected, the farther the interferer is away from the AUT, the better the possible improvements that can be achieved using the spatial filtering. However, these improvements require further oversampling. Furthermore, it is observed that the improvement threshold is reached for ESS radius slightly smaller than the actual echo source location. This means that there is no need to apply strong oversampling factors in order to include the interferer in the equivalent sphere and fully represent it with the spherical modes. Measurement time can thus be saved accordingly.

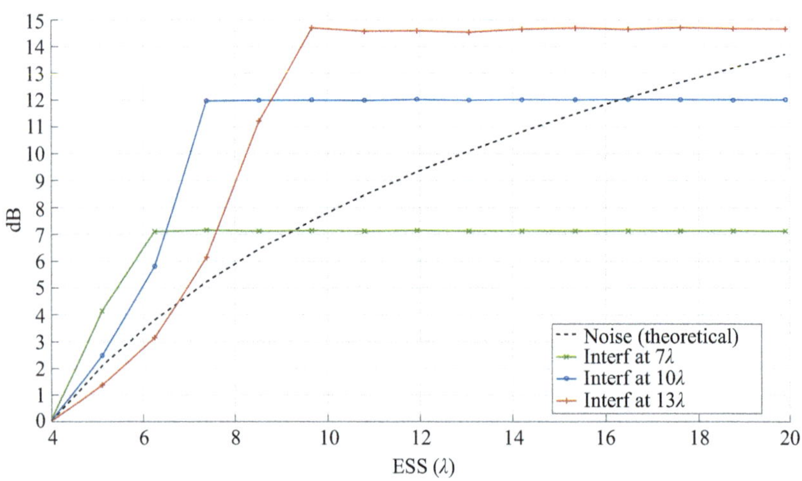

Figure 5.17 Improvement of the SIR and SNR due to the O_S in the case of emulated measurements of the SH600 at 3 GHz with error sources (interferences) at different distances

It should be noted that this modeled measurement scenario with echoes is of course not representative of any realistic measurement situations. Nevertheless, such simplistic assumptions allow to have a better understanding of the spatial filtering properties of the SWE against source of errors different from random noise.

Applicative experimental measurement scenarios with intentionally placed disturbances (e.g., reflecting plates) will be analyzed in the following paragraphs considering different types of NF ranges. As we will see, depending on the way the NF scanning is performed (e.g., by moving the probe as in PNF systems or by rotating the AUT as in many single probe SNF systems), the unwanted scattering signals have different impact in the different ranges and, consequently, they must be treated in a proper way. For instance, in the case of SNF measurements performed rotating the AUT along both scanning axes (such as the roll-over-azimuth systems), in order to separate the scattering contributions from the main antenna contribution, it is needed to mount the antenna in an offset position with respect to the center of rotation (see "Example—SNF with stationary probe and rotating AUT").

Example—PNF measurements (stationary AUT, moving probe)

In this paragraph, we will show how the spherical wave spatial filtering can be exploited to reduce the echoes and stray signals in PNF measurements. To do that, the measurement of an X-band standard gain horn (SGH) conducted in the presence of a reflecting plate, intentionally introduced in the anechoic chamber to create a perturbed environment, will be taken into account.

As described in [1] and shown Figure 5.18, the conventional PNF measurement setup is composed by a stationary AUT whose radiated NF is scanned at a certain distance by a probe which is moved in a rectangular x/y-plane. Once the PNF is measured, the FF is obtained applying the planar NF/FF transformation which, as also described in [1], can be simply performed with a 2D FFT.

The considered AUT is the MVG SGH820. The aperture dimension of the SGH820 is 195 × 148 mm and its directivity at 10 GHz is approximately 22 dB i.

The layout of the PNF setup used to measure the SGH820 is illustrated in Figure 5.18. As can be seen, a reflecting plate is introduced in the measurement

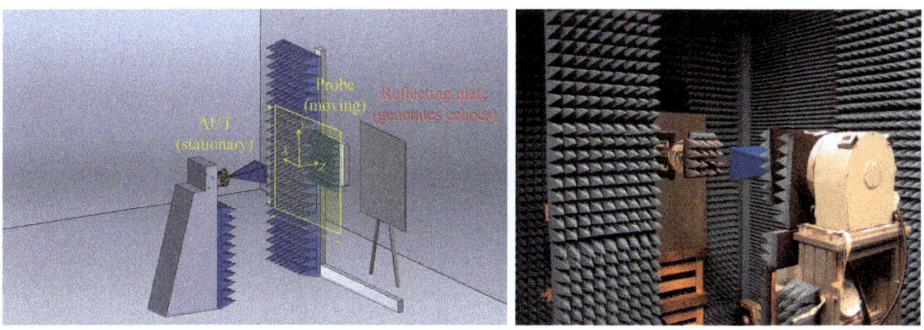

Figure 5.18 PNF measurement with reflecting plate to generate echoes: conceptual drawing (left); illustration of the measurement setup (right)

setup to intentionally perturb the measurement. Such reflecting plate has been placed behind the scanner parallel to the wall. The dimensions of the plate are 1.25 × 1.25 m (approximately $41\lambda \times 41\lambda$ at 10 GHz). A reference measurement has also been performed without the interferer. Both PNF measurements have been performed considering an AUT-to-scan distance of 0.375 m, a scanner size of 1.5 m × 1.5 m (view angle of approximately 60°) and using the conventional $\lambda/2$ sampling spacing. The probe used for the measurements is a rectangular open-ended waveguide (WR90 [17]). As in any typical PNF systems, the origin of the coordinate system is located on the probe aperture (when this is in the center of the scanning plane) and the z-axis is aligned with the AUT main axis.

To apply the spatial filtering on the perturbed measurement, the workflow shown in Figure 5.12 has been considered. First, from the NF data, the conventional planar NF/FF transformation has been applied. The probe pattern correction has also been considered during the transformation using the theoretical model of the probe [17].

To optimize the AUT minimum sphere, the coordinate system of the perturbed FF has been translated along the z-axis of 0.375 m so that the AUT aperture is centered in the origin of the modified coordinate system. The original and modified coordinate systems are illustrated in Figure 5.19 together with the corresponding AUT minimum spheres.

Since the FF coming from PNF measurements is defined only in the forward hemisphere, before applying the SWE, virtual field data have been added on the backward hemisphere in order to correctly perform the expansion. This has been done by simply padding the missing portion of the FF sphere with zeros (ZP) which, in the case of relativity high directly antennas (e.g., >15 dB i), do not introduce significant errors.

The 2D representation of the computed Q_{tot} coefficients of the perturbed PNF measurement after the optimization of the AUT minimum sphere is show in Figure 5.20 (bottom). The effect of the reflecting plate is clearly visible in the center of the spectrum where higher order modes emerge from the noise floor. In Figure 5.20 (top), the Q_{tot} coefficients of the perturbed measurement that would have been obtained without the

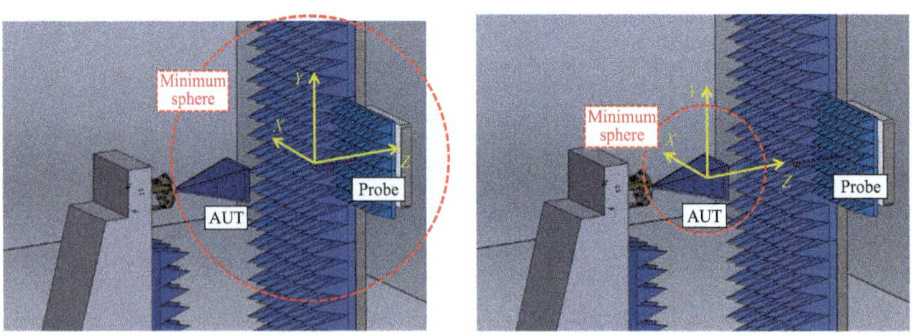

Figure 5.19 Illustration of original coordinate system (left) and modified coordinate system (right) and the corresponding AUT minimum spheres

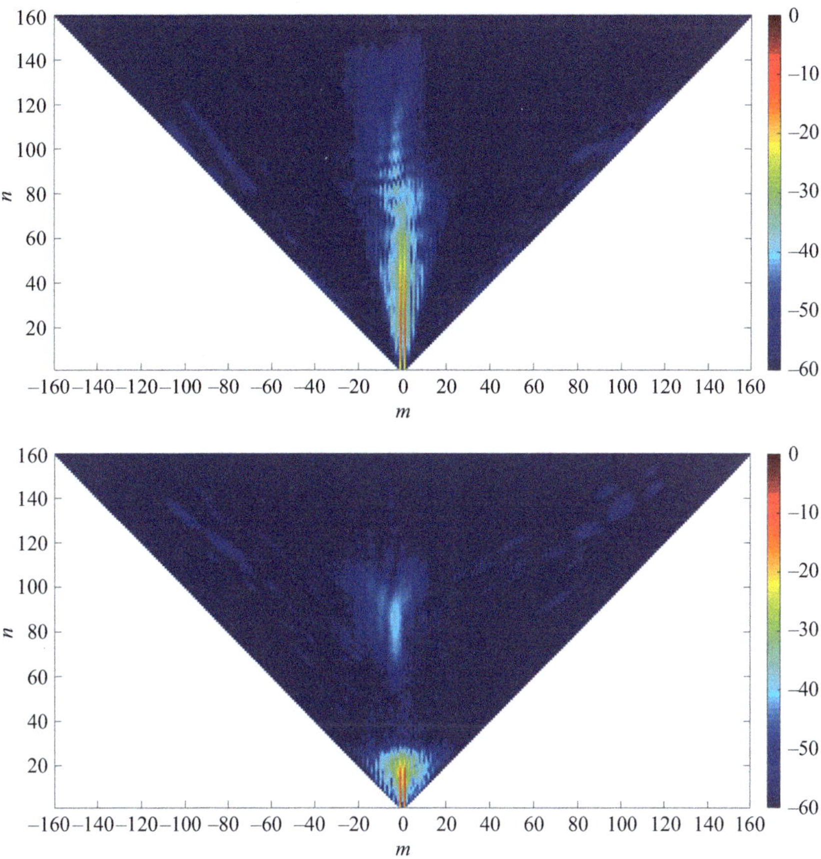

Figure 5.20 2D representation of the SWC computed from perturbed PNF measurement of the SGH820: without optimization of the AUT minimum sphere (top) and with optimization of the AUT minimum sphere (bottom)

optimization of the AUT minimum sphere are shown. It is evident that in such case, the AUT and echo signals would be entangled without any possibility to apply the modal filtering. It is thus remarked that, when the origin of the FF coordinate system is moved on the AUT aperture, the size of AUT minimum sphere is strongly reduced. This ensures that the spherical modal spectrum distribution of the AUT is concentrated in the lower order modes (i.e., low-pass behavior) allowing the application of low-pass filter and thus an effective suppression of echo pollution characterized by higher order modes.

The P_n power spectrum associated to the Q_{tot} coefficients shown in Figure 5.20 (bottom) is depicted by the orange trace shown in Figure 5.21. For comparison, the P_n power spectrum of the reference measurement, performed without the reflecting

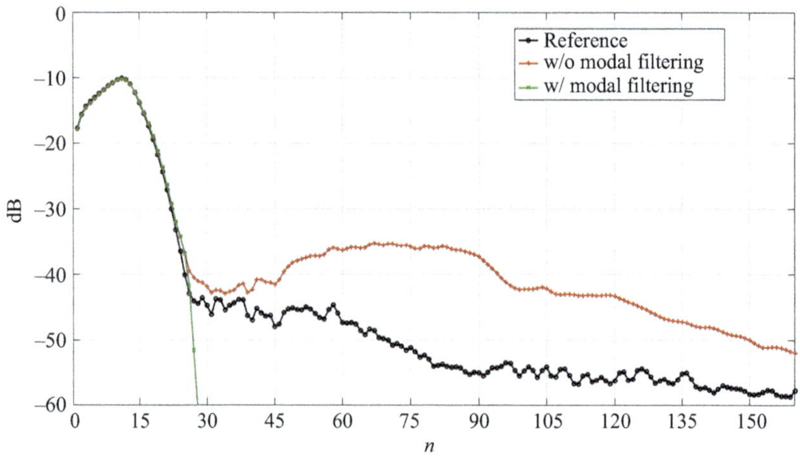

Figure 5.21 *P_n power spectrum comparison of the SGH820 measured in PNF range after optimization of the AUT minimum sphere: without the reflecting plate (black "o"), with reflecting plate (orange "+") and with reflecting plate plus filtering (green "x")*

plate, is shown by the black trace. The effects of the reflecting plate are confirmed and it affects modes with order >25.

A low-pass filtering has been applied considering an AUT minimum sphere of $R_{min} = 0.125$ m, which fully enclose AUT aperture. The P_n power spectrum after the modal filtering is illustrated by the green trace also shown in Figure 5.21.

The co-polar E-plane directivity pattern comparison between the reference FF (solid black trace), the FF without spatial filtering (orange dotted trace) and FF recomputed from the filtered spectrum (green dashed trace) is shown in Figure 5.22. As can be seen, the ripple generated by the presence of the reflecting plate is strongly attenuated applying the spherical wave spatial filtering. The deviation of the two patterns with respect to the reference is also illustrated by the pointwise ENL reported (in dB scale) in the same plot and defined by (5.21). As can be seen, the reduction of the errors coming from the presence of echoes is up to approximately 15 dB.

To further illustrate the effectiveness of the spherical wave spatial filtering, the holographic maps on the aperture plane of both the perturbed and echo reduced data have been computed. Results are shown in Figure 5.23 (top) and (bottom), respectively. The effect of the reflecting plate is clearly visible in the co-polar map on the left for negative y-axis values. The wide spot is supposedly associated to a re-irradiation of the field reflected by the plate from the antenna positioner. As seen in Figure 5.23 (bottom), the effect of the reflecting plate is strongly attenuated by the applied spatial filtering without impacting the AUT aperture field distribution.

Example—SNF with stationary probe and rotating AUT

In this paragraph, we analyze the spatial filtering properties of the SWE applied to echo polluted SNF measurement performed rotating the AUT and keeping the

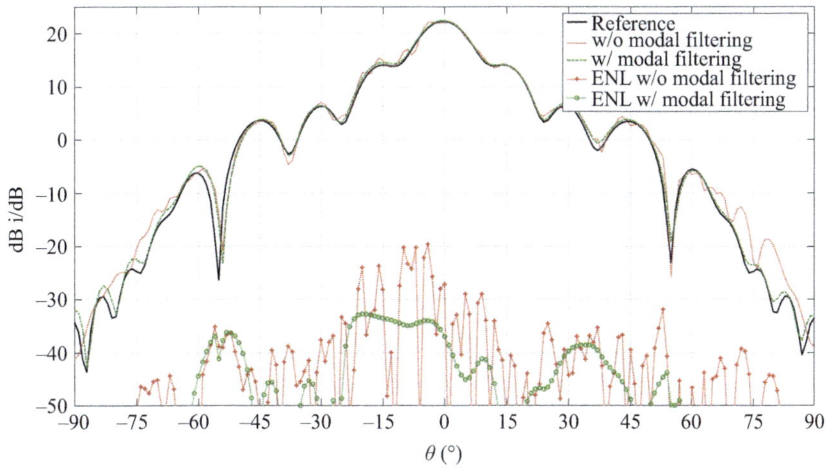

Figure 5.22 E-plane directivity pattern comparison for the SGH820 measured in the PNF system and associated equivalent errors. Reference pattern (black solid), pattern without spatial filtering (orange dotted), pattern with spatial filtering (green dashed), ENL without spatial filtering (orange dotted "+") and ENL with spatial filtering (green dotted "o")

probe fixed, such as conventional roll-over-azimuth, azimuth-over-elevation and elevation-over-azimuth ranges.

Let us consider that the echoes are generated by a scattering object located outside the measurement sphere (e.g., on the side wall). When the measurement of a relatively small AUT is performed with the AUT located in the origin of the measurement system (see Figure 5.24 top), the reflected paths (Path *i*) generated by the scattering object at any *i*-position of the AUT have a comparable length (Path 1≈Path 2≈…≈Path *n*). The spatial interference generated at the probe aperture creates a "slow" distortion of the pattern without generating higher order modes in the associated spherical modal spectrum. The spherical modal spectrum of AUT and the modal spectrum of echoes are thus overlapped at low modal frequencies and the unwanted contributions could not be removed by the spatial filtering.

Instead, when the AUT is displaced out of the center of rotation of the measurement system (offset position), as depicted in Figure 5.24 (bottom), the length of each reflected paths is different. In such situation, AUT and echo signals interfere with different phases at each *i*-position of the AUT generating a "fast" ripple that is added to the wanted signal. Such ripple can be isolated in the spherical wave domain after mathematically translating the origin of the measurement system back to the AUT. With such measurement setup and data processing, higher order modes associated to the echoes can be generated and thus filtered out considering the minimum sphere of the AUT. Unfortunately, to apply the above-described procedure, a higher number of measurement samples are required for the offset

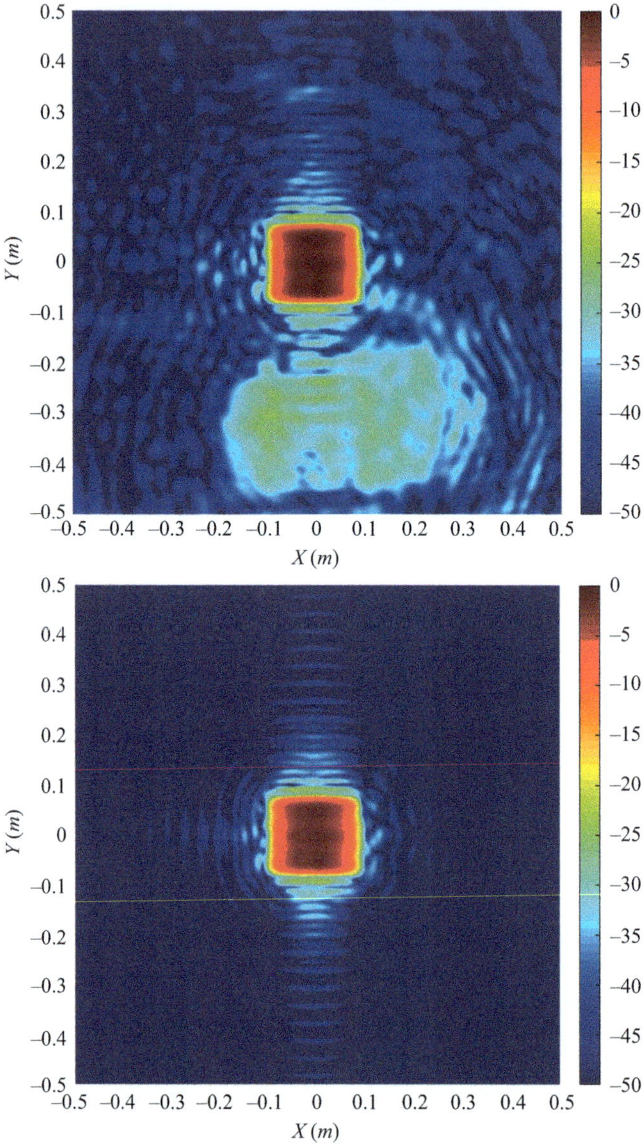

Figure 5.23 Holography on the AUT aperture plane. Computed from the FF without spatial filtering (top) and computed from the FF with spatial filtering (bottom)

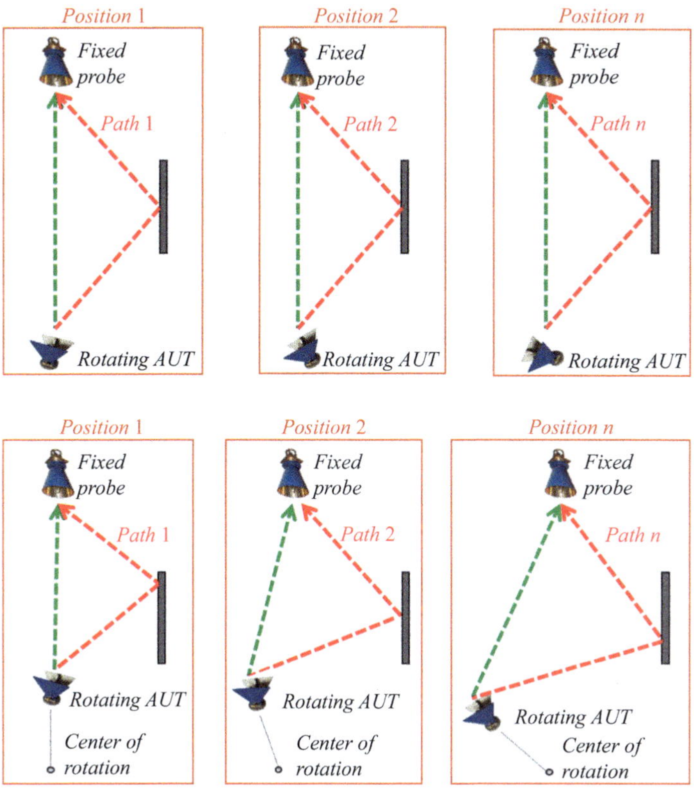

Figure 5.24 Measurement with rotating AUT, fixed probe and reflective plate on the side wall. Onset AUT (top); offset AUT (bottom)

measurement to ensure the same accuracy as of an onset measurement, with the drawback of an increased testing time.

The echo reduction properties of the SWE in such type of measurement scenario are verified considering the MVG SH2000 dual-ridge wideband horn as device under test. SNF measurements have been performed using a robotic arm system.[‡] As illustrated in Figure 5.25, the AUT has been placed on the robotic arm and the probe on a tower located in front of the robot. The alignment of the AUT and probe positioners is provided by sliding the probe tower in proximity of the robot and mating the AUT and probe interfaces with a precision mechanical adaptor. Unfortunately, due to mechanical constraints of the robot, only the forward hemisphere has been measured. Some errors due to the truncation of the spherical scanning area are thus expected.

The SH2000 maximum dimensions are 105 × 61 × 105 mm resulting in a minimum sphere radius of approximately 81 mm (considering the coordinate

[‡]https://www.kuka.com/.

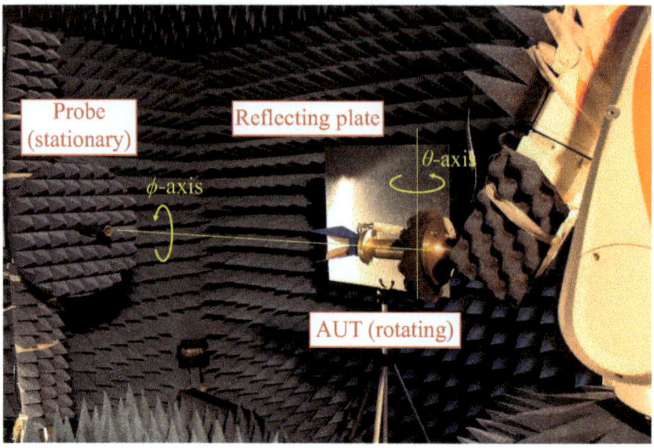

Figure 5.25 SNF measurement setup with offset antenna using a robotic arm. A reflecting plate is intentionally placed on the side wall to generate specular reflections

system on the geometrical center of the antenna). The frequency chosen for the test is 12 GHz. According to (5.4), the maximum sampling step along θ for this (onset) AUT is approximately $\Delta\theta = 6^\circ$ ($P_{tr} = -80$ dB).

The spherical NF setup used to measure the SH2000 is illustrated in Figure 5.25. A standoff of approximately 140 mm length has been used to mount the AUT on the robotic arm. As illustrated, the acquisition is performed by rotating the device under test around the θ- and φ-axis, while the probe is kept fixed. The considered probe is the MVG DOEW6000 first-order probe. Due to the relative low directivity of such probe (approximately 10 dB i at 12 GHz), probe correction has been neglected during the NF/FF transformations.

A scattering plate of size 0.5 m × 0.5 m ($20\lambda \times 20\lambda$ at 12 GHz) has been placed outside the measurement sphere at approximately 90 cm from AUT and probe. The plate is intended to generate a specular reflection like those schematized in Figure 5.24.

Onset measurements with and without the scattering plate have been performed allowing the robotic arm to make the θ-rotation around the geometrical center of the UAT. The onset measurement without the scattering plate has been considered as the reference.

An offset measurement has been performed moving the center of the θ-rotation at the bottom of the standoff where the antenna is mounted (as illustrated in Figure 5.25), generating an offset of 200 mm (or 8λ at 12 GHz) along positive z-axis. The obtained minimum sphere radius of the offset AUT is thus approximately 281 mm. According to (5.4), the maximum sampling step along θ of this offset measurement is approximately $\Delta\theta = 2^\circ$ ($P_{tr} = -80$ dB).

The onset measurements without and with the reflecting plate have been performed with a sampling along the θ-axis of $\Delta\theta = 5.0^\circ$ and $\Delta\theta = 4.0^\circ$, respectively. The one with the scatterer has been slightly oversampled ($O_{s,\theta} = 1.5$) with respect to the minimum required sampling in order to check if some improvements can be obtained by filtering out some higher order modes. The offset measurement with the scattering plate has been performed with the minimum requited sampling, $\Delta\theta = 2.0°$. It is observed that, since the offset is along the z-axis, the φ-sampling should not change. For this reason, each measurement has been performed with $\Delta\varphi = 10°$.

The P_n power spectrum associated to the SWE of the three different measurements is shown in Figure 5.26. The black trace marked with "o" is the P_n power spectrum of the onset measurement without the plate (reference). The orange trace marked with "+" is the one associated to the onset measurement with the plate. Despite the presence of the scatterer, the latter is very similar to the reference one and also the additional higher order modes are very low. This means that the stray signals are mixed up within the low order modes associated to AUT and, consequently, no improvements are expected by the application of the modal filtering. Instead, the green trace marked with "x" is the P_n power spectrum of the offset measurement with the plate, after the application of a mathematical translation of 200 mm along the z-axis to report the origin of the coordinate system on the geometrical center of the AUT. In this case, the effect of the scatter is clearly visible on the higher order modes and it can be mitigated by applying a filtering based on the dimension of the AUT minimum sphere.

The comparison of directivity co-polar patterns along the two main cuts, the E-plane and the H-plane, is shown in Figures 5.27 and 5.28, respectively. The black solid traces are the patterns obtained from the onset measurement without the reflecting plate (reference).

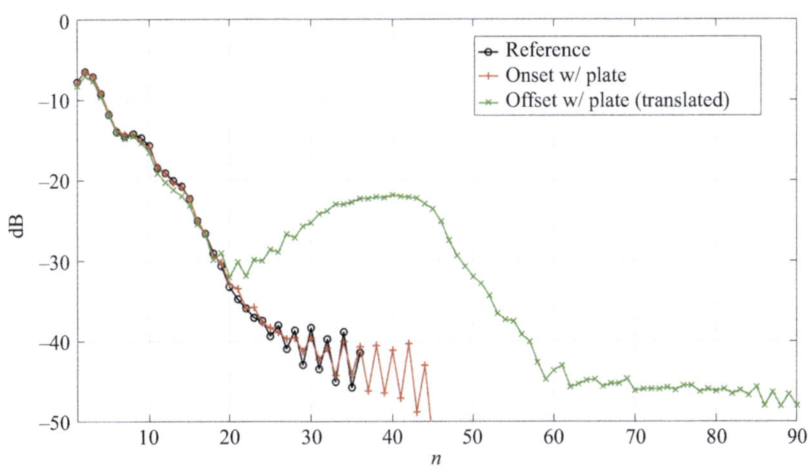

Figure 5.26 P_n power spectrum comparison of the SH2000 at 12 GHz measured in SNF range with the robotic arm: onset measurement without the reflecting plate (black "o"), onset measurement with reflecting plate (orange "+") and offset measurement with reflecting plate with applied mathematical translation (green "x")

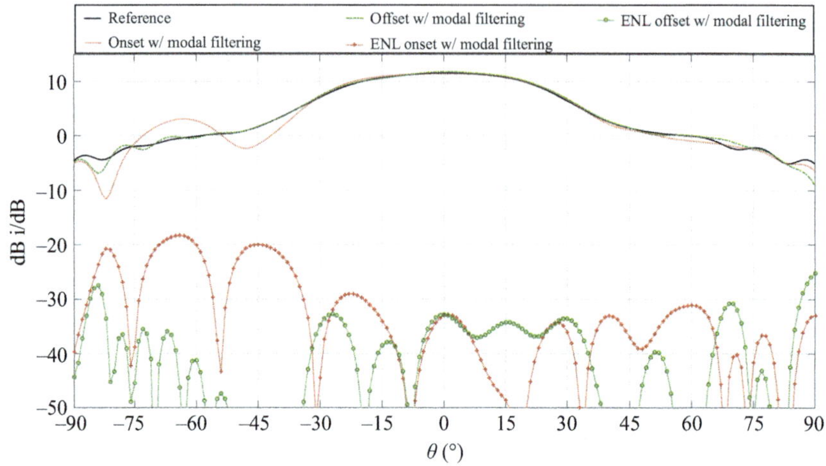

Figure 5.27 E-plane directivity pattern comparison. Reference (black solid), onset AUT with spatial filtering (orange dotted), offset AUT with spatial filtering (green dashed), ENL of onset AUT with spatial filtering (orange dotted "+") and ENL of offset AUT with spatial filtering (green dotted "o")

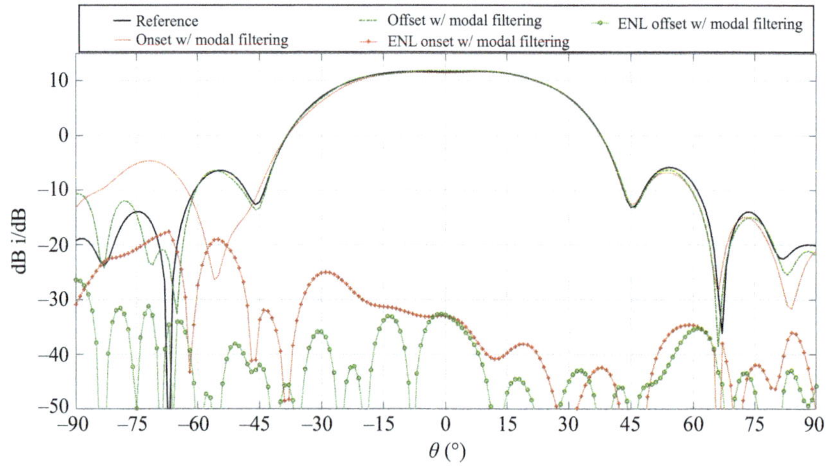

Figure 5.28 H-plane directivity pattern comparison. Reference (black solid), onset AUT with spatial filtering (orange dotted), offset AUT with spatial filtering (green dashed), ENL of onset AUT with spatial filtering (orange dotted "+") and ENL of offset AUT with spatial filtering (green dotted "o")

The orange dotted traces are the patterns of the onset data with the reflecting plate. A spatial filtering based on the AUT minimum sphere has been applied to such data but, as expected, it is unable to remove the effect of the scattering of the plate which is visible as a pattern distortion on the left side of both presented cuts. In fact, as explained above, the AUT and echo contributions are overlapped in the spherical wave spectrum resulting in a very poor effectiveness of the spatial filtering. This is also underlined by the pointwise ENL illustrated in both figures (in dB scale) by the dotted orange traces marked with "+".

The green dashed trace is the pattern obtained from the offset measurement with plate, where, as mentioned above, the spatial filtering is applied considering a mathematical translation of 200 mm along the z-axis to report the origin of the coordinate system on the geometrical center of the AUT. As observed in Figure 5.26, this operation, combined with the introduced offset, allows to generate higher order spherical modes associated to the echoes which can be filtered out applying a filtering based on the AUT minimum sphere. As can be seen, the agreement with the reference is very good almost all over the pattern. This is also confirmed by the ENL associated to these patterns (green dotted trace marked with "o") which are almost everywhere much lower than the ones corresponding to the onset measurement. Some relatively small deviations are observable at the edges of the patterns where some residual errors due to a not perfect rejection of the echoes are present. It shall also be highlighted that some differences between the presented patterns are partially due to the truncation of spherical scanning surface which is limited to only the forward hemisphere. This may create different truncation errors in the case of onset and offset measurements.

5.1.2.4 Reduction of truncation errors in SNF measurements

SNF measurements are often performed on only a portion of the scanning sphere for different reasons. For example, in the case of the electrically large antennas (e.g., high gain arrays or reflectors), the scanning time could be very high, especially if single probe systems are used. In such cases, the testing time can be significantly reduced by measuring the SNF only in the area where most of the radiated power is concentrated. The NF/FF transformation can then be performed by simply setting the missing samples to zero (ZP). In the case of high gain antennas, the ZP does not introduce significant errors, since is a good approximation of the actual radiated field being very low on angular regions further away from the main beam [18,19].

In other situation instead, a full spherical scan might be unfeasible due to mechanical constraints of the AUT positioner or due to the physical size of the test object which could be too big and or too heavy (e.g., vehicle, aircraft and satellite). In such cases, the conventional SWE-based NF/FF transformation applied to the zero-padded field can generate non-negligible errors especially at low frequency and in the case of low-gain antennas [19–23].

In the case of low directive devices, the ZP applied to truncated spherical NF dataset introduces an abrupt field discontinuity resulting in the generation of higher order modes in the SWE spectrum.

The SWE spectra shown in Figure 5.29 are relative to a z-oriented sleeve dipole measured at 1.9 GHz over a full sphere (top) and over a hemisphere (bottom). As can be seen, the discontinuity introduced by the ZP creates high-order modes and also modify the low-order modes distribution. The effect on the reconstructed pattern is the generation of unwanted ripples all over the pattern as illustrated by the directivity pattern comparison shown in Figure 5.30. It should be noted that, due to the fact that also the lower order modes are distorted by the truncation of the scanning area, a simple low-pass filtering would not be as effective as in the examples reported in the previous paragraphs relative to the spatial filtering of the echoes.

The errors caused by the truncation of the scanning area vary depending on the device under test. In the presented example (sleeve dipole), a truncated SNF acquisition is considered in the case of a very low directive device. Moreover, the NF truncation has been truncated at an angle corresponding to the maximum radiation of the antenna ($\theta = 90°$) generating a strong field discontinuity. Such

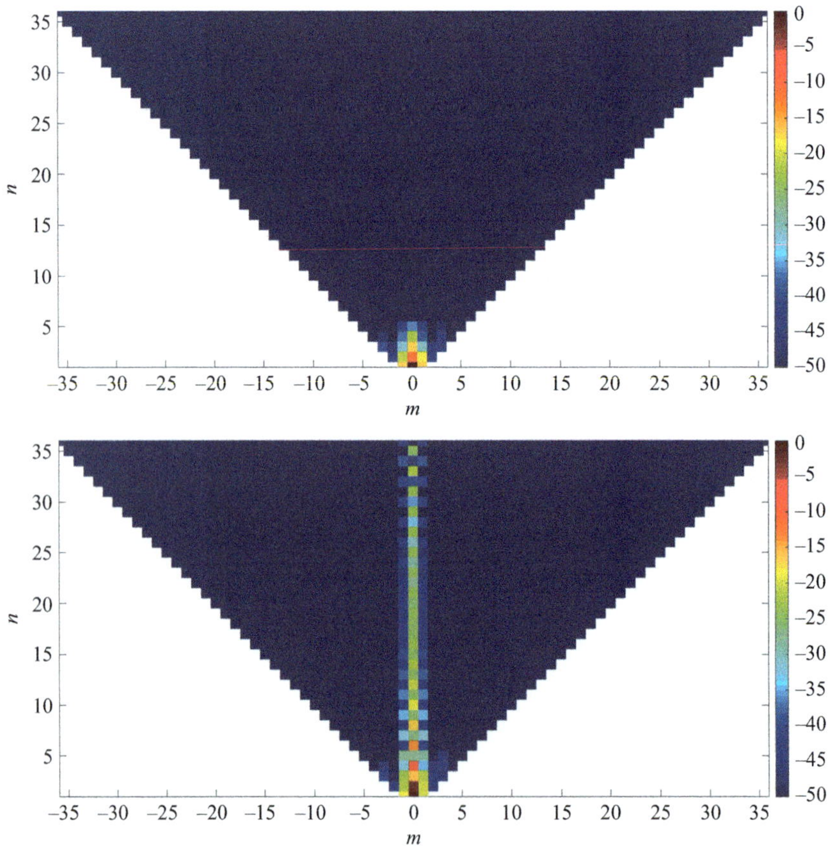

Figure 5.29 Q_{tot} *spectrum of a sleeve dipole at 1.9 GHz without SNF truncation (top) and with SNF truncation at* $\theta = 90°$ *(bottom)*

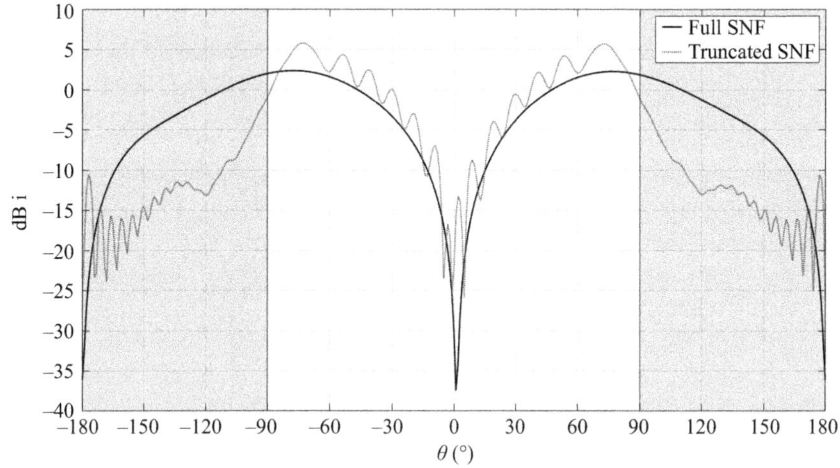

Figure 5.30 Directivity pattern comparison of sleeve dipole at 1.9 GHz without SNF truncation (solid back) and with SNF truncation at $\theta = 90°$ (dotted orange)

example represents by far a worst case scenario for the truncation of the SNF scanning area. Other measurement scenarios with higher directive antennas and/or smaller truncated areas at lower field levels are less sensitive to these types of errors.

It is also worth noting that, truncation errors are generally stronger at lower frequencies. In fact, as illustrated above, in order to represent the field discontinuity introduced by the truncation in the spherical wave domain, a very large number of SWF is needed (ideally an infinite number). Due to the cutoff properties of the SWE [7], the highest SWC order is given by (5.23):

$$N_{\max} = kR_{meas} = \frac{2\pi f}{c} R_M \tag{5.23}$$

where f is the frequency and R_M is the radius of the measurement sphere. For a given and finite R_M, the total number of computable spherical modes becomes bigger when the frequency is increased. Truncation errors are thus less important at higher frequencies because the discontinuity introduced by the zero-padding of the NF can be represented by a larger number of SWC. Instead, at lower frequencies, truncation errors are more pronounced because the field discontinuity is represented by a smaller number of modes. This concept will be illustrated with an example in "Reduction of truncation errors in automotive free-space antenna ranges" section.

To mitigate the errors caused by the truncation of the spherical scanning area, different advanced post-processing techniques have proposed in literature [18–24]. The technique described in the following paragraph is an iterative process where the same spatial/modal filtering presented in previous sections for echo/noise reduction purposes is applied at each iteration (iterative modal filtering, IMF) [20–23].

Iterative modal filtering technique

The IMF [20–23] is based on the SWE and spatial/modal filtering of the measured field, which are both repeated iteratively. The workflow of the IMF is shown in Figure 5.31. The ZP is first applied to the truncated area of the measured SNF in order to obtain full spherical dataset which is expanded in terms of spherical waves (SWE). The computed SWC are then low-pass filtered on the base the DUT physical dimensions as described in previous sections. In particular, a filtering at mode indexes defined by (5.2) and (5.3). With such filtering a first correction is performed canceling out the higher order modes caused by the truncation and illustrated in reference to Figure 5.29 (bottom).

From the filtered SWC, the SNF at the same measurement radius is computed on the full sphere by using (5.14). The computed NF samples in the truncated region are an improved estimate of the missing samples. In the last step of the procedure, these extrapolated samples are thus kept, while the recomputed NF samples on the un-truncated region are replaced with the real measured data, being not affected by errors due to the truncation.

This process is repeated until the convergence is reached. It is specified that the convergence of the algorithm can be controlled by computing at each iteration the global ENL (Equation (5.22)) between measured and reconstructed samples in

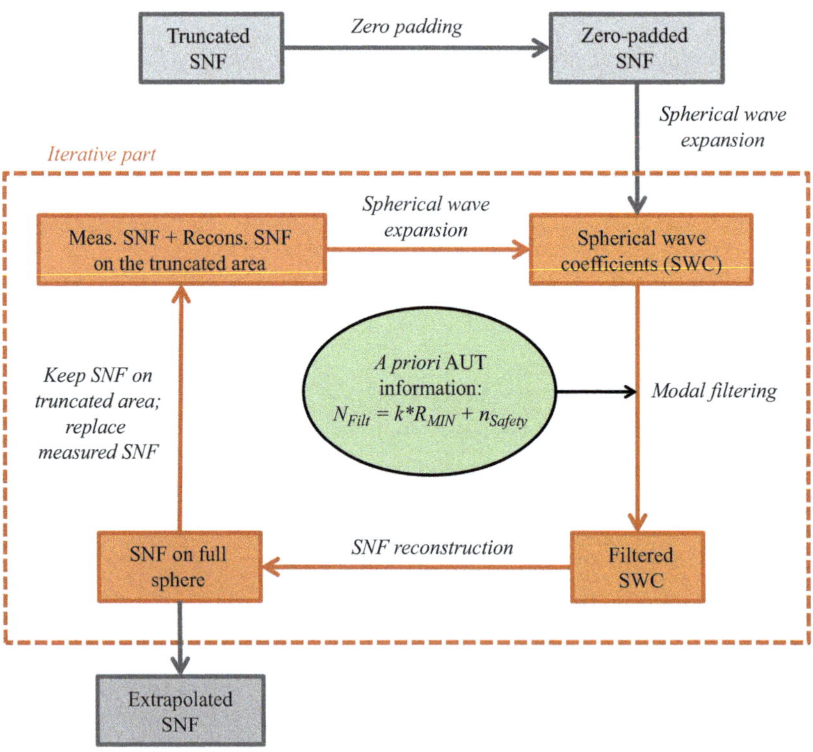

Figure 5.31 Block diagram of the IMF technique

the un-truncated region. When a certain threshold level defined by the user is reached (e.g., −40 dB), the iterative process is terminated. The output of this algorithm is the extrapolated SNF over the full sphere which can be NF/FF transformed as usually done.

The IMF technique has been compared with other truncation error-mitigation methods, such as equivalent currents technique, in [22,23] where it has been shown that they are both effective against such type of errors. The main advantage of the IMF against other techniques is its low computational cost. In fact, the main computational effort at each iteration resides on the SWE and on the field recomputation, which, as explained in Section 5.1.1.3, are both FFT-based operators. Furthermore, few iterations (usually 50–100 depending on the DUT and on the extent of the truncated area) are normally needed to reach the convergence keeping the overall computational effort relatively low.

Reduction of truncation errors in automotive free-space antenna ranges

Automotive antenna ranges are a typical example of truncated SNF systems [6,25,26]. In order to fully characterize the radiating properties of the vehicle-mounted antennas, a large measurement system capable of accommodating the full vehicle is required. A full spherical scanning around the car is desirable; however, due to vehicle size, weight and/or economic factors, the scanning over a whole spherical surface is often unfeasible. For this reason, truncated spherical scanners are typically preferred.

One solution is to combine hemispherical scanning with a metallic ground plane. The latter is then assumed to be a perfect electric conductor (PEC) in the NF/FF transformation [25,26], avoiding the truncation errors. However, PEC ground planes are far from being representative of realistic automotive environments such as asphalt that is strongly dielectric. A further drawback is the interaction of the large metallic ground plane with the ambient structure, compromising the quality of the measurements especially at lower frequencies. For this reason, NF measurements are rarely performed (Figure 5.32) below 400 MHz in PEC-based systems.

A widely involved alternative solution for automotive testing is a multi-probe system placing the vehicle in a condition like free-space by using absorber materials on the floor. Such solution enables to perform accurate NF measurements even below 70 MHz [6]. An example of such system is shown in (multi-probe automotive range provided by MVG sited at Laird, MI, USA).

Typical implementations of such automotive multi-probe systems have a measurement radius ranging from 4 to 6 m, and they can measure up to 10°–20° below the horizon, resulting in a conical truncated area of 70°–80° in elevation. While the elevation scan is performed by the probe array, the azimuthal scanning is performed by rotating a turntable over full 360°. Systems like this are currently involved for automotive measurement in the 70 MHz–6 GHz frequency range. To achieve this, a 70–400 MHz probe array is typically installed on one side of the arch and a second probe array working in the 400–6,000 MHz range is installed on the other side of the arch.

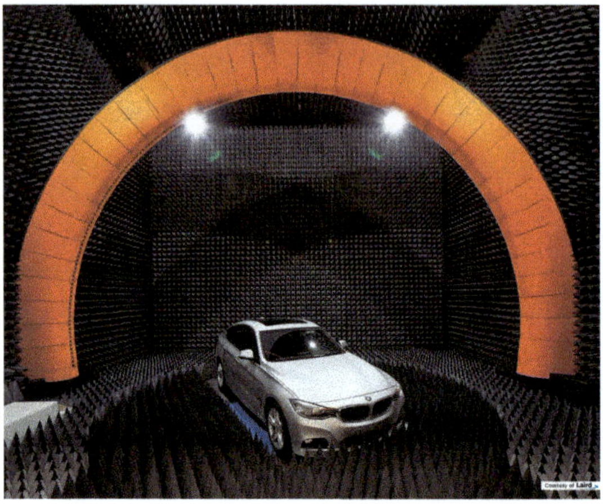

Figure 5.32 Example of truncated spherical NF automotive range. MVG multi-probe system at Laird, MI, USA

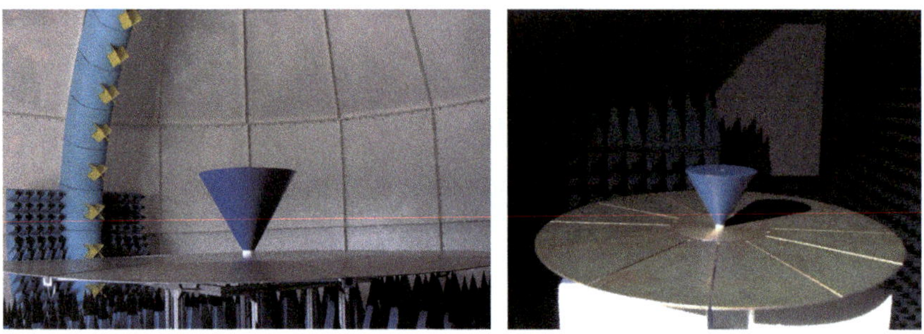

Figure 5.33 SMC70 monocone antennas with 4 m circular ground plane working in the 70–220 MHz range (left); SMC700 monocone antennas with 40 cm circular ground plane working in the 700–2,200 MHz range (right)

In order to study the effect of the truncation of scanning area in free-space automotive ranges, and also to illustrate the effectiveness of the introduced IMF technique, monocones antennas like the ones shown in Figure 5.33 are considered. Such devices are often used as reference gain antennas in order to calibrate the systems (i.e., substitution or gain transfer method [1]) and, as can be seen, they are composed by the monocone itself and a circular metallic ground plane.

As explained above, truncation errors are expected to be higher at lower frequencies. It is thus interesting to investigate such errors at lower frequencies, such as 70–700 MHz, where also several telecommunication services are operating. Monocones antennas at such frequencies require large ground planes (e.g., 4 m diameter) making a full spherical measurement very challenging or even impractical. For this reason, the scaled model techniques have been adopted for this test.

As described in [27–29], such antenna measurement technique can be involved to test antennas placed in complex environments such as ships, aircrafts, satellites or other structures that can have an important effect on the properties of the antennas but that are physically too large to be measured. To overcome these limitations, the technique uses scaled models as similar as possible to that of the actual device, which can be measured in smaller systems. The scale model technique is based on the basic concept that the electromagnetic performance of a system depends on its dimensions in terms of wavelengths (electrical size). In other words, if the frequency is multiplied by a factor N and the dimensions are divided by the same factor N, the electromagnetic behavior is the same. As an example, an antenna working at 70 MHz in a 5 m length vehicle has the same performance of a ten times smaller antenna working at 700 MHz in a 50 cm scaled vehicle. Other than the physical size and the frequency, the only other quantity that should be scaled (multiplied) by the same factor N is the conductivity. This is in general the most problematic parameter as is not easy to find materials that properly fit such scaling. However, this is not a problem in the case of low loss devices (e.g., fully metallic devices) whose high conductivity allows to obtain negligible errors by using the same material in the actual and in the scaled model.

The scaled-model technique has been applied to investigate and mitigate the truncation errors in automotive ranges at lower frequencies considering two monocones antennas provided by MVG: the SMC700 (see Figure 5.33, right) and the SMC2200 working respectively in the 700–2,200 and 2,200–7,000 MHz frequency range. Both antennas have been mounted on a circular ground plane of 40 cm diameter. Such antennas are the ten times scaled version of other two MVG monocone antennas with a 4 m circular ground plane: the SMC70 (see Figure 5.33, left) and the SMC220 working, respectively, in the 70–220 and 220–700 MHz frequency range. Both the scaled and the full-size devices are almost fully metallic with very low ohmic losses, making the scaling of the conductivity unnecessary.

The radiation pattern of the SMC700 and SMC2200 has been measured in a quasi-full spherical multi-probe system with a measurement radius of 1.6 m (MVG SG64). The entire bandwidth of the first device has been measured with a 100 MHz sampling step, while the second device has been measured in the 2,200–6,000 MHz range with the same sampling step. The FF has been computed with the standard spherical NF/FF transformation, and it has been assumed to be the reference datasets, not being affected by truncation errors. From the SWC computed during the NF/FF transformation process, a partial spherical NF in the $\theta = [0^\circ, 100^\circ]$ and $\varphi = [0^\circ, 360^\circ]$ range and at 60 cm from the origin of the coordinate system have been computed with (5.14). Such generated SNF datasets are equivalent to the truncated SNF measurements of the SMC70 and SMC220 (respectively in the 70–220 MHz and 220–600 MHz frequency range) in a free-space system like the ones described above, able to measure up to 10° below the horizon and having a measurement radius of 6 m. The SNF datasets have been computed with an angular sampling step of $\Delta\theta = \Delta\varphi = 2.5^\circ$. According to (5.4) and (5.5), and considering the radius of the minimum sphere of the monocones to be 2 m (the ground plane is 4 m diameter), such sampling steps result in an oversampling factor of approximately 9 at the lowest frequency of the SMC70 (70 MHz) and an oversampling factor of approximately 2 at the highest frequency of the SMC220 (600 MHz).

First, the truncated SNF datasets have been NF/FF transformed simply considering ZP on the truncated area without applying any type of filtering. The obtained FF directivity pattern has been compared to the reference one (not affected by truncation errors) at each frequency using the global ENL (see (5.22)) as shown by the orange dotted trace with "+" markers shown in Figure 5.34. The ENL has been evaluated only on the measured portion of the sphere ($\theta = [0°, 100°]$). As expected, the truncation errors are higher at lower frequencies and gradually decrease as soon as we reach higher frequencies. As explained above, this phenomenon is due to the number of SWC used to represent the scenario which become smaller and smaller reducing the frequency.

As shown in Figure 5.29, the effect of the truncation of the scanning area on the spherical wave spectrum when ZP is applied is the generation of higher order modes. A simple modal filtering based on the minimum sphere of the monocones devices ($R_{\min} = 2$ m) has thus also been applied. The results in terms of global ENL over the full band are illustrated by the yellow dashed trace with "x" markers reported in Figure 5.34. As can be seen, at lower frequencies the ENL is almost at the same level of the unfiltered FF, meaning that the modal filtering alone is not capable of improving the accuracy results. This is due to the fact that the truncation of the scanning area also modifies the low-order modal distribution associated to the antenna. The improvements are instead more pronounced at higher frequencies.

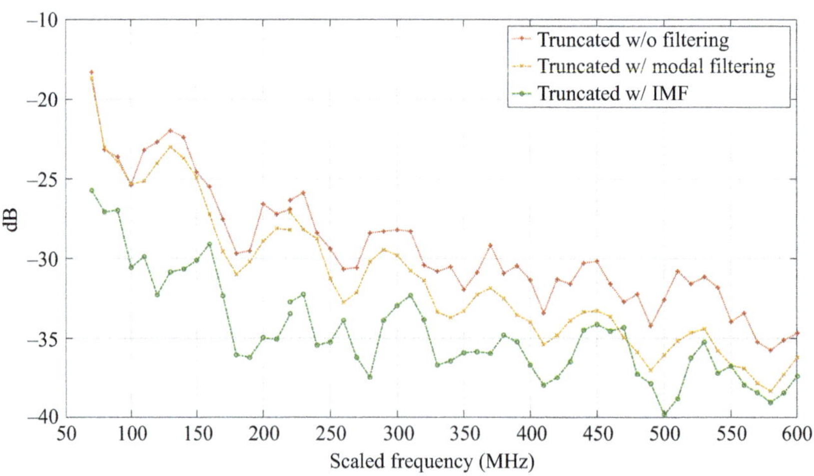

Figure 5.34 Global ENL over frequency of the truncated SNF measurements of the SMC70 and SMC220: transformed with zero-padding without filtering (dotted orange with "+"); transformed with zero-padding with modal filtering (dashed yellow with "x"); transformed with IMF (dashed green with "o")

The IMF technique described in previous paragraph has finally been applied to the truncated datasets. The applied modal filtering is again based on the minimum sphere of the SMC70 and SMC220 ($R_{\min} = 2$ m). The iteration processes have been stopped when the ENL computed on the reconstructed NF on the measured portion of the sphere reached the threshold of −40 dB. This resulted in 45 iterations for the SMC70 and 80 iterations for the SMC220. The results in terms of global ENL over the full band are illustrated by the green dashed trace with "o" markers reported in Figure 5.34. As can be seen, by applying the IMF technique up to 10 dB, improvement of the ENL is obtained at lower frequencies. At the high end of the band, the accuracy of the results is instead comparable to the one obtained with simple modal filtering.

Results in terms of directivity pattern for the SMC70 are shown in Figure 5.35, respectively, at 70 MHz (top) and 120 MHz (bottom). The shadowed areas highlight the truncated and unreliable angular regions where samples have not been taken. The comparison is between the reference (solid black trace), standard NF/FF with ZP and without any filtering (dotted orange trace) and IMF technique (dashed green trace). The IMF can mitigate the ripple introduced by truncation in the measured portion of the sphere allowing to improve the measurement of peak directivity and gain. The associated pointwise ENL are reported in the same plot (in dB scale) and further underlines how the IMF is capable of reducing the error caused by the truncation.

A similar directivity pattern comparison is reported in Figure 5.36 for the SMC220 at 220 MHz (top) and 500 MHz (bottom). Due to the higher frequencies, the equivalent error obtained by the standard NF/FF without any filtering is lower with respect to the previous case. Nevertheless, the accuracy of the results is further improved by the application of the IMF technique.

5.2 Exploiting cylindrical mode expansion for fast single-cut measurements with improved accuracy

This second part of the chapter explains cylindrical NF/FF transformation and the use of cylindrical modes for improving the results. Cylindrical NF/FF transformation using cylindrical mode decomposition has been widely studied, and the basic theory is presented in Section 5.2.1. Basically, similar application to the ones shown in Section 5.1 of this chapter can be done working with these modes. In particular, a detailed study can be found in [30]. In this chapter, instead of focusing in classical cylindrical NF/FF transformation, we focus in single-cut measurements, and the application to specific cases. Although, there are different approaches to single-cut in NF/FF transformation [31], we are focused on the one that use CWE. The theory of this method is explained in Section 5.2.2, where some application examples are detailed. Finally, Section 5.2.3 focuses on the use of cylindrical waves to filter out the reflections, as an example of post-processing techniques, although this same technique can be used to reduce some other measurement errors, in the same way it is explained in the chapter for spherical waves or chapter four for plane waves. Some examples are included to validate the method.

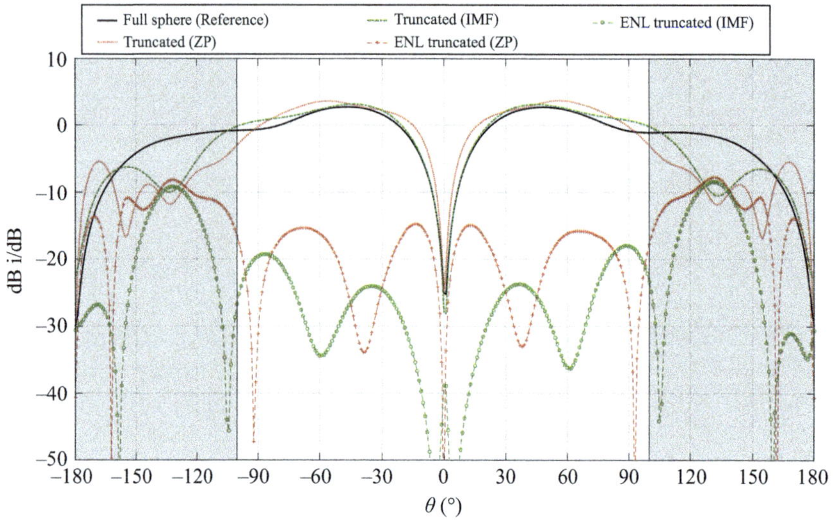

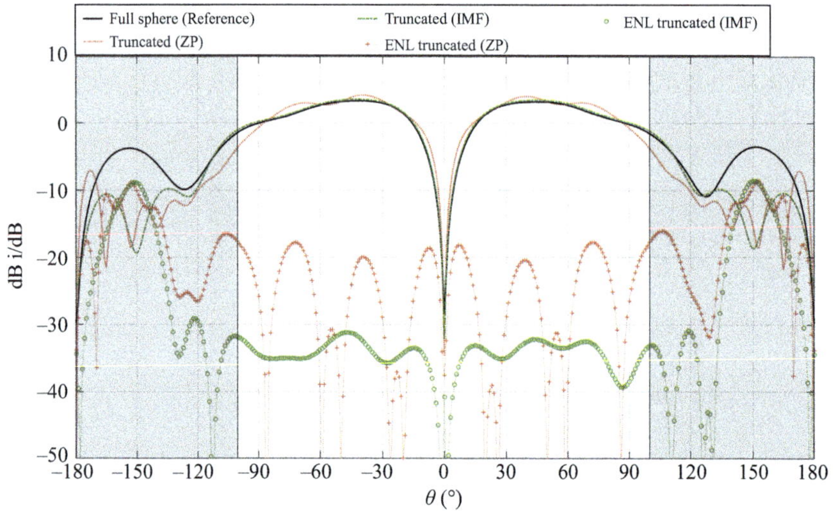

Figure 5.35 Directivity elevation pattern comparison of the SMC70 at 70 MHz (top) and at 120 MHz (bottom). Pattern of the SNF measurement without truncation (solid black); pattern of the truncated SNF measurement transformed with zero-padding (dotted orange) and with the IMF (dashed green); pointwise ENL of the truncated SNF measurement transformed with zero-padding (dotted orange with "+") and with the IMF (dashed green with "o")

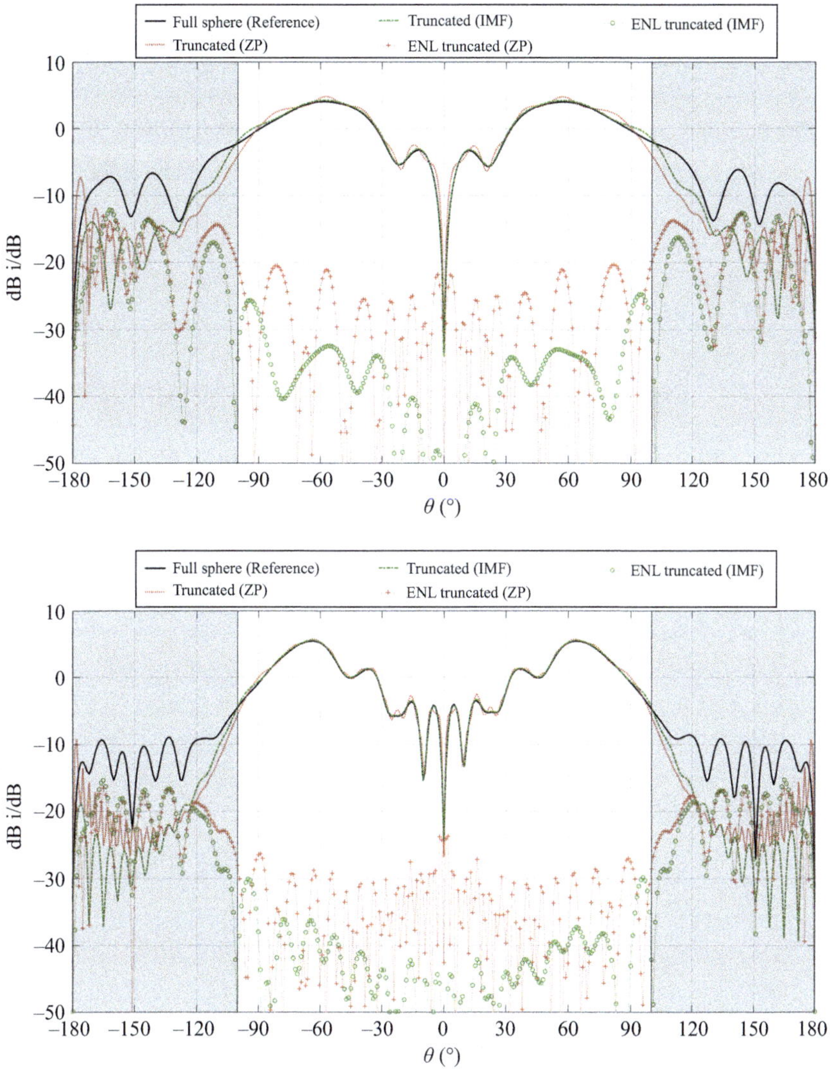

Figure 5.36 Directivity elevation pattern comparison of the SMC220 at 220 MHz (top) and at 500 MHz (bottom). Pattern of the SNF measurement without truncation (solid black); pointwise pattern of the truncated SNF measurement transformed with zero-padding (dotted orange) and with the IMF (dashed green); ENL of the truncated SNF measurement transformed with zero-padding (dotted orange with "+") and with the IMF (dashed green with "o")

5.2.1 Theory of cylindrical waves

The solution to the vector Helmholtz equation in cylindrical coordinates has been widely studied by different authors [13,32,33]. For time variations of the form $e^{j\omega t}$, the function that represents an outward traveling wave is the Hankel function of second kind, obtaining the following solution to the scalar Helmholtz equation:

$$f(\rho,\phi,z) = H_n^{(2)}(\kappa\rho)e^{jn\phi}e^{-jk_z z} \tag{5.24}$$

The solution to the electromagnetic equations with this function is shown in (5.25), representing the well-known CWE of the electric field:

$$\vec{E}(\rho,\phi,z) = \sum_{-\infty}^{\infty}\int_{-\infty}^{\infty}\left(a_n(k_z)\vec{F}_{1,nk_z}(\rho,\phi,z) + b_n(k_z)\vec{F}_{2,nk_z}(\rho,\phi,z)\right)dk_z \tag{5.25}$$

where the two polarization vectors can be expressed as

$$\vec{F}_{1,nk_z}(\rho,\phi,z) = \hat{r}\frac{jn}{r}H_n^{(2)}(\kappa\rho) - \hat{\phi}\frac{\partial H_n^{(2)}}{\partial r}(\kappa r)\bigg|_{r=\rho} \tag{5.26}$$

$$\vec{F}_{2,nk_z}(\rho,\phi,z) = -\hat{r}\frac{jk_z}{k_0}\frac{\partial H_n^{(2)}}{\partial r}(\kappa r)\bigg|_{r=\rho} + \hat{\phi}\frac{nk_z}{k_0\rho}H_n^{(2)}(\kappa\rho) + \hat{z}\frac{\kappa^2}{k_0}H_n^{(2)}(\kappa\rho) \tag{5.27}$$

where $a_n(k_z)$ and $b_n(k_z)$ are the coefficients of the CWE, which can be calculated from the field measured over a cylindrical surface as

$$a_n(k_z) = \frac{1}{4\pi^2\left(\partial H_n^{(2)}/\partial r\right)(\kappa r)\big|_{r=\rho_0}}\left(\frac{nk_z}{\rho_0\kappa^2}\int_{-\infty}^{\infty}\int_{-\pi}^{\pi}E_{meas,z}(\rho_0,\phi,z)e^{-jn\phi}e^{jk_z z}\,d\phi dz - \int_{-\infty}^{\infty}\int_{-\pi}^{\pi}E_{meas,\phi}(\rho_0,\phi,z)e^{-jn\phi}e^{jk_z z}d\phi dz\right) \tag{5.28}$$

$$b_n(k_z) = \frac{k_0}{4\pi^2 H_n^{(2)}(\kappa\rho_0)\kappa^2}\int_{-\infty}^{\infty}\int_{-\pi}^{\pi}E_{meas,z}(\rho_0,\phi,z)e^{-jn\phi}e^{jk_z z}d\phi dz \tag{5.29}$$

where $E_{meas,\phi}(\rho_0,\phi,z)$ and $E_{meas,z}(\rho_0,\phi,z)$ are the ϕ- and z-components of the field measured on a cylinder or radius ρ_0. Once the cylindrical wave coefficients are obtained from the measured data, the field can be calculated in FF applying expression (5.30):

$$\vec{E}_{far}(r,\theta,\phi) = -2k_0\sin(\theta)\frac{e^{-jk_0 r}}{r}\sum_{-\infty}^{\infty}j^n\left(a_n(k_z)\hat{\phi} + jb_n(k_z)\hat{\theta}\right)e^{jn\phi} \tag{5.30}$$

The calculation of these coefficients can be done using FFT in a very efficient way, as it is shown in [13].

Figure 5.37 (left) explains cylindrical acquisition. AUT rotates on an azimuth axes, while the probe is displaced on a vertical line. The distance between probe and AUT axis is ρ_o, while R_o is the minimum radio of AUT cylinder, that is the maximum radial distance from the AUT rotation axes to the extreme of the AUT (in a plane). The acquired samples are shown in Figure 5.37 (right). The distance between two probe positions defines the step in z (Δz), while the step in azimuth defines the distance between two angular samples of the acquisition cylinder ($\Delta\phi$).

The number of required acquisition samples depends on the electrical dimensions of the AUT. The minimum number in azimuth is $N_\phi = 4\pi R_o/\lambda$, although usually a security factor equal to 10 is added to that value (as in the spherical case). Therefore, the angular step must fulfill the following: $\Delta\phi \leq 360/N_\phi$. For the vertical dimension, z, the sampling theorem fixes the vertical step. Usually, this is fixed to $\Delta z \leq \lambda/2$, although some under-sampling could be applied for some specific cases.

Cylindrical setup is used for antennas with narrow main beam in elevation and broad main beam in azimuth. As in the case of planar acquisitions, the AUT is not fully enclosed by the acquisition surface, so the radiation pattern is not completely reconstructed, and a truncation error will appear after the FF transformation. In this case, the truncation is only present in elevation, since in azimuth, the AUT is fully enclosed by the cylindrical acquisition surface as shown in Figure 5.37. Therefore, there is only an angle that limits the validity in elevation and it is determined as shown in the following equation:

$$\theta = atan\left(\frac{L - A}{2\rho_o}\right) \tag{5.31}$$

where L is the vertical dimension of the cylinder, A is the vertical dimension of the AUT and ρ_o is the distance between both axes. In this case, the theoretical reliable spectral region is mathematically defined (5.32) that represents a horizontal strip within the visible range:

$$|k_z| = k \sin\theta \tag{5.32}$$

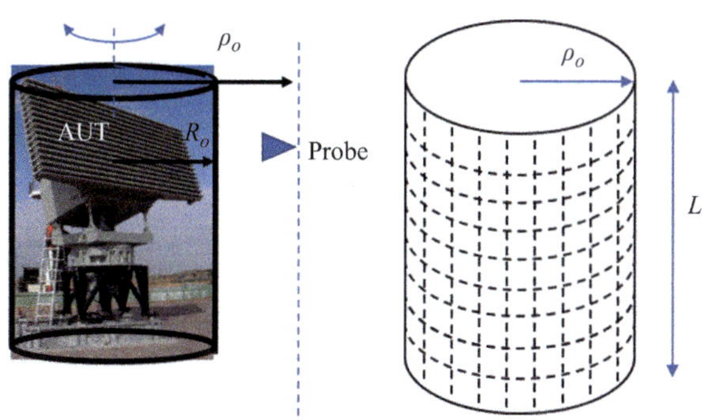

Figure 5.37 AUT and probe (left) and acquisition samples (right)

5.2.2 Single-cut measurements

In this section, we are processing one cut measurements using the cylindrical NF technique [31,34,35]. With the aim of having a generic procedure that is valid when measuring with an arbitrary probe, probe-corrected cylindrical modal coefficients (CMCs), calculated from only one cut (ring) is used. The ring is considered as a zero-height cylinder (Figure 5.38).

The method to obtain this probe compensated cylindrical modes coefficients is the modification of [13] for only one cut, considering each cut as a cylinder of zero height. This is equivalent to aggregate all the cylindrical modes in the axis, having a large aliasing (infinite step in the linear slide). However, if the antenna patterns are separable in both principal planes, the information of the pattern in that cut can be extracted. Obviously, any array is separable, but this can be a good approximation for rectangular array antennas, horns or simple reflectors. Figure 5.39 shows the mode representation for a cylindrical acquisition. The information of the

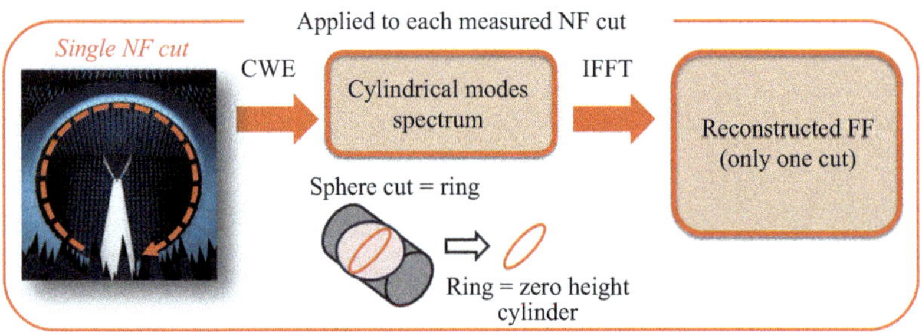

Figure 5.38 One ring acquisition near field transformation

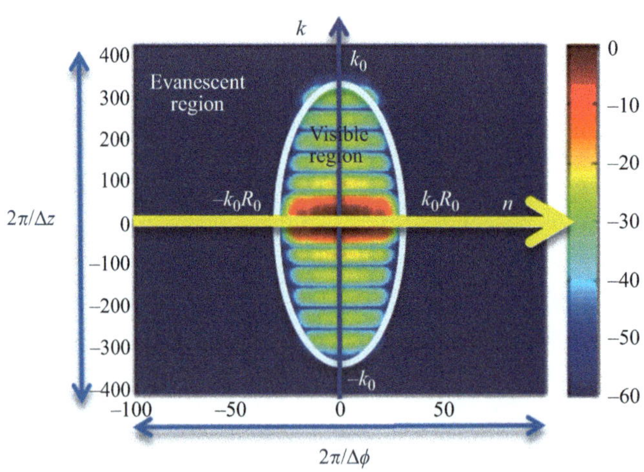

Figure 5.39 Mode representation of cylindrical expansion

radiation pattern is concentrated in the visible region, defined by the wavenumber and the radius of the minimum cylinder containing the AUT. In the case of one-ring transformation, all the modes are collapsed on the horizontal axis (yellow line).

The procedure employed for this case can be summarized in four steps:

1. Calculation of the transmitting modal coefficients of the AUT: The algorithm employed is [13], but instead of performing the 2D FFT that would be needed in a general case, a 1D FFT along ϕ is applied, since the height of the cylinder is zero:

$$\begin{aligned} I_n(h) &= \int_{-\pi}^{\pi} E_z(r_0, \phi) \cdot e^{-jn\phi}\, d\phi \\ I_n'(h) &= \int_{-\pi}^{\pi} E_\phi(r_0, \phi) \cdot e^{-jn\phi}\, d\phi \end{aligned} \tag{5.33}$$

2. Calculation of the probe CMCs for horizontal and vertical polarization from the probe FF pattern (*c* and *d* modes).
3. Calculation of the cylindrical mode coefficients: Both of the previous coefficients are combined, and the Hankel functions are evaluated to calculate the probe-corrected CMCs, with equations shown in [13] for only one k line. The modes a and b are calculated solving (5.34) for both polarizations:

$$a_n \sum_{m=-\infty}^{\infty} c_m \cdot H_{n+m}^{(2)}(k_o R_o) + b_n \sum_{m=-\infty}^{\infty} d_m \cdot H_{n+m}^{(2)}(k_o R_o) = \int_{-\pi}^{\pi} v(R_o, \varphi) e^{-jn\varphi} d\varphi \tag{5.34}$$

4. Once the CMCs are calculated, it is easy and fast to reconstruct the FF through an inverse FFT [32], as it is expressed in [36].

This algorithm has been applied to different antennas, as it was shown in [31,34] to show its validity. Here, it is applied to a linear polarized horn antenna (MVG SGH820) shown in Figure 5.40. The antenna has been measured at 10 GHz, at a 1-m distance when FF distance is 8.3 m.

Figure 5.41 shows the results for this antenna. As it is observed, the comparison with the conventional spherical near to FF transformation is excellent for

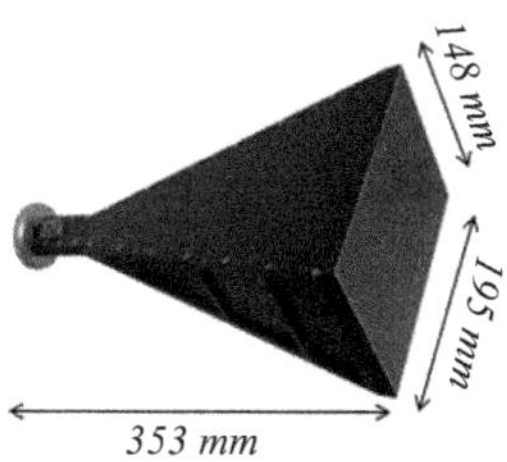

Figure 5.40 MVG SGH820 horn antenna

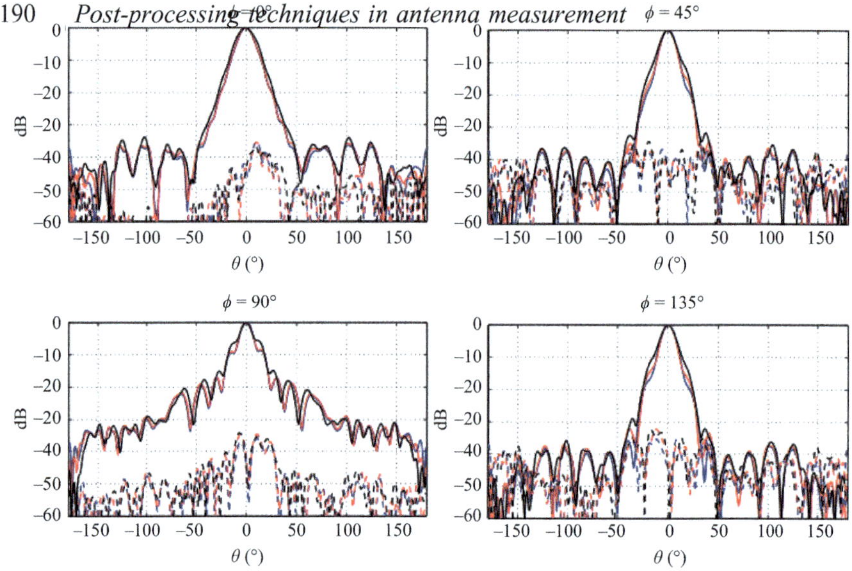

Figure 5.41 Comparison of results: in black measured near field, in blue processed with a conventional spherical near-to-far-field transformation and in red processed only one cut using cylindrical modes

the horizontal and vertical patterns, where the antenna pattern is separable. However, for the diagonal patterns, some discrepancies are found. As it was explained in the theory, this algorithm works well for this kind of antennas. The main advantage is the measurement time, since the FF is reconstructed using only the information of one single cut. Other examples are shown in [31,35] obtaining the same conclusions.

Therefore, the main application of this algorithm is the rapid measurement of commodity antennas as horns, conventional reflectors, separable rectangular arrays among others.

5.2.3 Applications to echo reduction in single-cut measurements

Cylindrical mode coefficients can be applied to the reduction of some errors, as it was done in the previous section for spherical case. If only the n-modes within that interval are considered in the NF/FF transformation, all the spurious contributions coming from other directions (as echoes or reflections) are filtered out. In a cylindrical general case, the AUT modes are the included in the visible region (Figure 5.42) [13].

In this case, since the information of the AUT is concentrated in the n values for $k_z = 0$, the valid interval is $(-k_oR_o, k_oR_o)$ for $k_z = 0$ (Figure 5.39), where all the visible modes are concentrated.

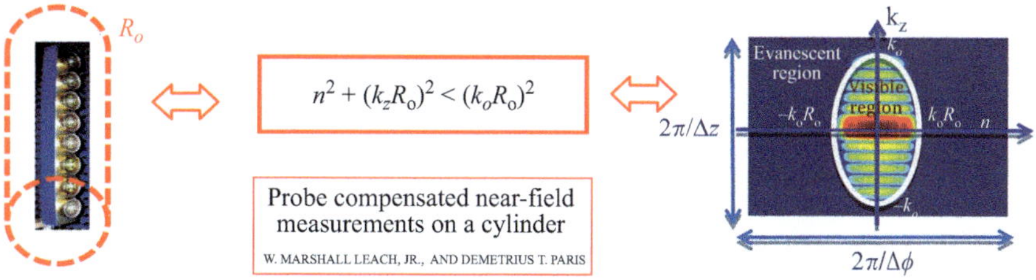

Figure 5.42 Representation of modes in spherical near field transformation

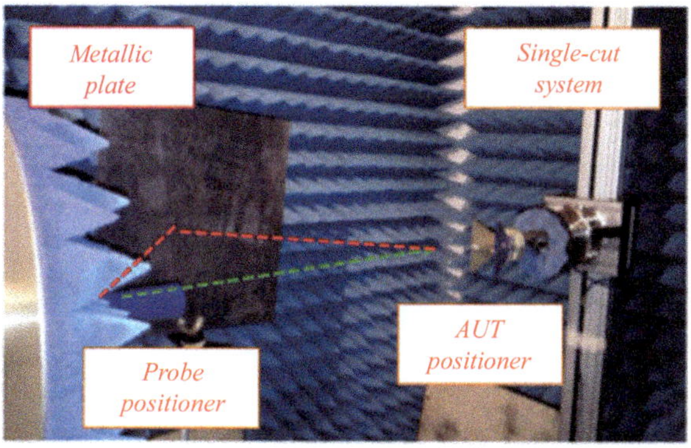

Figure 5.43 Single-cut measurements setup for the SH2000 wideband horn

As an application example, the MVG SH2000 horn antenna is measured in a single-cut NF setup able to perform single-cut data rotating on azimuth (Figure 5.43) at X-band. In this case, the AUT is offset 50 cm from the origin of the measurement system. To add a reflection to be reduced later by a post-processing, a metallic plate was collocated into the chamber. In order to reduce the number of cylindrical modes, and allowing a better filtering, before applying any processing, the AUT is mathematically translated to the origin. In this example, two echo-reduction techniques are compared: cylindrical modal filtering and time-domain filtering based on FFT.

Figure 5.44 shows a representation of the cylindrical (one-ring) modes for this particular case. As it is observed, the echo can be perfectly canceled since the modes associated to this reflection appear out of the AUT modes. As explained also in "Example—SNF with stationary probe and rotating AUT" section, this is due to the fact that, since the AUT is not centered in the setup, the spatial interference created on the probe strongly varies with the antenna position generating higher order modes associated to the echoes. Those modes can be separated from the ones of the AUT after the mathematical translation of the AUT to origin of the coordinate system allowing to filter them out. In this particular case, time gating does not work very well since the distance of the direct and reflected ray is not different

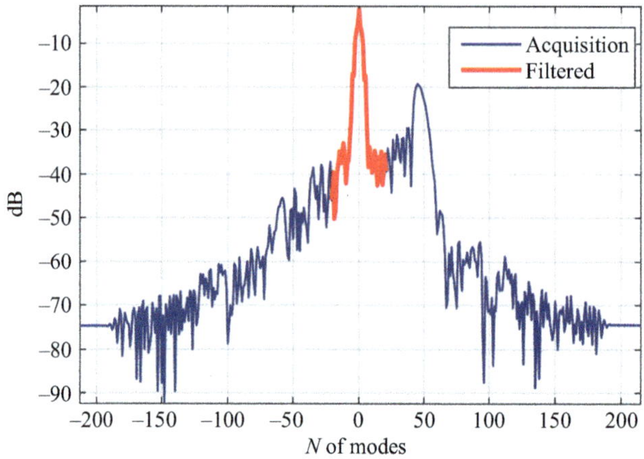

Figure 5.44 Single-cut spectral domain of the Wideband Horn (SH2000)

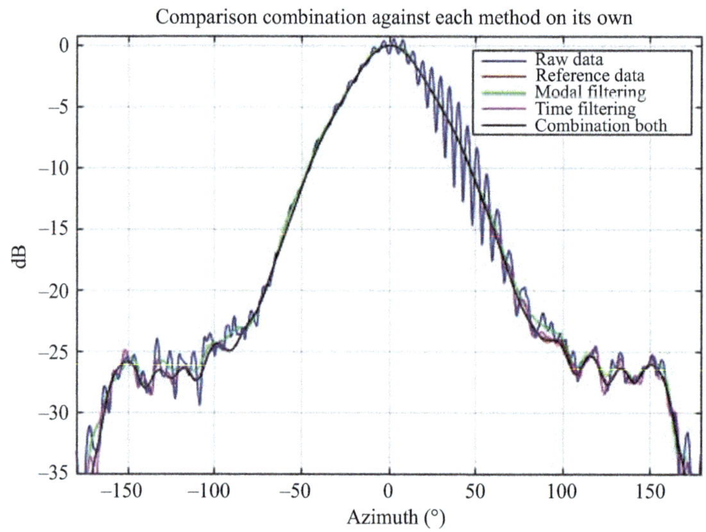

Figure 5.45 Acquired data including reflection, reference data and results applying different echo-reduction techniques

enough. In [36], other situations where time gating works better than modal filtering can be found.

Figure 5.45 shows the effect in the radiation pattern of the modal filtering, time filtering and combination of both techniques for this particular setup at 8 GHz. It is observed that the contribution from the echo can be almost perfectly filtered using cylindrical mode filtering technique, since it is enough separated from the contribution of the AUT. This is due to the offset of the AUT, as it observed in Figure 5.44.

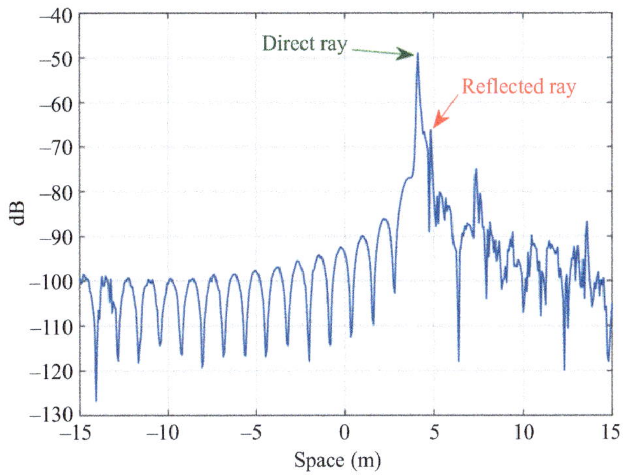

Figure 5.46 Time domain signal for the SH2000 antenna

With the measurements performed, this method allows us to obtain a great agreement between the reference pattern at 8 GHz and the pattern with the reflection after filtering (Figure 5.45). The red pattern is the reference one, the blue one contains the effects of the reflection and the green one is the pattern after modal filtering. The magenta one is the same antenna applying time-gating techniques and black one when both techniques are applied simultaneously. As explained above, time gating applied to this specific setup is not able to produce accurate results because the direct signal and the contribution from the echo are very close as also shown Figure 5.46. The effectiveness of the time filtering could be improved by increasing the measurement bandwidth, but sometimes this is not possible. This is also one of the advantages of the spatial/modal filtering which allow for accurate echo reduction using only one frequency point.

This section was focused on the echo-reduction techniques using one-ring cylindrical mode expansion and filtering. However, similar to the holography/planar technique shown in Chapter 4 and the SWE shown in the first part of this chapter, this technique also can improve the measurement result reducing the noise, filtering the leakage and other spurious contributions. This chapter has explained this particular case of one-ring NF/FF transformation, but the results can be extended to the general case of fully cylindrical acquisition and relative NF/FF transformation.

5.3 Conclusions

This chapter analyzes the use of spherical or cylindrical wave expansion for the improvement of antenna measurements. The source of errors analyzed in this chapter are noise, reflections and truncation errors, presenting different algorithms to mitigate their effects.

The chapter explains in a more detailed way the use of spherical wave expansion, beginning by a detail theoretical analysis of the meaning of the spherical modes and the relation with the antenna dimensions and measurement setup; and then, using the proposed algorithms for different applications in noise, echo reduction and reduction of the truncation effect, due to incomplete measurements.

In the case of cylindrical wave expansion, in order to avoid redundancy, we decided to show only a special case: the echo reduction for single cut measurements, explaining the theory and limitations of this special case. In any case, this technique can also be applied for general cases and for all the previous applications.

Therefore, Chapter 5 completes the other approaches for the same purpose explained from Chapters 2–4. While Chapters 2 and 3 focus on the potential of equivalent currents and Chapter 4 in the direct use of plane wave spectrum, Chapter 5 exploits the signal filtering through the spherical or cylindrical wave coefficients. These are the classical domains used for the improvement of antenna measurements through post-processing techniques using a priori information of the dimensions of the AUT.

References

[1] IEEE Std 1720-2012, "Recommended Practice for Near-Field Antenna Measurements". 2012.

[2] L. J. Foged, L. Scialacqua, F. Mioc, *et al.*, "Echo Suppression by Spatial Filtering Techniques in Advanced Planar and Spherical NF Antenna Measurements", AMTA Symposium, October 2012, Seattle, WA, USA.

[3] L. J. Foged, J. Estrada, and P. O. Iversen, "Spherical Near Field Testing of Small Antennas from 800 MHz to 18 GHz", IEEE International Symposium on Antennas and Propagation, 2007, June 10–15, Honolulu, HI, USA.

[4] L. J. Foged, and A. Scannavini, "Efficient testing of wireless devices from 800 MHz to 18 GHz", Radio Engineering Magazine, vol. 18, no. 4, pp. 460–466, 2009.

[5] A. Scannavini, L. J. Foged, A. Giacomini, L. Duchesne, N. Gross, and F. Herbiniere, "Efficient Testing of Antennas for 5G mm-Wave Applications", EuCAP, 2018, London, UK.

[6] P. Noren, P. Garreau, and L. J. Foged, "State of the Art Spherical Near-Field Antenna Test Systems for Full Vehicle Testing", 6th EuCAP, March 2012, Prague, Czech Republic.

[7] J. E. Hansen (ed.), Spherical Near-Field Antenna Measurements, Peter Peregrinus Ltd., on behalf of IEE, London, United Kingdom, 1988.

[8] F. Jensen, and A. Frandsen, "On the Number of Modes in Spherical Wave Expansions" AMTA 26th Annual Symposium (AMTA2004), 2004.

[9] F. Saccardi, A. Giacomini, and L. J. Foged, "Probe Correction Technique of Arbitrary Order for High Accuracy Spherical Near Field Antenna Measurements", AMTA, 2016 October 30–November 4, Austin, TX, USA.

[10] F. Saccardi, A. Giacomini, and L. J. Foged, "Comparative Investigation of Spherical NF Measurements with Full and First Order Probe Correction Using Calibrated or Simulated Probe", EuCAP, 2017 19–24 March, Paris, France.

[11] T. Laitinen, S. Pivnenko, J. M. Nielsen, and O. Breinbjerg, "Theory and practice of the FFT/matrix inversion technique for probe-corrected spherical near-field antenna measurements with high-order probes", IEEE Transactions on Antennas and Propagation, vol. 58, no. 8, pp. 2623–2631, 2010.

[12] C. Ludwig, "The definition of cross polarization", IEEE Transactions on Antennas and Propagation, vol. AP-21, no. 1, pp. 116–119, 1973.

[13] W. M. Leach, and D. T. Paris, "Probe compensated near-field measurements on a cylinder", IEEE Transactions on Antennas and Propagation, vol. 21, no. 4, pp. 435–445, 1973.

[14] A. G. Repjar, A. C. Newell, and M. H. Francis, "Accurate determination of planar near-field correction parameters for linearly polarized probes", IEEE Transactions on Antennas and Propagation, vol. 36, no. 6, pp. 855–868, 1988.

[15] C. A. Balanis, Antenna Theory Analysis and Design. Hoboken, NJ: John Wiley & Sons, Inc., Publication, 3rd edition, 2005.

[16] L. J. Foged, M. Faliero, and P. O. Iversen, "Random Noise in Spherical Near Field Systems", AMTA, November 2009, Salt Lake City, UT.

[17] A. D. Yaghjian, "Approximate formulas for the far field and gain of open-ended rectangular waveguide", IEEE Transactions on Antennas and Propagation, vol. AP-32, no. 4, pp. 378–384, 1984.

[18] F. J. Cano-Fácila, S. Pivnenko, and M. Sierra-Castañer, "Reduction of truncation errors in planar, cylindrical and partial spherical near-field antenna measurements", International Journal of Antennas and Propagation, vol. 2012, Article ID 438727, 2012.

[19] L. J. Foged, L. Duchesne, P. Garreau, P. O. Iversen, and J.-C. Bolomey, "Truncation Impact on Measured Radiation Pattern in Spherical Near Field Antenna Test Ranges", IEEE APS Symposium and USNC-URSI National Radio Science Meeting, 2001 8–12 July, Boston, USA.

[20] E. Martini, S. Maci, and L. J. Foged, "Spherical Near Field Measurements with Truncated Scan Area", 5th European Conference on Antennas and Propagation, EuCAP 2011, 2011 11–15 April, Rome, Italy.

[21] E. Martini, S. Maci, and L. J. Foged, "Reduction of Truncation Errors in Spherical Near Field Measurements", AMTA, 2010, 10–15 October, Atlanta, Georgia, USA.

[22] L. J. Foged, L. Scialacqua, F. Saccardi, *et al.*, "Comparative Investigation of Methods to Reduce Truncation Errors in Partial Spherical Near-Field Antenna Measurements", EuCAP, 2012 26–30 March, Prague, Czech Republic.

[23] F. Saccardi, F. Rossi, L. Scialacqua, and L. J. Foged, "Truncation Error Mitigation in Free-Space Automotive Partial Spherical Near Field Measurements", AMTA, 2017, 15–20 October, Atlanta, GA, USA.

[24] R. C. Wittmann, C. F. Stubenrauch, and M. H. Francis, "Using truncated data sets in spherical scanning antenna measurements", International Journal of Antennas and Propagation, vol. 2012, 2012.

[25] R. Cornelius, D. Heberling, and D. Pototzki, "Spherical Near-Field Far-Field Transformation with Infinite Ground Plane At Arbitrary Height z", EuCAP, 2017 19–24 March, Paris, France.
[26] R. A. M. Mauermayer, and T. F. Eibert, "Spherical Field Transformation for Hemispherical Antenna Measurements above Perfectly Conducting Ground Planes", AMTA, October 2016, Austin, TX, USA.
[27] ANSI/IEEE Std 149-1979, "IEEE Standard Test Procedures for Antennas". 1979.
[28] G. Sinclair, E. C. Jordan, and E. W. Vaughan, "Measurement of aircraft-antenna patterns using models", Proceedings of the IRE, vol. 35, pp. 1451–1462, 1947.
[29] G. Sinclair, "Theory of models of electromagnetic systems", Proceedings of the IRE, vol. 36, pp. 1364–1370, 1948.
[30] S. Gregson, A. Newell, and G. Hindman, "Reflection Suppression in Cylindrical Near-Field Antenna Measurement Systems-Cylindrical MARS", Antenna Measurement Techniques Association, AMTA 2009, 2009, Salt Lake City, UT, pp. 119–125.
[31] R. Cornelius, T. Salmerón-Ruiz, F. Saccardi, L. Foged, D. Heberling, and M. Sierra-Castañer, "A comparison of different methods for fast single-cut near-to-far-field transformation", IEEE Antennas and Propagation Magazine, vol. 56, no. 2, pp. 252–261, 2014.
[32] A. D. Yaghjian, "An overview of near-field antenna measurements", IEEE Transactions on Antennas and Propagation, vol. AP-34, no. 1, pp. 30–44, 1986.
[33] O. M. Bucci, and C. Gennarelli, "Use of sampling expansions in near-field-far-field transformations: the cylindrical case", IEEE Transactions on Antennas and Propagation., vol. 36, no. 6, pp. 830–835, 1988.
[34] T. Salmerón-Ruiz, M. Sierra-Castañer, F. Saccardi, S. Burgos, F. J. Cano-Fácila, and L. J. Foged, "A Fast Single Cut Spherical Near-Field-to-Far-Field Transformation Using Cylindrical Modes", Proceedings of 8th European Conference on Antennas and Propagation, EuCAP 2014, April 2014, Den Haag (Netherlands).
[35] M. Sierra Castañer, F. Saccardi, and L. J. Foged, "Gain Antenna Measurement using Single Cut Near Field Measurements", Proc. of the 38th Annual AMTA Symposium, Austin (Texas), 2016, October 30–November 4, pp. 309–312.
[36] M. J. López Morales, F. Saccardi, M. Sierra Castañer, and L. J. Foged, "Comparison of Echo Reduction Techniques for One-Single Cut Antenna Measurements", Proceedings of the 10th European Conference on Antennas and Propagation, 2016, 10–15 April, Davos, Switzerland.

Chapter 6

Phase recovery in antenna measurements

Ana Arboleya[1], Yuri Álvarez[2], Jaime Laviada[2], and Fernando Las-Heras[2]

Antenna characterization based on amplitude-only acquisitions is described in this chapter. This kind of characterization can be necessary either due to budget constraints or when only scalar equipment is available at the considered working frequency. In those cases, a variety of phaseless techniques can be applied to recover, directly or indirectly, the complex near-field (NF), radiated by the antenna. After that, full characterization of the antenna can be made by resorting to conventional NF to far-field (NF–FF) transformations or diagnostics approaches.

In particular, this chapter focuses on two specific techniques, both derived from optics. In the first technique, an interferometry between the unknown radiated field and a known radiated field is used. As it will be illustrated, if certain conditions are met, then it is possible to directly retrieve the radiated field with a negligible error by means of simple operations in the spatial spectral domain, also known as k-space.

In the second technique, the phase recovery requires measuring the amplitude of the radiated field in at least two independent surfaces. After that, a postprocessing step consisting of an iterative algorithm is performed to find a phase distribution so that the propagation between planes matches the measured amplitudes.

In addition, several enhancements are formulated for each approach, enabling a simplification of the interferometric setup and reduced requirements on sampling criteria as well as an improvement of the convergence of the iterative schemes by means of the use of the sources reconstruction method (SRM) and a recast of the problem as a nonlinear optimization one.

[1]Área de Teoría de la Señal y Comunicaciones, Universidad Rey Juan Carlos, Madrid, Spain
[2]Área de Teoría de la Señal y Comunicaciones, Department of Electrical Engineering, Universidad de Oviedo, Gijón, Spain

6.1 Introduction

The amplitude of the radiated field provides the most insightful information about an antenna for most of applications as many parameters such as directivity or secondary lobe level can be directly computed if the acquisition is performed in the FF. However, it is not always the case. For instance, an antenna for imaging applications is usually required to have a constant phase along the antenna main lobe. Otherwise, the field impinging in the area illuminated by the spot will not be phase uniform, and, consequently, it will not be added coherently in reception as it is usually assumed.

Furthermore, FF measurements are not always feasible due to the required dimension of the facilities. In those usual cases, FF is indirectly computed by performing an NF acquisition followed by an NF–FF transformation. This transformation requires the full knowledge of the NF in terms of not only amplitude but also phase.

Despite being the fact that the phase of the field plays a key role in many cases, its acquisition is much more challenging than amplitude acquisition. Although amplitude can be measured by relatively simple devices sensitive to the power of the radiated field such as bolometers, phase acquisition requires a significantly more complex and precise circuitry, which might not be available at cutting-edge high frequencies. In addition, approaches for direct phase acquisition required in antenna metrology entail to guide a reference signal, phase locked to the radiated signal, from the transmit unit to the receiver unit. This electrically large path is subject to different adverse conditions, which have a negative impact on the phase acquisition. First, this path must be implemented by some flexible parts as the transmitter or the receiver will be moved and/or rotated to perform planar, cylindrical, or spherical acquisitions. Thus, the movement of the transmission lines, usually coaxial cables conveying a fraction of the transmitted signal,* is subject, up to some degree, to flexing, which affects the received phase. Unless this impact can be correctly characterized and compensated, which is also challenging, the accuracy of the result will be degraded. In a similar fashion, if the temperature of the acquisition setup components varies, a phase drift will happen distorting the result. Although the aforementioned effects can be negligible at low frequencies, they become very relevant at high frequencies.

Finally, it is worth noting that *vector* instrumentation is only available up to certain frequency bands, although the limit of these bands keeps raising thanks to technological advances. Moreover, vector instrumentation, if available, is usually much more expensive than scalar instrumentation. Consequently, *scalar* equipment is the only choice for antenna characterization at very high frequencies or in cost-effective facilities.

*In high-frequency regime (W-band and beyond), in which the transmitted signal is generated by using frequency multipliers, this signal is usually a fraction of the pre-multiplied signal.

6.1.1 Main phase recovery techniques

Phaseless techniques, also known as scalar or amplitude-only acquisitions, have been mostly based on approaches already available in optics. These techniques pursue to retrieve the phase of the field from one or several amplitude-only acquisitions either to evaluate the phase in the NF (e.g., in the case of NF-focused antennas), as a previous step to perform an NF–FF transformation or for diagnostics purposes.

Despite the large amount of techniques, in general, they can be classified into two large sets: (1) interferometric techniques and (2) iterative techniques. A general overview of these techniques is given next, before providing an extensive description of two specific approaches, one for each set of techniques, in Sections 6.2 and 6.3.

6.1.2 Interferometric techniques

Interferometric techniques are based on acquiring at least the amplitude of the sum of two or more signals. If one of these signals is known (i.e., it is a reference signal) or the phase shift between the acquired signal is known, then the amplitude and phase of the desired signal can be retrieved by means of relatively simple math.

A first subset of these techniques is based on using *modified probes*. These probes usually contain two or more antennas, spaced by a known distance. The received signals are usually split into several branches, phase shifted and combined. One or more power meters can be used to acquire the field at the different stages of the modified probe.

One of the pioneering works was the *differential phase measurement method*, which originally required three power meters [1]. The phase retrieval, based on vector sum equivalences, was able to retrieve the phase difference between the fields acquired by the two antennas in the probe. The method was later modified so that only one power meter would be required [2]. This differential method was initially applied to cylindrical acquisitions with a constant angular step, which depended on the distance between both probe antennas. The method has been successfully tested at the terahertz regime [3] .

A similar method, also based on a modified probe to measure the phase difference, was later proposed with significant differences. In particular, the proposed method was based on acquiring the field at different sets of samples [4]. The same sampling step was used in all the sets but with a different spatial offset. After the acquisition, the phase offset between the independent set of acquisitions was found by searching a potential solution among a reduced set of possibilities by resorting to *nonredundant sampling theory* [5]. Consequently, arbitrary scanning surfaces could be used without the requirement of using a sampling step given by the separation of the probe antennas [6]. Further studies have also proved the possibility of using the approach for wideband acquisitions [7] as well as for imaging applications [6].

The second largest subset of interferometric techniques is derived from *Gabor holography*. The word *holography* comes from the Greek words *hólos* (whole) and *gráphō* (written or represented) and was coined by Gabor in 1948, who developed a technique for reconstructing the complex field distribution of the recorded intensity of a coherent wave disturbance with a properly characterized reference field in the optics field [8]. However, the quality of the complex field retrieved by the original approach, in which the reference source and the unknown source were placed along the same axis, was degraded by several aliasing terms. This limitation was bypassed by including an off-axis reference. This technique was developed by Leith and Upatnieks [9], which is why it is often called *Leith–Upatnieks Holography*. Indirect off-axis holography[†] was first used in the late 1960s for imaging applications in the microwave and acoustic bands [10] and subsequently adapted for antenna metrology applications by Napier [11] and Bennett [12]. Figure 6.1 shows the typical layout of a measurement setup for indirect off-axis holography antenna measurement applications [13].

The most relevant limitations of indirect off-axis holography are due to the nature of the interferometric signal built by adding the unknown field and the reference field. In particular, although spectra overlapping is much smaller than in the original on-axis holography approach, it is still not negligible for antennas with a large spatial bandwidth [14]. For this reason, the use of synthetic reference waves by means of a variable phase shifter, which enables further control of the spatial spectrum terms, was proposed by several authors [15,16]. In addition, the latter technique simplifies the characterization of the reference wave. Another drawback associated to the involved spectra terms is that the spatial bandwidth is significantly larger than the that of the original signal. In order to mitigate this effect, the use of nonredundant sampling theory has also been proposed [17,18].

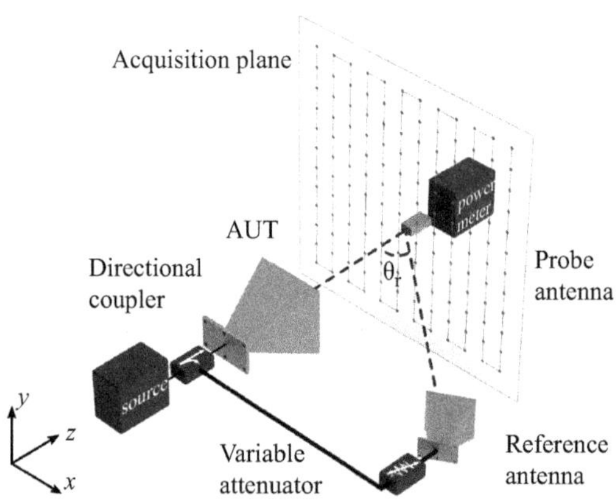

Figure 6.1 An overview of a typical setup for antenna measurement using amplitude-only information and interferometric techniques

[†]The term indirect holography is often used to refer to these techniques as they allow one to acquire the “whole” information (i.e., amplitude and phase) by means of an indirect process. For further details on *direct holography* schemes for antenna metrology, please see Chapter 4.

In Section 6.2, together with a review of off-axis indirect holography, several enhanced approaches are presented. The first approach, based on a simple modification of the acquired interferometry pattern, enables a significant reduction of the sampling requirements. Next, an approach based on including mechanical shifts to synthesize the reference wave is presented. Thus, the error due to the aliasing in the spectral domain is removed *without the need of a phase shifter*. Finally, a third approach is described as an alternative method to enable the acquisition of *wide-band signals* in contrast to the single frequency formulation. This latter approach also allows the use of conventional $\lambda/2$ spatial sampling criteria.

6.1.3 Iterative techniques

Iterative phase retrieval techniques rely on numerical methods, which exploit the information contained in the amplitude of the radiated field collected at two or more acquisition surfaces.

The main advantage of iterative phase retrieval techniques is hardware simplicity, as only a single power sensor is required to collect the radiated field, as depicted in Figure 6.2. However, the main disadvantage is the iterative nature of these techniques, making them slower with respect to interferometric techniques. Besides, they are quite dependent on the conditions of the problem, namely, antenna under test (AUT) and acquisition surfaces characteristics (such as AUT electric size, directivity, and radiated field acquisition distance). Another disadvantage is the need of at least two acquisition surfaces, doubling measurement time.

Different iterative phase retrieval techniques have been developed, depending on the kind of AUT measurement range. For example, for planar antenna measurement techniques, a well-known method is the iterative Fourier technique (IFT), described in [19,20]. In this method, the NF amplitude of the AUT is measured at two different

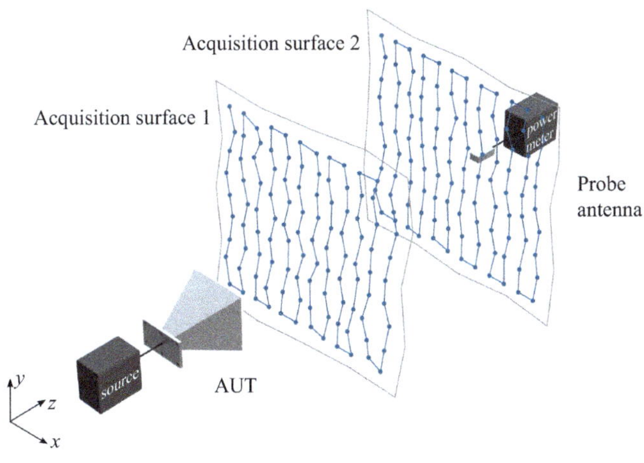

Figure 6.2 An overview of a typical setup for antenna measurement using amplitude-only information and iterative techniques

surfaces. First guess of the aperture field of the AUT is assumed based on its aperture shape. This initial guess (IG) is propagated to the first measurement surface, replacing the calculated NF amplitude by the measured one, and keeping the calculated NF phase from the initial first guess. Next, the field on the first acquisition surface is propagated toward the second acquisition surface, replacing again the calculated amplitude by the measured NF. After that, the field is propagated back to the aperture plane of the AUT. Several metrics can be set to define the stopping criteria of the algorithm, for example, variation of the root mean square (RMS) value of the aperture fields, RMS error between the measured and calculated NF amplitude in the acquisition planes, and maximum number of iterations.

One of the advantages of the IFT is the ease and fast calculation of the forward and backward transformation thanks to the use of fast Fourier transforms (FTs) [19,20]. IFT has been successfully extended to spherical measurement domains in [21], based on the well-known spherical mode expansion of the fields of the AUT.

Among the main limitations of the IFT (also common to all iterative phase retrieval techniques) is the fact that convergence of the IFT greatly depends on the first guess of the aperture field, which can condition the convergence of the iterative procedure, prematurely ending in a local minimum. The woks presented in [22,23] have proposed the use of global search algorithms such as differential evolution (DE), real coded genetic algorithm (RCGA), and particle swarm optimization (PSO) to improve the choice of the IG, thus minimizing the risk of stagnation of the iterative procedure.

Another approach is based on the minimization of a nonlinear cost function (CF) relating the amplitude of the measured field radiated by the AUT and the field calculated from an electromagnetic model of the AUT. This methodology is described in [24] for planar domains, then extended in [25] for cylindrical measurement ranges. The SRM for antenna diagnostics with NF–FF transformation (described in Chapters 2 and 3) has been modified to handle phaseless NF measurements. The resulting nonlinear CF is minimized by means of well-known nonlinear optimization techniques such as Levenberg–Marquardt algorithm (LMA) [26], as well as by means of a global search algorithm [27]. Linearization of the nonlinear CF has been recently proposed in [28], together with the use of regularization techniques to improve convergence.

In brief, iterative phase retrieval techniques can be classified into two main groups: (1) iterative schemes based on forward–backward propagation of the field between the two acquisition surfaces and the aperture of the AUT, (2) minimization of a nonlinear CF relating the amplitude of the measured NF and the NF estimated from an equivalent electromagnetic model of the AUT (equivalent currents, aperture fields, and modal expansion).

6.2 Indirect off-axis holography

This section is devoted to the description of interferometric phase-retrieval techniques based on indirect off-axis holography approaches. It focuses on the conventional method and on a recently presented method adapted for broadband phaseless antenna characterization. Two application examples are included in order to show the capabilities of interferometric techniques.

6.2.1 Method

Indirect off-axis holography is a two-step technique based on (1) recording the amplitude of the interference pattern of a reference field and the field radiated by the AUT, also called *hologram*, and (2) retrieving the complex field of the AUT by postprocesing the acquired hologram. The second step is performed by filtering part of the spectrum of the hologram in the spatial frequency domain or the k-space and removing the effect of the reference field, which has to be previously known in amplitude and phase.

Conventional setups are implemented as shown in Figure 6.1 using a radiated reference field [14,15,29], obtained from a sample of the radio-frequency (RF) source by means of a directional coupler, and radiated toward the acquisition plane with a certain off-axis angle. A variable attenuator is often employed in order to balance the power level between the AUT and the reference branches and maximize the dynamic range of the hologram [14,18].

More recent approaches, which will be explained later, substitute the radiated reference with a synthesized reference generated by means of a phase shifter [16,30] as shown in Figure 6.3(a). There are other implementation options such as using a shaped mirror as collimator in compact antenna ranges to create a plane reference wave [31] or resorting to mechanical shifts to avoid the use of the phase shifter as shown in Figure 6.3(b) [18,32].

In the conventional approach, the hologram is acquired at each point of the acquisition plane as

$$H(\vec{r}) = |E_{aut}(\vec{r}) + E_r(\vec{r})|^2, \qquad (6.1)$$

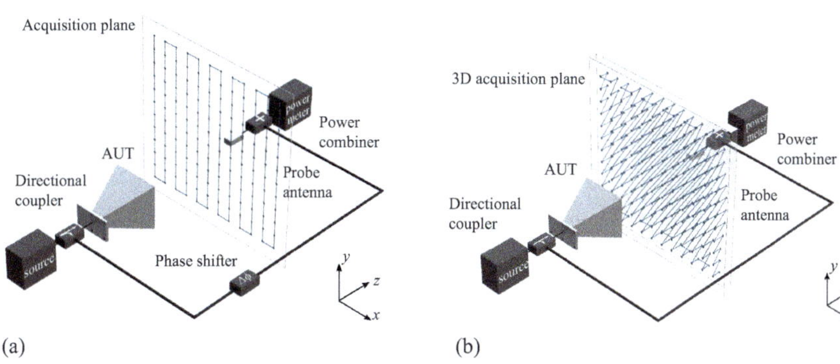

Figure 6.3 Setups for indirect off-axis holography with synthesized reference wave for antenna measurement on a planar range: (a) electronic phase shifts, (b) mechanical phase shifts [13]

where $\vec{r}$ is the position vector defining the position of the probe antenna in the acquisition plane, and E_{aut} and E_r are the components of the fields radiated, respectively, by the AUT and the reference antenna received by the probe.

Further development of (6.1) yields to

$$H(\vec{r}) = |E_{aut}(\vec{r})|^2 + |E_r(\vec{r})|^2 + E_{aut}(\vec{r})E_r^*(\vec{r}) + E_{aut}^*(\vec{r})E_r(\vec{r}), \tag{6.2}$$

where the asterisk is used to denote complex conjugate.

The spectrum of the hologram in the k-space can be obtained by means of an FT of (6.2) as

$$h(\vec{k}) = |e_{aut}(\vec{k})|^2 + |e_r(\vec{k})|^2 + e_{aut}(\vec{k}) \otimes e_r^*(-\vec{k}) + e_{aut}^*(-\vec{k}) \otimes e_r(\vec{k}), \tag{6.3}$$

where e_{aut} and e_r are the FT of E_{aut} and E_r, respectively, and $\otimes$ the convolution operator.

The usual representation of the spectrum can be seen in Figure 6.4. The central term corresponds to the sum of the two zero-frequency harmonics defined by the squared amplitudes of the AUT and the reference antenna and is known as autocorrelation term, whereas the other two terms contain information about the complex field of the AUT shifted and distorted by the convolution with the reference field and are known as cross-correlation or image terms.

If there is a negligible overlap between the image terms and the autocorrelation terms, then band-pass filtering of the spectrum yields:

$$h_{bpf}(\vec{k}) \simeq e_{aut}(\vec{k}) \otimes e_r^*(-\vec{k}), \tag{6.4}$$

and, since the complex reference field is known, the complex field of the AUT can be retrieved in the spatial domain as

$$E_{ret}(\vec{r}) = \frac{H_{bpf}(\vec{r})}{E_r^*(\vec{r})} \simeq E_{aut}(\vec{r}), \tag{6.5}$$

where H_{bpf} is the inverse FT of the filtered image term h_{bpf}.

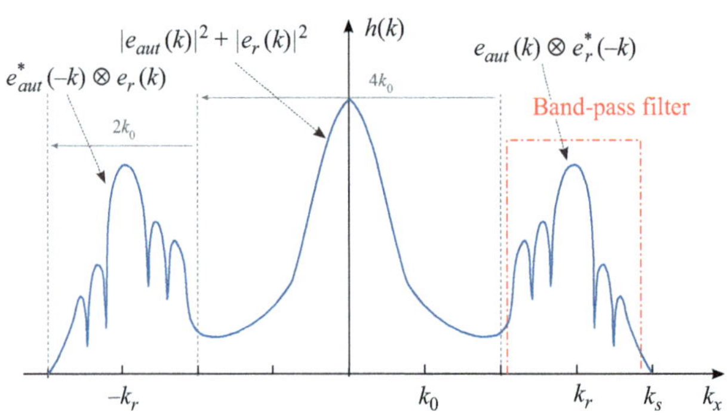

Figure 6.4 Spectrum of a hologram for conventional indirect off-axis holography. Ideal scenario with no off-axis angle limitation

Two important aspects need to be noted at this point:

- The complex reference field must be fully characterized in amplitude and phase. This can be achieved by means of an initial vector calibration, which does not need to be repeated unless the reference antenna is modified, or by an analytical characterization.
- Low-directivity antennas such as open-ended waveguides (OEWG) are recommended as reference antennas because they provide uniform illumination of the acquisition plane [14]. Moreover, the lack of nulls in the amplitude distribution of the reference field guarantees that (6.5) can be evaluated without the risk of divisions by zero.

If the reference antenna is far enough from the acquisition plane, its radiated field can be approximated by a plane wave. Thus, position of the image terms, which are centered at $\pm k_r$, and, therefore, degree of overlap between the image terms and the autocorrelation terms will be determined by the off-axis angle θ_r formed by the normal to the aperture of the reference antenna and the normal to the acquisition plane, and the free-space wavenumber $k_0 = 2\pi/\lambda$:

$$k_r = \pm k_0 \cos(\theta_r) \tag{6.6}$$

This off-axis angle can be introduced either in the x-axis, in the y-axis, or in both, but for the sake of simplicity, the formulation of the problem will be developed considering the first option (see Figure 6.1).

On the other hand, following the *Nyquist* criteria, and in order to avoid aliasing, sampling step must be selected according to the extension of the spectrum k_s as [33]:

$$\Delta x < \frac{\pi}{k_s}. \tag{6.7}$$

According to [34], the maximum spatial bandwidth of a radiated field in a planar acquisition is $2k_0$. This bandwidth is doubled, for the case of considering the squared amplitude fields. Therefore, as graphically shown in Figure 6.4, the no-overlapping condition is defined by

$$k_r > 3k_0, \tag{6.8}$$

which implies a plane wave in the non-visible part of the spectrum. Consequently, the conventional approach cannot guarantee an aliasing-free spectrum. If the maximum value of k_r is assumed (maximum off-axis angle for a radiated reference field must be lower than 90°), then the central position of the image terms is $\pm k_0$, and overlapping, though moderate, always occurs (see Figure 6.5(a)). In this case, the extension of the spectrum is reduced to $k_s = 2k_0$ and the sampling step can be relaxed up to $\lambda/4$.

6.2.2 Improved schemes

As aforementioned, several improved schemes focused on reducing the degree of overlap and increasing the quality of the phase retrieval have been developed. The most known approaches are described in this section.

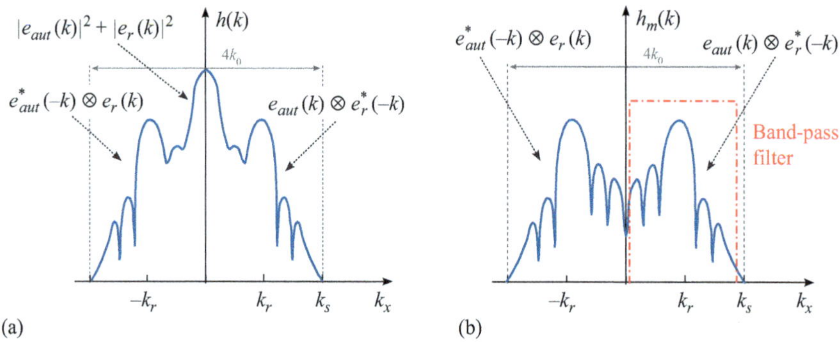

Figure 6.5 Spectrum of a hologram for indirect off-axis holography for $\theta_r < 90°$: (a) conventional hologram, (b) modified hologram

6.2.2.1 Modified hologram

The modified hologram technique is a simple overlapping reduction technique, which consists in the removal of the autocorrelation terms of the hologram. The simplest implementation involves an extra scan in order to record the amplitude of the field radiated by the AUT (the amplitude of the reference field is supposed to be acquired in a calibration stage) [14,35]. The other common implementation, known as *opposite-phase* technique, employs two different power meters and an Hybrid-T that provides the complete hologram in the sum port and the sum of the auto-correlation terms in the difference port [36].

The modified hologram can, therefore, be expressed as

$$H_m(\vec{r}) = H(\vec{r}) - |E_{aut}(\vec{r})|^2 - |E_r(\vec{r})|^2 = E_{aut}(\vec{r})E_r^*(\vec{r}) + E^*_{aut}(\vec{r})E_r(\vec{r}), \tag{6.9}$$

being the expression of the spectrum as follows:

$$h_m(\vec{k}) = e_{aut}(\vec{k}) \otimes e_r^*(-\vec{k}) + e^*_{aut}(-\vec{k}) \otimes e_r(\vec{k}). \tag{6.10}$$

For off-axis angles smaller than 90°, the image terms will be centered at $k_r < k_0$ as shown in Figure 6.4, thus considering a maximum width of $2k_0$ for the image terms:

$$\Delta x = \frac{\pi}{k_s} < \frac{\lambda}{4}. \tag{6.11}$$

In practice, grazing angles of 90° are clearly not feasible. However, even considering smaller off-axis angles, this technique significantly reduces the overlap, specially for the case of directive antennas, with narrower spectra. In these cases, the off-axis angle can be further reduced yielding larger sampling steps. For the case of low-directive antennas, some overlap will necessarily exist [36] but much more moderate than when using the standard hologram.

Although this approach needs of an extra scan, or the modification of the measurement setup for the implementation of the *opposite-phase* technique, the use of the modified hologram has several advantages.

- The amplitude of the field radiated by the AUT is directly acquired; thus, the final field of the AUT can be composed by the retrieved phase and the measured amplitude, reducing the error of the phase retrieval.
- Overlapping is reduced, with a further reduction of the error of the phase retrieval.
- If the spatial bandwidth of the antennas is small (i.e., directive antennas), the reduction of the extension of the k-space allows using of larger sampling steps; thus, acquisition time can be reduced, compensating the time needed for the extra acquisition.
- If aliasing is tolerable, lower off-axis angles can be used, and the reference antenna can be placed closer to the AUT yielding similar paths for the AUT and reference fields, which implies more robust setups for scanning errors and RF source instabilities [14,37].

6.2.2.2 Synthesized reference-wave indirect holography

Measurement setups employed for conventional off-axis holography can be modified in order to avoid the use of radiated reference waves. The use of synthesized reference waves allows the displacement of the image terms to the non-visible part of the spectrum, overcoming main limitations in conventional setups related to the position and off-axis angle of the reference antenna [14] and the spectrum overlapping. There are two different approaches for the implementation of this type of setups based on the use of electronic or mechanical shifts.

Synthesized reference field by means of electronic shifts

Plane reference waves can be synthesized from a sample of the RF source by means of a phase shifter that introduces constant increments of the phase $\Delta\varphi$ between consecutive samples in the acquisition plane. Then that signal is added to the field from the AUT acquired by the probe by means of a power combiner yielding an interference pattern that can be expressed as

$$H_{synth}(\vec{r}) = |E_{aut}(\vec{r}) + E_0 e^{-jk_r x}|^2, \tag{6.12}$$

where $E_0 e^{-jk_r x}$ represents the plane wave synthesized as the probe moves in the x-axis direction.

The spectrum of the hologram can be computed by means of an FT as

$$h_{synth}(\vec{k}) = |e_{aut}(\vec{k})|^2 + |e_0|^2 + e_0^* e_{aut}(k - k_r) + e_0 \, e_{aut}^*(-k - k_r). \tag{6.13}$$

From (6.13) the field of the AUT can be retrieved in the same way than that in the conventional approach by applying (6.4) and (6.5).

The spectrum of the hologram will be similar to the spectrum for the ideal scenario depicted in Figure 6.2 with some differences. The image terms are not distorted by the convolution with the reference field, as it happened in the conventional case. In this case, they are only displaced. Furthermore, the central position of the image terms is no longer defined by the layout of the elements but by the following expression:

$$k_r = \pm\frac{\Delta\varphi}{\Delta x}. \tag{6.14}$$

Therefore, the use of synthesized references removes the limitation imposed in the case of using radiated reference waves and allows the total control of the overlap of the spectrum by displacing the image terms to the non-visible part of the spectrum, defined as $|k| > k_0$, at the expenses of using denser spatial sampling.

Typical values for the phase shifts are usually between $\pi/2$ and $3\pi/4$; thus, sampling step ranges from $\lambda/4$ to $\lambda/8$ [16,18,36] in order to fulfill the no-overlapping condition described in (6.8). However, sampling step requirements can be relaxed if the modified hologram technique is implemented or while characterizing directive antennas.

Synthesized reference field by means of mechanical shifts

An alternative way of implementing synthesized reference-wave indirect holography avoiding the use of phase shifters is by means of mechanical displacements of the probe in order to obtain an interference pattern similar to the hologram described in the previous cases [13,18].

In this way, a constant sample of the RF source is added by means of a power combiner to the signal received by the probe, which acquires the amplitude of the AUT in a three-dimensional (3D) grid (see Figure 6.6 for an example).

The 3D grid is computed specifically for each setup by adding small mechanical displacements, defined as $\vec{d} = d\vec{r}/\|\vec{r}\|_2$, between the original points of a conventional

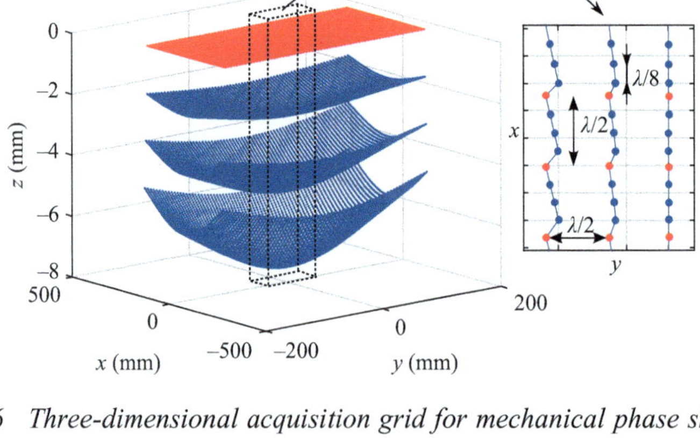

Figure 6.6 Three-dimensional acquisition grid for mechanical phase shifts indirect off-axis holography in the K_a-band and detailed view of a vertical column of points

rectangular acquisition grid. Disregarding amplitude variations of the field of the AUT, the field of the AUT can be expressed, in those new added points, as

$$E_{aut}(\vec{r} + \vec{d}) \approx E_{aut}(\vec{r})e^{-jk_0 d}. \tag{6.15}$$

In that case, taking into account the constant amplitude C from the reference branch added to the recorded field of the AUT, the following expression is obtained:

$$H'(\vec{r} + \vec{d}) = |E_{aut}(\vec{r} + \vec{d}) + C|^2 \tag{6.16}$$

If (6.16) is further developed, it can be clearly seen that the obtained hologram is equivalent to the hologram obtained by means of the use of a phase shifter described in (6.12):

$$\begin{aligned} H'(\vec{r} + \vec{d}) &= |E_{aut}(\vec{r})e^{-jk_0 d} + C|^2 = |E_{aut}(\vec{r})|^2 + |C|^2 \\ &\quad + CE_{aut}(\vec{r})e^{-jk_0 d} + CE^*_{aut}(\vec{r})e^{jk_0 d} \approx H_{synth}(\vec{r}). \end{aligned} \tag{6.17}$$

Thus, if the mechanical displacements are appropriately selected, the use of the phase shifter can be avoided.

The 3D grid is a layered one in which the number of layers equals the number of considered phase shifts defined as $N_\varphi = 2\pi/\Delta\varphi$. Those new points increase the *equivalent sampling* in the sweep direction with respect to the original sampling employed in the initial regular grid so that, position of the image terms is defined as in (6.14) according to the *equivalent sampling* and phase shifts. Figure 6.6 shows the measurement grid computed for the application example. In this case, the initial grid (red points) is a regular rectangular grid with $\lambda/2$ sampling in the x- and y-axes. Then, the mechanical displacements are added to introduce 90° shifts, in a four-layer grid, yielding an equivalent sampling of $\lambda/8$ in the x-axis.

The use of synthesized reference waves, either with electronic or mechanical shifts, has two main advantages with respect to conventional indirect holography schemes with radiated reference waves.

- The image terms can be displaced to the non-visible part of the spectrum; and thus, spectrum overlap can be totally controlled by selecting the adequate sampling rate and phase shifts.
- Since the reference wave is a plane wave, its expression can be obtained by measuring the amplitude of the signal at the output of the directional coupler and adding the phase behavior according to the electronic or mechanical displacements. In this way, the need of vector calibration is bypassed.

However, synthesized reference wave holography exhibits two big drawbacks for antenna measurement:

- At high-frequency bands, where the setup has to be implemented by means of waveguide components, connection of the phase shifter to the receiver's end might be difficult. An alternative approach in which a radiated reference field is employed and two holograms are multiplexed in order to displace the image terms to the non-visible part of the spectrum can be found in [13,32].

- Availability of phase shifters at high-frequency bands might suppose a limitation, increasing the cost of the measurement setup. The use of phase shifters can be replaced with mechanical shifts [13,18]; however, not all the positioning systems in antenna measurement ranges allow for noncanonical surface acquisition.
- Sampling requirements might also be a limitation at high-frequency bands; the modified hologram technique can be implemented in order to relax those requirements. Furthermore, nonredundant sampling techniques can also be applied in order to reduce the number of acquisition points [17].

Application example

To illustrate the capabilities of interferometric phaseless techniques, a measurement example based on indirect off-axis holography is shown next. The setup has been implemented with synthesized reference wave by means of mechanical shifts.

The measurement setup, implemented on the planar measurement range described in [38], mimics the layout shown in Figure 6.3(b). The acquisition plane is a 3D grid with a rectangular section of 40 cm × 34 cm placed at 20 cm of the aperture of the AUT shown in detail in Figure 6.6. The AUT is a 15 dB i gain horn antenna at 30 GHz.

Equivalent sampling has been selected to be $\lambda/2$ in the y-axis and $\lambda/8$ in the x-axis, in which the spectrum displacement will be introduced. The mechanical phase shifts have been calculated to introduce a phase shift of 90° yielding $N_\varphi = 4$, that is, four different layers in the acquisition plane, as shown in Figure 6.6.

As described in (6.14), 90° phase shifts and spatial sampling of $\lambda/8$ will produce image terms centered at $k_r = \pm 2k_0$.

Figure 6.7(a) shows the acquired hologram, where the interference pattern is clearly observed, while the spectrum of the hologram is shown in Figure 6.7(b). Since no overlap occurs between the image terms and the autocorrelation term, the spectrum is filtered from $k_x = 0.5k_0$ to $k_x = 4k_0$, and the field is retrieved in the spatial domain after compensating for the effect of the synthesized reference wave as described in (6.5).

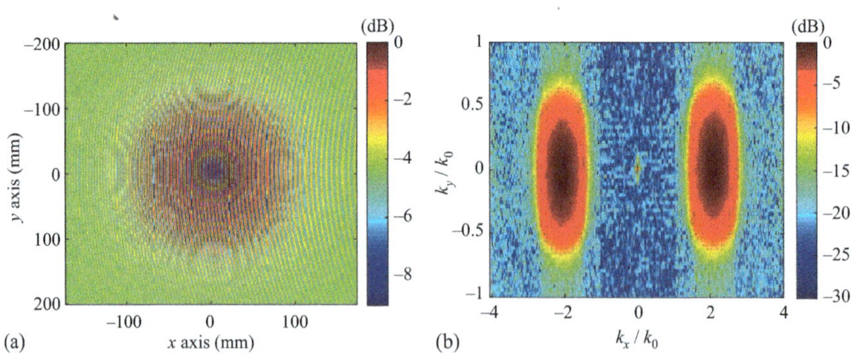

Figure 6.7 Acquired hologram: (a) spatial domain, (b) k-space

Main cuts of the retrieved NF amplitude and phase of the AUT in the acquisition plane are shown in Figure 6.8 and compared to the measured field acquired directly with amplitude and phase, showing very good agreement.

In order to quantify the quality of the method, the following error metric is employed:

$$error[\%] = 100\frac{\|E_{measured}(\vec{r}) - E_{retrieved}(\vec{r})\|_2}{\|E_{measured}(\vec{r})\|_2} \tag{6.18}$$

where $E_{measured}$ is the reference field obtained from the complex acquisition with amplitude and phase and $E_{retrieved}$ is the retrieved field obtained from the scalar measurement of the hologram. $\|\cdot\|_2$ defines the Euclidean norm. The error for the presented application example is 3.02%, proving the validity of the proposed method. In this case, the complete hologram has been acquired, and no extra measurements for the application of the modified hologram technique have been performed; thus, both amplitude and phase have to be retrieved. Error is decreased if the amplitude of the field is measured independently.

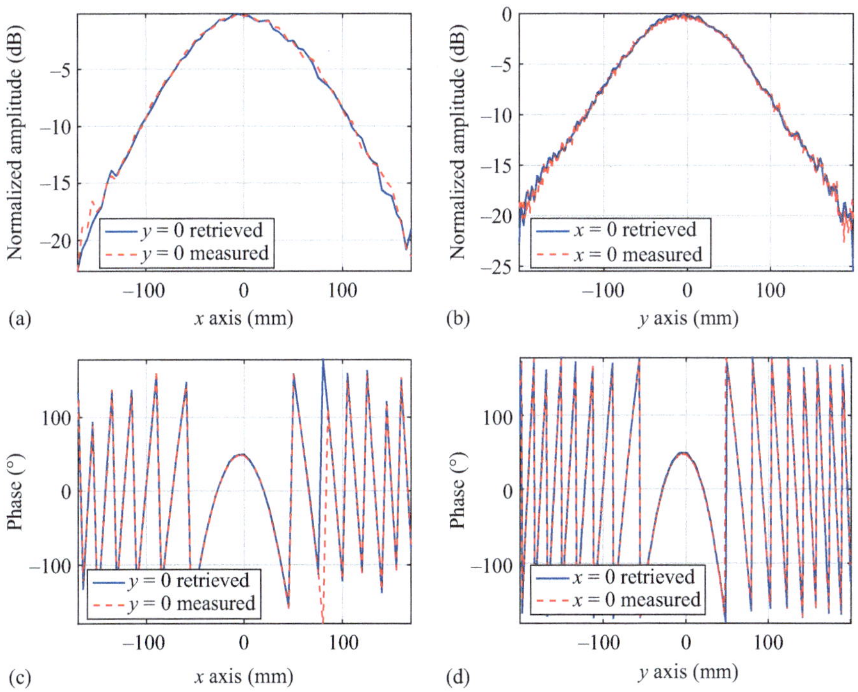

Figure 6.8 Comparison between measured and retrieved fields: (a) amplitude cut for x = 0, (b) amplitude cut for y = 0, (c) phase cut for x = 0, and (d) phase cut for y = 0

6.2.3 Broadband indirect off-axis holography

Conventional indirect off-axis holography and the presented improved schemes in Section 6.2.2 are monochromatic techniques. Nowadays, advances in fabrication technologies have contributed to the development of high-performance antennas, particularly at high-frequency bands, contributing to the growth of new applications demanding broadband antennas such as high-resolution synthetic aperture radar [39] or high-rate wireless communications [40].

Accurate and efficient characterization of this type of antennas requires novel and efficient techniques, normally in the time domain (TD) [41], since the application of conventional monochromatic techniques would be very time consuming. The proposed method is a broadband phaseless method based on extrapolation of conventional off-axis holography.

6.2.3.1 Method

The proposed broadband phaseless technique is based on the same setup as conventional techniques, as shown in Figure 6.9, although different data acquisition and phase-retrieval procedures are employed. For this technique, a frequency sweep is performed at each point of the acquisition plane, and the hologram is recorded as a frequency interference instead of a spatial one. As in conventional techniques, if the amplitude of the AUT and the reference antenna are independently characterized, the modified hologram technique for overlapping and sampling reduction can be applied; and, thus, the hologram at each spatial point can be expressed as

$$\begin{aligned} H_m(\vec{r},\omega) &= |E_{aut}(\vec{r},\omega) + E_r(\vec{r},\omega)|^2 - |E_{aut}(\vec{r},\omega)|^2 - |E_r(\vec{r},\omega)|^2 \\ &= E_{aut}(\vec{r},\omega)E_r^*(\vec{r},\omega) + E_{aut}^*(\vec{r},\omega)E_r(\vec{r},\omega). \end{aligned} \tag{6.19}$$

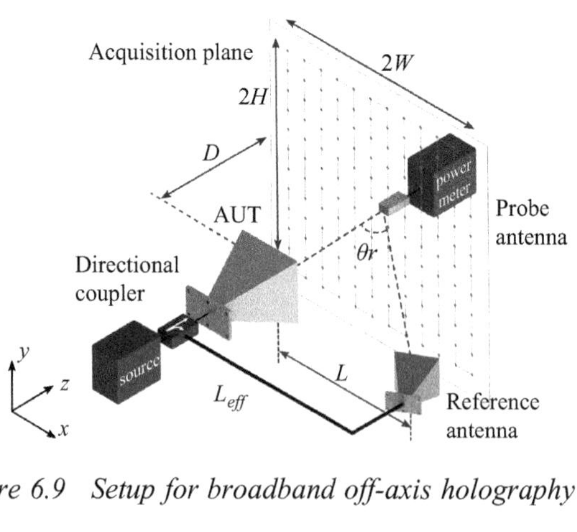

Figure 6.9 Setup for broadband off-axis holography [13]

The spectrum of the modified hologram is computed by means of an inverse FT in the TD as

$$h_m(\vec{r},t) = e_{aut}(\vec{r},t) \otimes e_r^*(\vec{r},-t) + e_{aut}^*(\vec{r},-t) \otimes e_r(\vec{r},t). \tag{6.20}$$

Providing that no overlapping exists, the desired image term can be filtered by means of a time window, yielding the following expression:

$$h_{filt}(\vec{r},t) \approx e_{aut}(\vec{r},t) \otimes e_r^*(\vec{r},-t). \tag{6.21}$$

And, finally, the complex field of the AUT is retrieved independently at each acquisition point simultaneously for all the frequencies of the working band after removing the effect of the reference field as

$$E_{ret}(\vec{r},\omega) = \frac{H_{filt}(\vec{r},\omega)}{E_r^*(\vec{r},\omega)} \approx E_{aut}(\vec{r},\omega), \tag{6.22}$$

where $H_{filt}(\vec{r},\omega)$ is the FT of the filtered image term in (6.21).

The procedure described by (6.19)–(6.22) allows the retrieval of one the components of the tangential field received by the probe; in order to acquire the orthogonal component, the AUT needs to be rotated 90°, while the reference and probe antennas keep the same orientation. A complete description of the acquisition process can be found in [35].

As mentioned above, the quality of the phase retrieval will depend mostly on the overlap of the image terms in the TD. This overlap is determined by the time required from the signals in the AUT and reference branches, t_{aut} and t_r, respectively, to travel from the RF source to the probe and, therefore, will be determined by the frequency sampling and the layout of the elements and the size and distance D to the acquisition plane. The parameters of the measurement setup, which influence the overlap of the image terms, are defined in Figure 6.9.

From further analysis of the spectrum, schematically depicted in Figure 6.10(a), the no-overlapping condition can be defined as $t_{aut} - t_r > 0$. This condition can be expressed in terms of the distance as

$$\sqrt{D^2 + W^2 + H^2} > \sqrt{D^2 + (W-L)^2 + H^2} + L_{eff}, \tag{6.23}$$

meaning that $L > L_{eff}$, that is, physical separation between the AUT and the reference antenna should be larger than the length of the transmission line connecting the directional coupler to the reference antenna, L_{eff}. Unfortunately, due to the layout of the setups, particularly at high-frequency bands, in which connections are made by means of waveguide sections, this condition is usually tough to fulfill.

Nevertheless, the length of the transmission line connecting the reference antenna and the directional coupler L_{eff}, which controls t_r, can be increased until the image terms of the hologram are swapped. In this case, as shown in Figure 6.8(b),

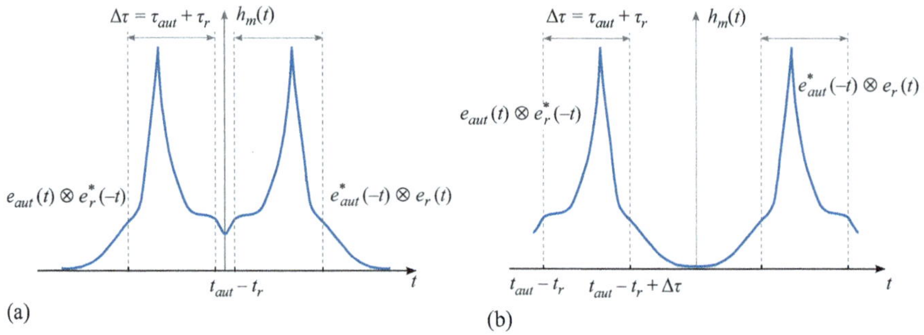

Figure 6.10 Scheme of the spectrum of the hologram (a) original, (b) with swapped terms

the nonoverlapping condition is achieved if $t_{aut} - t_r < 0$, which, as a function of distance, can be expressed as

$$\sqrt{D^2 + W^2 + H^2} - \left(\sqrt{D^2 + (W - L)^2 + H^2} + L_{eff}\right) < c\Delta\tau, \tag{6.24}$$

where c is the speed of light in vacuum, and $\Delta\tau$ the width of the image term.

The previous condition yields that, to avoid overlapping, the difference between L_{eff} and the distance between antennas D has to be larger than the dispersion of the image terms, that is, $L_{eff} - L > c\Delta\tau$.

The frequency sampling must be selected according to the Nyquist rule and (6.24) as

$$\Delta f < \frac{1}{2(t_r - t_{aut})}. \tag{6.25}$$

This technique presents several advantages with respect to the conventional indirect off-axis holography:

- The phase is retrieved independently at each point of the acquisition grid (simultaneously for all the frequencies). Consequently, standard sampling of $\lambda/2$ as for complex acquisitions (amplitude and phase) can be employed, in contrast to conventional off-axis holography or iterative techniques where the sampling needs to be increased at least up to $\lambda/4$.
- Conventional nonredundant sampling techniques [24] can be directly applied to further reduce the number of acquisition points with high accuracy in the phase retrieval [42].
- Position of the image terms and, consequently, overlapping can be simply controlled with the physical length of the signal paths in the AUT and reference branches, particularly, with L_{eff}; and, therefore, neither mechanical nor electrical phase shifts are needed, with the consequent setup complexity reduction.
- The phase is retrieved coherently for all the frequencies in the working band. Therefore, the retrieved data can be employed for group delay computation, very relevant for broadband antennas and high-rate wireless communications.

Main disadvantages of the technique are summed up below:

- The reference branch still needs to be characterized in amplitude and phase, as in conventional approaches, in order to remove the effect of the reference field as indicated in (6.22).
- All the components of the setup are required to be broadband.
- The frequency sweep at each acquisition point may require to stop the probe antenna at each acquisition point or to highly reduce the speed of on-the-fly acquisition systems.

6.2.3.2 Application example

The broadband off-axis holography technique is applied in this example for the characterization of a double-reflector antenna in the W-band from 75 to 110 GHz. The reflector, shown in Figure 6.11 in the measurement setup, has been designed *ad hoc* in order to model a directive antenna with directivity close to 50 dB. The main reflector is a 30 cm diameter parabolic dish, while the secondary one, at 13 cm of the vertex of the former, is a 4.8 cm planar dish. The antenna feed is a 10 cm length WR-10 OEWG implemented by means of two 5 cm straight waveguide sections, whose aperture is 2 cm below the secondary reflector.

The reference antenna is a 3 cm length tapered-end OEWG antenna, with the center of its aperture placed at $L = 17$ cm from the vertex of the reflector in the x-axis, pointing toward the center of the acquisition plane ($\theta_r \approx 15^\circ$). A 3D-printed support has been designed in order to hold the main structure of the reflector and fix the position of the reference antenna. The AUT is installed on a rotary positioner in order to achieve 90 rotations for the acquisition of the second field component (see [35] for an exhaustive description of the complete acquisition procedure). Connection between the directional coupler (inside the 3D-printed support) and the reference antenna is made by means of a flexible cable with $L_{eff} = 30$ cm.

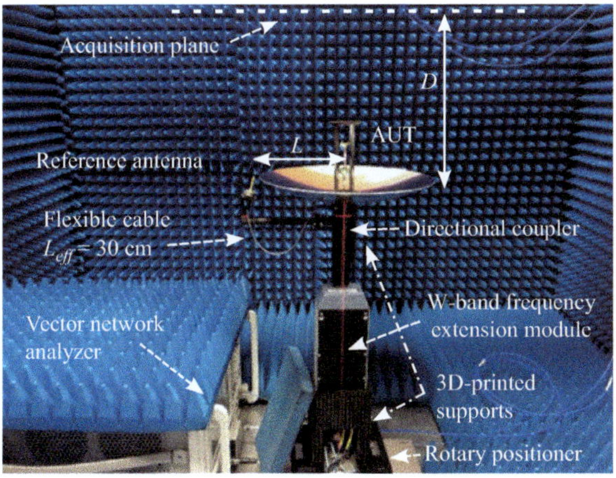

Figure 6.11 Measurement setup for the double-reflector characterization in the W-band

The acquisition plane is a square plane with $W = H = 25$ cm placed at $D = 39.9$ cm from the aperture of the AUT. Spatial sampling has been set to $\Delta x = \Delta y = \lambda_{\min}/2$, where $\lambda_{\min}$ is the wavelength at 110 GHz, the upper frequency of the band. A frequency sweep with 201 points in the complete W-band is made at each of those spatial points of the acquisition grid.

With the described setup, t_{aut} and t_r are 1.1 and 5.5 ns, respectively, for the most restrictive points, which are the corners of the acquisition plane on the reference antenna side [35], right corners in Figure 6.8. The -30 dB width or dispersion of each of the signals is $\Delta\tau_{aut} = 0.34$ ns and $\Delta\tau_r = 0.5$ ns yielding a total duration of the image term of approximately $\Delta\tau = 0.8$ ns. The minimum frequency sampling, as described in (6.25), should be 625 MHz; thus, the employed sampling, $\Delta f = 175$ MHz, largely satisfies the Nyquist rule.

The other nonoverlapping condition described in (6.24) is also fulfilled with the implemented setup; and therefore, as can be seen in Figure 6.12(a), the image term corresponding to the convolution of e_{aut} with e_r^* can be filtered for both the x and y components of the hologram considering the negative half-space of the spectrum. Figure 6.12(b) shows the retrieved phase for the central point of the acquisition plane compared to the phase from a direct acquisition (considered as the reference measurement) in the central part of the frequency band.

Figure 6.13 shows a comparison of the main cuts of the retrieved phase and the reference measurement for the central frequency of the band of both NF components E_x and E_y. Apart from some 180° phase shifts, there is a good agreement between them.

In order to quantify the behavior of the method in the complete frequency band, Figure 6.14 shows the error obtained at each frequency of the band as defined in (6.18). An average error of 1.78% is obtained for the complete frequency band. Due to the windowing effect of the TD filtering process, error is very large in the extremes of the band; and thus, it is advisable to discard the frequencies at the extremes.

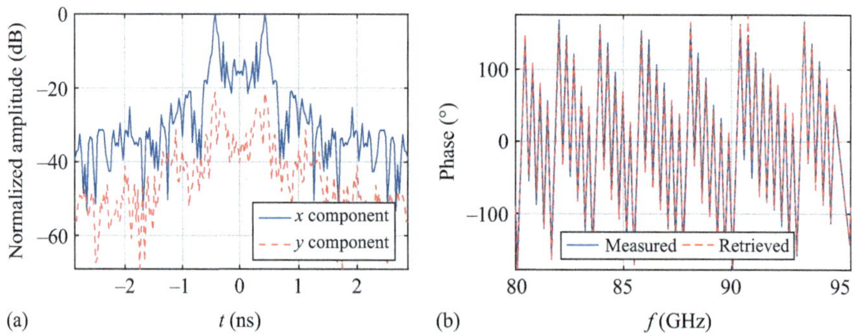

Figure 6.12 (a) Spectrum of the hologram and (b) retrieved phase for E_x at the central point of the acquisition plane in the central part of the frequency band

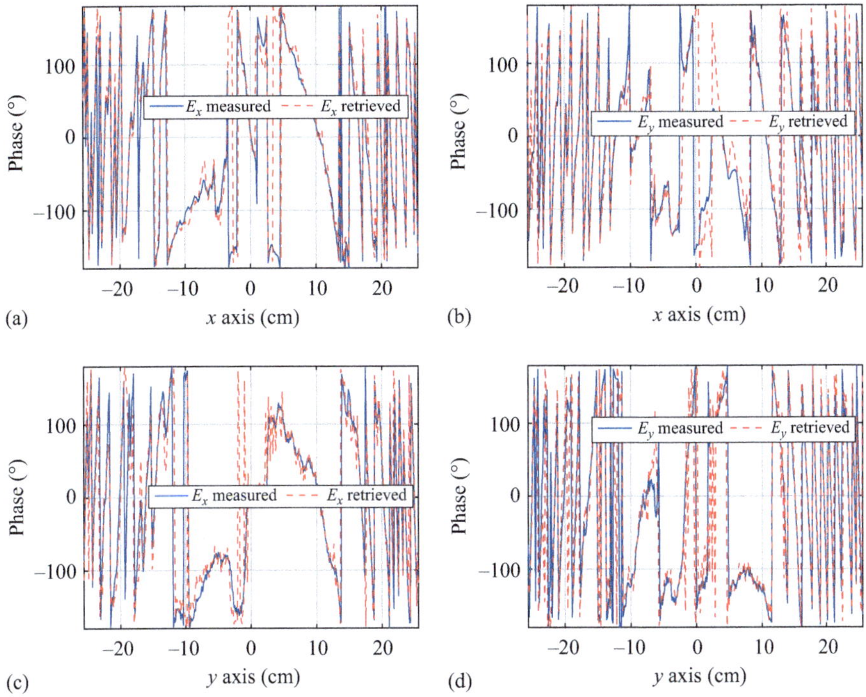

Figure 6.13 Main cuts of the retrieved phase of the E_x and E_y near-field components for the central frequency of the band, (a) y = 0 cut for E_x, (b) y = 0 cut for E_y, (c) x = 0 cut for E_x, and (d) x = 0 cut for E_y

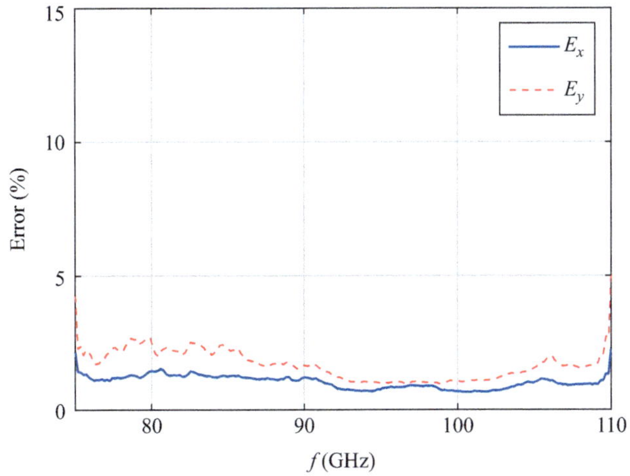

Figure 6.14 Mean error of the phase retrieval in the working band

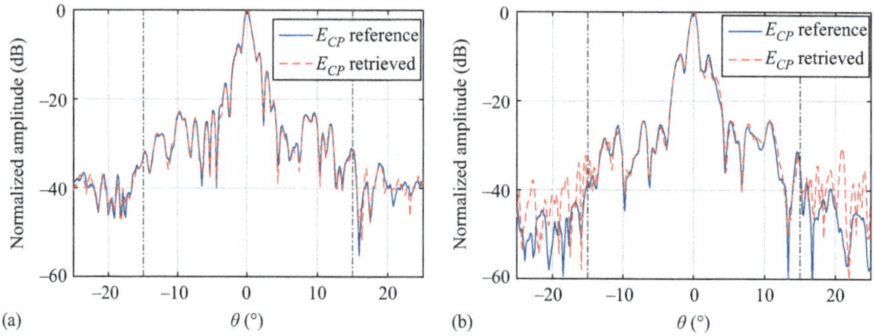

Figure 6.15 Main cuts of the copolar far-field pattern: (a) $\phi = 0°$ and (b) $\phi = 90°$

Finally, Figure 6.15 shows the main cuts of the copolar FF pattern after NF–FF transformation compared to those obtained from the reference measurements. With the implemented setup, the valid margin of the transformation is $\pm 15°$. Agreement within the valid margin of the transformation is good, being the maximum difference between the reference pattern and the pattern from the retrieved phase -87 dB. The crosspolar component computed from the retrieved fields exhibits good agreement with the reference pattern as well; however, it is not shown here due to the low level of its main cuts.

6.2.4 Error analysis

Main sources of error in indirect off-axis holography are the errors in the characterization of the reference antenna and errors in the measurement system regarding positioning accuracy and electrical errors such as noise and drift of the RF instrumentation.

For the characterization of the reference antenna, there are two main approaches: (1) using an initial vector calibration to characterize the antenna with amplitude and phase [32,35], and (2) placing the reference antenna far enough from the AUT, with the subsequent losses in the reference path, so its radiation pattern can be approximated by a plane wave [14].

The former requires complex equipment, especially at high-frequency bands, but it is the option that provides better results. If the reference antenna is kept in the same position, the vector calibration only has to be carried out once. It is also possible to characterize the reference antenna with no off-axis angle and then introduce the tilt in the field, according to the reference antenna position and off-axis angle, during the postprocessing; this will require the accurate knowledge of the reference antenna position and off-axis angle.

The latter yields a fully scalar process. However, placing the reference antenna far from the AUT will decrease the dynamic range of the hologram, since the power level of the AUT and reference branches will be unbalanced, requiring larger acquisition planes and increasing the system sensitivity to positioning errors [14].

The use of the modified hologram allows for a reduction of the separation between antennas with the consequent error reduction [14].

In setups with synthesized reference waves, the reference field can be analytically computed from the amplitude measurements of the reference branch and the known phase shifts, also yielding a purely scalar approach [18,36,43].

The effect of errors arising during the acquisition process due to mechanical and electrical errors has been widely studied in [14,42,44]. In [44], the effect of positioning and alignment errors is studied for the conventional off-axis holography approach, and in [14] the study is repeated for the modified hologram method. While errors in the scan axes ($x - y$ for planar, $z - \rho$ in cylindrical, or $\phi - \theta$ in spherical acquisitions) have a similar effect on indirect holography than on complex acquisitions, schemes based on off-axis holography are much less sensitive to planarity errors (z-axis for planar and $r-$ for cylindrical and spherical acquisitions). In [42], the study is extended for the case of broadband holography and a set of nonredundant points, proving that broadband holography has the same behavior, regarding those errors, than the conventional approach. Regarding the effect of noise and drift, holographic techniques are slightly more sensitive to noise [42], while being very robust to drift in the RF instrumentation; relative receiver channel drift better than 1% level is enough for adequate performance. The smearing effects during on-the-fly scans have a considerable larger influence on conventional off-axis holography than on complex measurements due to the oscillatory nature of the interference pattern; fortunately, the effect of this type of error is low with modern instrumentation and can be reduced by adjusting the acquisition speed.

6.3 Iterative phase retrieval techniques

6.3.1 Fundamentals

As stated in Section 6.1.3, iterative phase retrieval techniques aim to recover the phase of the radiated field from amplitude-only information acquired in, at least, two acquisition surfaces. While NF–NF and NF–FF transformation techniques using complex NF information require just one acquisition surface, in the case of phaseless methods at least two acquisition surfaces are needed. The reason is that the NF pattern exhibits angular dependence with the distance from the AUT, whereas the angular dependence of the FF pattern is not a function of the distance. This feature of the NF pattern contains sufficient information to recover the phase of the NF field.

Iterative phase retrieval techniques can be classified into two main groups, as described in Section 6.1.3: (1) iterative schemes based on two separate measurement surfaces, where the fields are alternately propagated from one measurement plane to another until reaching an stationary solution, and (2) minimization of a nonlinear CF relating the amplitude of the NF measured and the NF estimated from the reconstructed AUT aperture fields.

Phase retrieval techniques based on methods (1) or (2) require solving an electromagnetic inverse problem. The study of the posedness of this inverse problem has been carefully conducted in [19,24] and recently in [27,28]. In brief, iterative phase retrieval techniques have to solve an ill-posed problem, where the solution space is often populated with local minima. The two aforementioned groups of iterative phase retrieval techniques address this problem under two different approaches: iterative schemes (1) simplify the nonlinear CF relating the measured and estimated field by adding phase information iteratively calculated, whereas nonlinear minimization algorithms (2) are based on an iterative search for the minimum of the nonlinear CF.

Both groups of iterative phase retrieval techniques are greatly influenced by the choice of the IG, which is eventually based on *a priori* information about the AUT. For example, [22,27,28] discuss about the impact of the main beam steering (broadside or scanned beam) of the AUT, improving the choice of the IG by means of global search techniques as PSO, DE, and RCGA.

Iterative phase retrieval techniques are eventually based on the minimization of a CF relating NF amplitude acquired at two different surfaces, $|E_1|, |E_2|$ with the NF amplitude calculated from an electromagnetic model of the AUT (I), $|(Z_{E_1,I}\ Z_{E_2,I})(\ I\ )|$, where $Z_{E_{1,2},I}$ denotes the formulation relating the electromagnetic model of the AUT with the NF on the acquisition surfaces 1 and 2.

AUT model (I) can be constructed, for example, from AUT aperture fields or from a 3D equivalent currents model of the AUT. Besides, depending on the geometry of the NF acquisition surfaces (planar, cylindrical, spherical, and arbitrary geometry), different formulation relating the electromagnetic model of the AUT and the NF calculated in the acquisitions surfaces can be used. These formulations are summarized in Table 6.1.

Table 6.1 Classification of the iterative phase retrieval techniques

Related work	Acquisition surface	AUT model	Formulation relating AUT model and NF on the acquisition surface	Iterative phase-less retrieval method
[19,20,24] [22,28]	Two planes	Aperture fields	Plane wave spectrum Fourier Transform	Iterative scheme
[25]	Two cylindrical surfaces	Aperture fields	Hankel functions	Nonlinear CF minimization
[23]	Two cylindrical surfaces	Aperture fields	Hankel functions	Iterative scheme
[21,27]	Two spherical surfaces	Spherical wave modes	Spherical wave expansion	Nonlinear CF minimization
[26,45]	Two arbitrary geometry surfaces	Equivalent currents (on arbitr. surf.)	EM integral equations	Nonlinear CF minimization

Among the iterative phase retrieval techniques cited in Table 6.1, this book chapter will focus on the one based on AUT characterization by means of equivalent currents, as it covers all possible geometries of the NF acquisition domain, at the expense of increasing the complexity of the inverse problem to be solved. Besides, this technique is based on the SRM explained in Chapters 2 and 3.

6.3.2 Phaseless sources reconstruction method

As explained in Chapters 2 and 3, the SRM has become a powerful antenna characterization technique, overcoming traditional limitations of Fourier-based NF–NF and NF–FF transformation methods [46,47]. SRM allows handling arbitrary-geometry NF measurement and aperture field domains, which is of special interest for diagnostics of antennas of complex geometry, such as antenna arrays [26] or base station mobile network antennas [48]. SRM has been successfully applied for radome analysis as well [49].

SRM provides an electromagnetic equivalent currents model of the AUT, which can be used to calculate the field radiated by the AUT at any point of the space outside the surface enclosing the AUT [46,47]. AUT equivalent currents model is calculated by solving the integral equations relating these equivalent currents and the fields radiated by the AUT (as explained in Chapter 2).

SRM formulation can be extended to recover such equivalent currents model of the AUT from amplitude-only NF measurements. Following the notation presented in Section 6.3.1, in the case of the SRM, formulation relating the AUT model based on equivalent currents (I) with the amplitude of the NF acquired in two acquisition surfaces ($|E_1|, |E_2|$), $Z_{E_{1,2},I}$, is based on the electromagnetic integral equations (Chapter 2).

6.3.2.1 Description of the method

Phaseless SRM (pSRM) can be used in both phase retrieval techniques based on iterative schemes and minimization of a nonlinear CF. The flowchart of these two methodologies is depicted in Figures 6.16 and 6.17, respectively.

In the case of the pSRM based on iterative scheme (Figure 6.16), the inputs to the first step of the iterative algorithm are the measured amplitude at the first acquisition surface ($|E_1|$) and an IG of the phase of the field on this surface. This IG can be based on *a priori* knowledge of the AUT: the better the IG of the phase, the faster the convergence will be. In the next step, a linear CF F relating the measured amplitude and the estimated phase of the NF with the field radiated by an equivalent currents model of the AUT, (I_u, I_v), is minimized. This formulation is based on the SRM explained in Chapter 2 (u, v denote two orthogonal components of the equivalent currents). Once an equivalent currents model of the AUT is recovered from the minimization of CF F, these currents are used to calculate the NF at the second acquisition surface: the calculated phase is kept, while the calculated amplitude is replaced by the acquired one ($|E_2|$). Next, again a linear CF F is minimized to refine the equivalent currents model of the AUT using, in this case, the NF information at the second acquisition surface. The equivalent currents

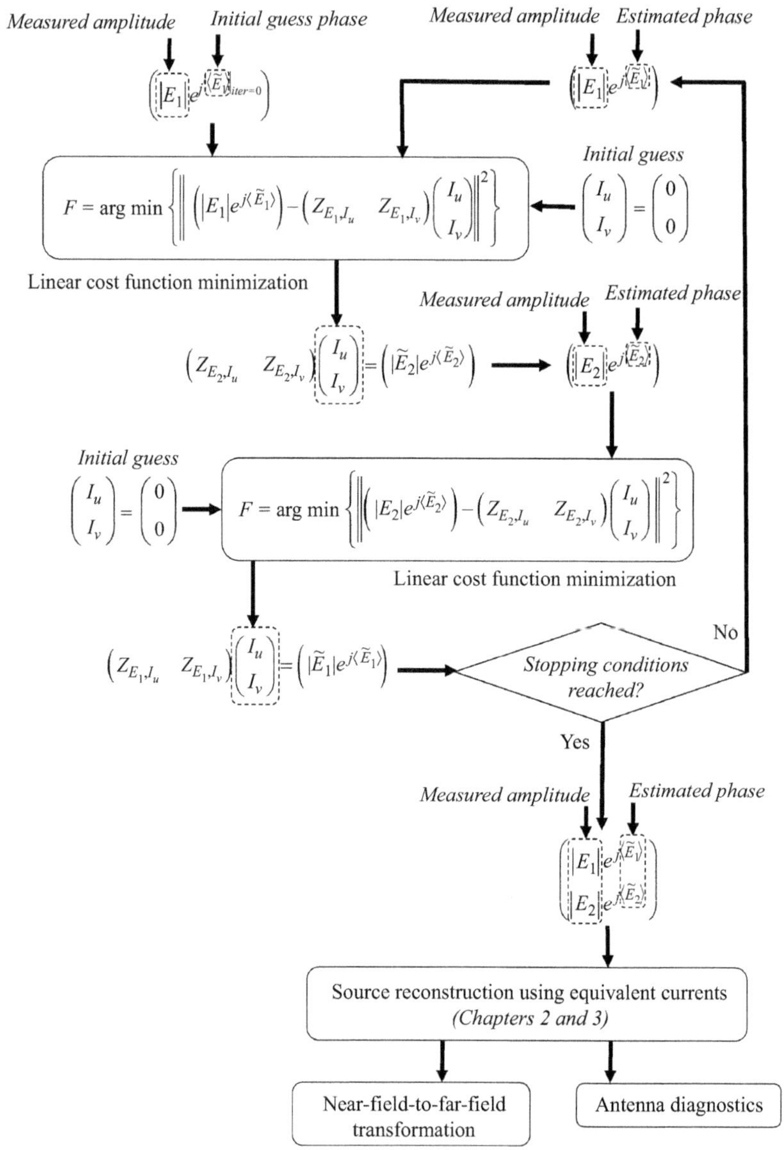

Figure 6.16 Phase retrieval methods based on an iterative scheme

model is refined again, using it to estimate the phase of the NF back at the first acquisition surface.

The described algorithm is repeated until a stopping condition is reached. In general, this condition can be (1) the difference between the RMS value of the amplitude of the NF estimated from the equivalent currents model of the AUT

(RMS_{est}) and the RMS value of the acquired NF amplitude (RMS_{acq}), denoted as $\delta = |RMS_{est} - RMS_{acq}|$, is less than a certain threshold $\delta < \delta_{threshold}$, (2) δ decreases below a certain tolerance for a certain number of consecutive iterations, and (3) a maximum number of iterations, N_{iter}, are reached. Convergence of this iterative scheme to a global or to a local minimum is directly related to the stopping condition reached. In general, stopping condition (1) is related to convergence to the global minimum, whereas stopping conditions (2) and (3) may be due to algorithm stagnation in a local minimum.

The iterative scheme finishes when one of the three stopping conditions is met. Then, the measured amplitude and the phase estimated from the retrieved equivalent currents model of the AUT, (I_u, I_v), is introduced as input in the SRM (see Chapter 2) to recover an equivalent currents model of the AUT, from which antenna diagnostics and radiation pattern calculation can be conducted.

Figure 6.17 shows the flowchart of the pSRM based on the minimization of a nonlinear CF relating the amplitude of the NF measured at the two acquisition surfaces, $|E_1|, |E_2|$, and the amplitude of the NF radiated by an equivalent currents model of the AUT, (I_u, I_v). Nonlinear CF F of Figure 6.17 can be analyzed as a nonlinear least squares problem.

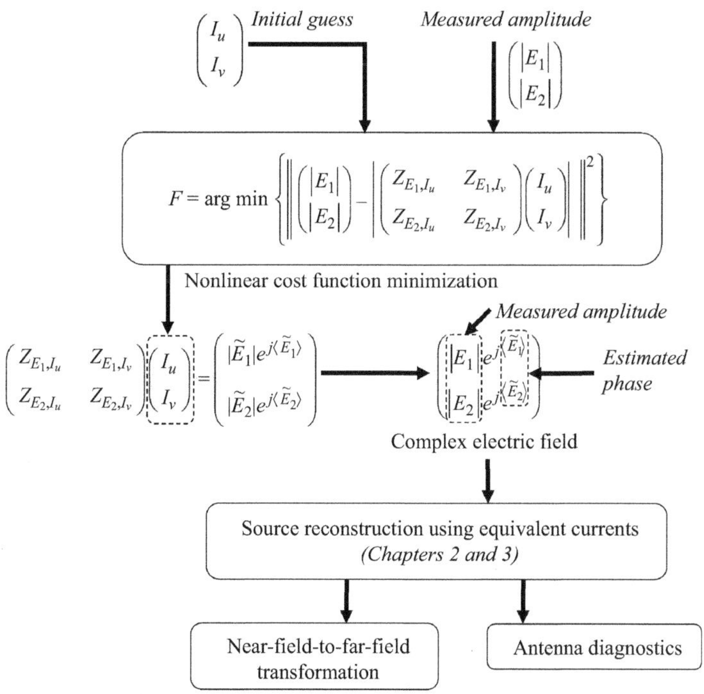

Figure 6.17 Phase retrieval methods based on nonlinear cost function minimization

Before discussing the algorithms for nonlinear CF minimization, main differences between linear (e.g., SRM) and nonlinear (e.g., pSRM) least squares problem are briefly revised next:

- Under noise-free conditions, a linear CF does not have local minima (i.e., the solution is unique), as opposed to nonlinear CFs. Thus, failure of the minimization algorithm to find a minimum is a common problem in the case of nonlinear CF problems.
- As a consequence of the previous fact, under noise-free conditions, linear CF minimization reaches convergence for any IG, though its convergence speed will be faster for IGs close to the solution. In the case of noisy measurements, it must be remarked that the choice of the IG can eventually determine the convergence to a global or local minimum.
- Nonlinear least squares problems cannot be solved by means of direct solvers, thus requiring an iterative procedure based on the linearization of the problem, then refining the parameters of the linearized approach by successive iterations.

The most well-known method for nonlinear CF minimization is the Gauss–Newton algorithm (GNA), which requires the calculation of the Jacobian of the CF. In the case of ill-posed problems, a more robust algorithm is the LMA. While GNA provides faster convergence to the global minimum than LMA for well-behaved functions, LMA is capable of finding the global solution even when the IG is far off such global minimum [50].

LMA algorithm introduces a damping factor Λ, which is refined at each iteration depending on the rate of decay toward the global minimum. For example, if the residual of the CF does not decrease from one iteration to another, Λ value is increased to avoid getting trapped in a local minimum. If the residual decreases, then Λ is decreased as well to steadily approach to the global solution [51].

Concerning pSRM based on nonlinear CF minimization, LMA has been proved to be more successful than GNA for minimizing the nonlinear CF F shown in Figure 6.17, as detailed in [26]. As LMA is an iterative technique, the same three stopping conditions defined for the pSRM based on the iterative scheme can be applied as well. Once an stopping condition is achieved, the reconstructed equivalent currents model of the AUT, (I_u, I_v), are used to estimate the phase of the NF on both acquisition surfaces. Together with the measured NF amplitude, these fields are used as inputs for the SRM for antenna diagnostics and characterization, as depicted in Figure 6.17.

Depending on the field components acquired, the flowchart of the pSRM based on nonlinear CF minimization depicted in Figure 6.17 can be modified. For example, in [26], the NF amplitude is measured on two spherical surfaces, acquiring $|E_\theta|$ and $|E_\phi|$ field components. If the AUT is characterized from its aperture fields (equivalent magnetic currents, M_x, M_y), then $|E_x|, |E_y|$ are calculated from the acquired spherical field components. According to the Second Electromagnetic Equivalence Principle, working with linear field components E_x, E_y allows decoupling integral equations relating NF and aperture fields, so that two nonlinear CFs relating $|E_x|$ with M_y, and $|E_y|$ with M_x are set. Decoupling of field components reduces the complexity of the nonlinear problem, although it requires an additional IG about the phase of the spherical field components.

If the spherical acquisition system is configured so that it allows the direct acquisition of copolar and crosspolar components ($|E_x|, |E_y|$ according to the Ludwig's third definition), then the intermediate step of transforming between spherical and Cartesian field components is avoided. This was the case of the truncated spherical acquisition system described in [40].

6.3.2.2 Application example

To illustrate the methodology explained in Section 6.3.2.1, a simulation-based application example for phaseless antenna array diagnostics and characterization is presented. This example is based on planar acquisition and reconstruction domains, to make easier the analysis of the results and the comparison of phase retrieval methods based on an iterative scheme and on nonlinear CF minimization.

A 2×2 antenna array working at 15 GHz is chosen as AUT. Aperture field amplitude and phase distribution (E_y component) is depicted in Figure 6.18(a) and (b), respectively, noticing that the amplitude of the upper right element is 6 dB, lower than the other. Elements are shifted 180° following a chessboard-line pattern (Figure 6.18(b)). Overall AUT aperture size is $D = 5.5 \times 5.5\lambda$, so FF distance is $R_{FF} = 2D^2/\lambda = 2.42$ m.

From the AUT aperture fields, NF has been calculated on two 4×4 m planar acquisition domains placed 1 and 2 m away from the aperture of the AUT. In order to highlight the fact that the SRM is capable of handling arbitrary-geometry acquisition surfaces, and in connection with recent advances in *in situ* antenna measurement systems [45], planar acquisition domains have been irregularly samples, as observed in Figure 6.18(c). The number of sampling points on each acquisition plane is 17,956, that is, 0.11 samples/cm^2 on average. Noise has been added according to a signal-to-noise ratio of 40 dB.

Next, the pSRM has been applied to retrieve the phase from the NF amplitude. An aperture field domain of 10×10 cm has been defined, discretized in 729 sub-domains. Thus, for each field component, E_x, E_y, the number of equations is 35,912. From $|E_x|, |E_y|$, equivalent magnetic currents components M_y, M_x (i.e., aperture fields) are recovered, resulting in 729 unknowns. Phase retrieval techniques based on iterative scheme (Figure 6.16) and nonlinear CF minimization (Figure 6.17) have been evaluated.

The same stopping conditions are set to both phase retrieval methods, being stopped when one of the following conditions is reached: (1) $\delta = |RMS_{est} - RMS_{acq}|$ is $\delta < 10^{-4}$, (2) δ decreases less than 10^{-7} for 10 consecutive iterations, and (3) $N_{iter} > 1{,}000$. In this example, pSRM based on the iterative scheme stopped due to condition (1) after 455 iterations, with $\delta = 2.4 \times 10^{-2}$. In the case of nonlinear CF minimization, nonlinear LMA minimization stopped after 51 iterations due to condition (2) as well, being $\delta = 8.6 \times 10^{-3}$. pSRM algorithms were implemented in MATLAB® and run on a conventional laptop equipped with an Intel Core i7 processor and an 8 GB RAM. Calculation time was 325 s for the pSRM based on iterative scheme (0.7 s per iteration), and 75 s for the pSRM based on nonlinear CF

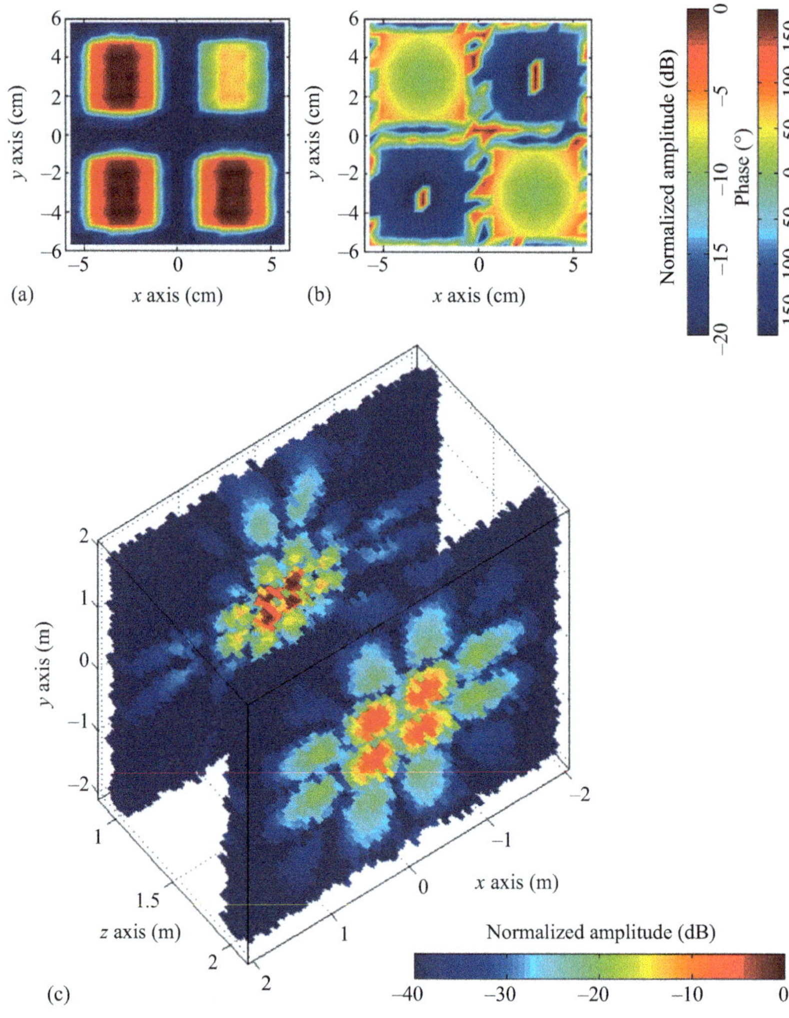

Figure 6.18 Application example: (a) amplitude of the aperture fields of the AUT; (b) phase of the aperture fields of the AUT; (c) amplitude of the NF evaluated at two 4 m × 4 m acquisition planes placed at $z_1 = 1$ m and $z_2 = 2$ m

minimization (1.5 s per LMA iteration), requiring in both cases less than 1,500 MB RAM. Note that the calculation time per iteration is lower in the case of the iterative scheme, as a linear system of equations (instead of a nonlinear one) is solved at each iteration.

The SRM is applied to the acquired NF amplitude and the retrieved phase on the two planar acquisition surfaces. Reconstructed aperture fields are depicted in Figure 6.19. Here, it must be pointed out the capability of the pSRM to recover the

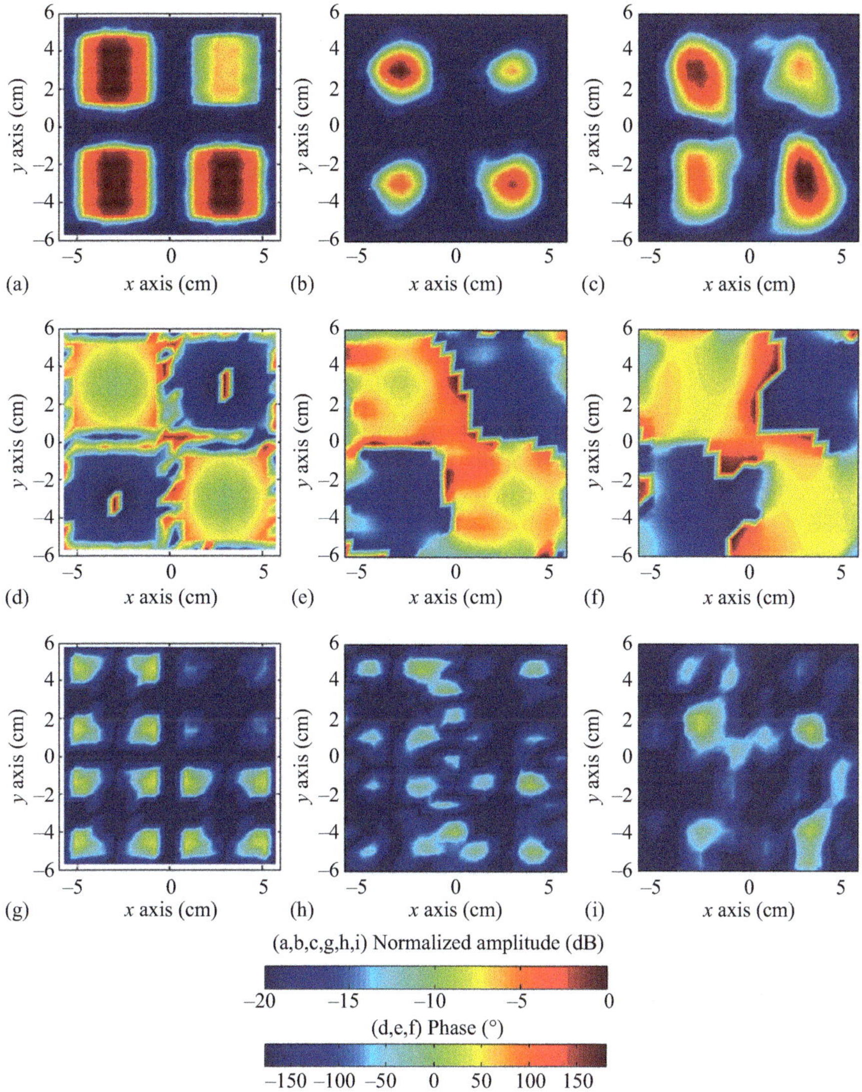

Figure 6.19 AUT aperture fields: (a, b, c) E_y, amplitude. (d, e, f) E_y, phase. (g, h, i) E_x, amplitude. (a, d, e) Reference aperture fields. (b, e, h) Reconstructed aperture fields, nonlinear cost function minimization. (c, f, i) Reconstructed aperture fields, iterative scheme

phase distribution of the antenna array, as shown in Figure 6.19(e) and (f), where it can be observed that the chessboard-like phase pattern (Figure 6.19(d)) is reconstructed. Concerning the amplitude distribution, the 6 dB attenuation of the upper right element can be fairly estimated as observed in Figure 6.19(b) and (c).

Crosspolar component (E_x) has been reconstructed as well, observing that nonlinear CF minimization provides better pattern results (Figure 6.19(h)) than the iterative scheme (Figure 6.19(i)).

AUT radiation pattern can be calculated from the aperture fields. A comparison of the FF pattern calculated from reference and reconstructed aperture fields (Figure 6.19) is shown in Figures 6.20 and 6.21. Size and position of the farthest planar acquisition surface determine the margin of validity of the FF pattern, which is $\pm 45°$ for this example, as depicted in Figure 6.21. Within this margin, maximum error between the reference FF pattern, and the FF pattern calculated from the reconstructed aperture fields from phaseless measurements is around 2 dB at $\phi = 0°$. The two-dimensional representation of the FF pattern depicted in Figure 6.20 shows that FF pattern calculated from pSRM based on nonlinear CF minimization (Figure 6.20(c) and (d)) is more similar to the reference FF pattern (Figure 6.20(a) and (b)) than the one calculated from pSRM based on iterative scheme (Figure 6.20(e) and (f)). This result is coherent with the fact that, when the pSRM stopping condition was reached, δ value for pSRM based on iterative scheme was greater than δ value for pSRM based on nonlinear CF minimization.

As mentioned in Section 6.3.2.1, pSRM based on nonlinear CF optimization is sensitive to the choice of the IG. In the previous example, the IG consisted of a uniform aperture field, $IG_1 : M_x = M_y = 1$. Several IGs for M_x, M_y have been tested as well, aiming to evaluate the convergence of the pSRM. Figure 6.22 shows the FF pattern calculated considering a sinc-like distribution of the aperture fields as IG ($IG_2 : M_x = M_y = f(sinc(x)sinc(y))$), and an IG distribution where the aperture fields are set to zero except at the center of the aperture ($IG_3 : \{M_x = M_y = 0\} \cup \{M_x(x = y = 0) = M_y(x = y = 0) = 1\}$). Differences in the visible margin of the FF field pattern due to the choice of the IG are within a 2–3 dB range.

LMA convergence has been also evaluated for the considered IG. As observed in Figure 6.23, IG_3 converges slower than IG_1 and IG_2, as the former IG is the most different among these three with respect to the nominal aperture fields distribution. In spite of that, LMA is able to find the global solution [51].

The influence of the damping factor Λ in the LMA for nonlinear CF minimization has been assessed as well. In particular, the scaling of the damping factor Λ depending on the rate at which the residual decreases, denoted as *Sf*, has been tested. The damping factor is updated as $\Lambda(n+1) = Sf\,\Lambda(n)$ if the residual does not decrease for two consecutive iterations (n denotes the nth LMA iteration). If the residual decreases, then the damping factor is decreased as well according to $\Lambda(n+1) = \Lambda(n)/Sf$. The impact on the LMA convergence is shown in Figure 6.24: results labeled as Λ_1 correspond to a scaling factor $Sf = 5$. A scaling factor $Sf = 2$ is considered in Λ_2, and a scaling factor $Sf = 1.2$ in Λ_3. As observed in Figure 6.24, Λ_1 converges faster but as the step is larger, LMA cannot reach the global minimum as in the case of Λ_2 or Λ_3. The latter exhibits the slowest convergence rate, as it requires more iterations to reach the region of the global minimum. The impact on the LMA damping parameter can be also observed in the FF pattern, in Figure 6.24. Largest differences between estimated FF pattern and

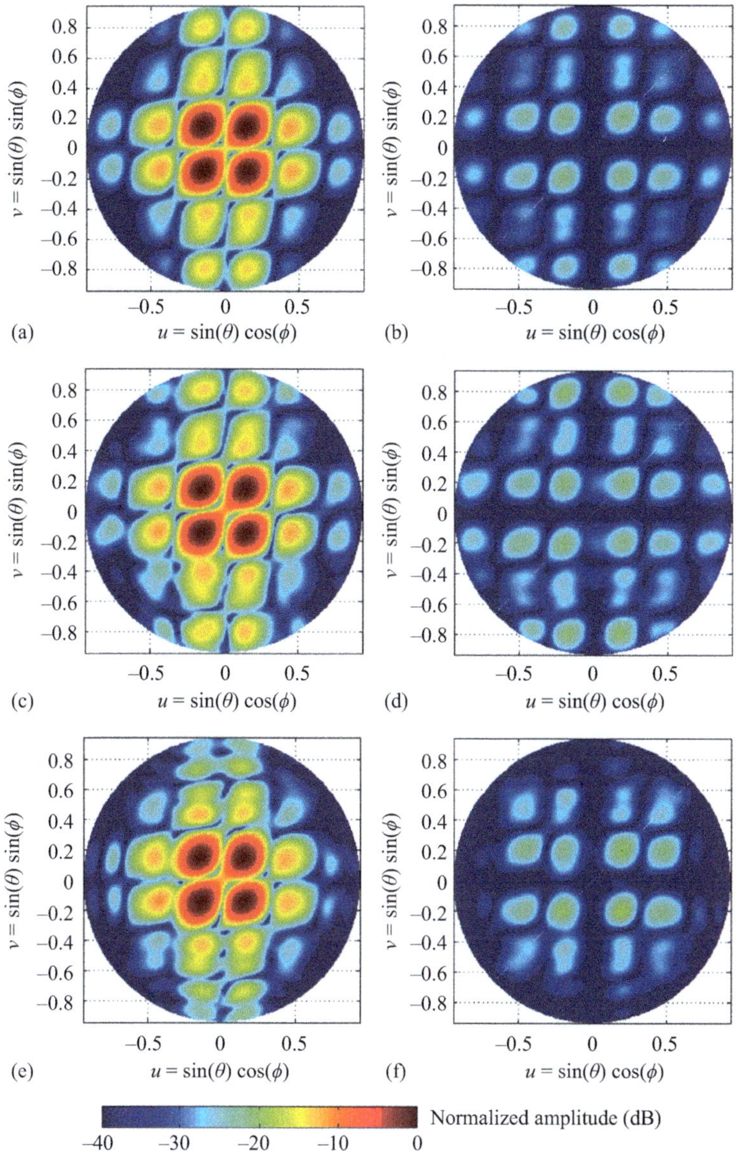

Figure 6.20 AUT FF pattern: (a, c, e) copolar component; (b, d, f) crosspolar component; (a, b) FF pattern calculated from reference aperture fields; (c, d) FF pattern calculated from reconstructed aperture fields (nonlinear cost function minimization); (e, f) FF pattern calculated from reconstructed aperture fields (iterative scheme)

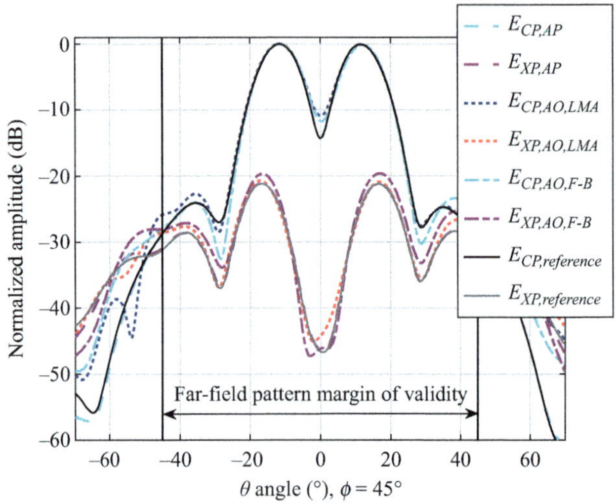

Figure 6.21 AUT FF pattern. From reconstructed aperture fields using NF amplitude and phase information ($E_{CP,AP}$: copolar, $E_{XP,AP}$: crosspolar). From reconstructed aperture fields using NF amplitude-only information, nonlinear cost function minimization ($E_{CP,AO,LMA}$: copolar, $E_{XP,AO,LMA}$: crosspolar). From reconstructed aperture fields using NF amplitude-only information, iterative scheme ($E_{CP,AO,F-B}$: copolar, $E_{XP,AO,F-B}$: crosspolar). Reference FF pattern ($E_{CP,reference}$: copolar, $E_{XP,reference}$: crosspolar). FF pattern margin of validity is 90°

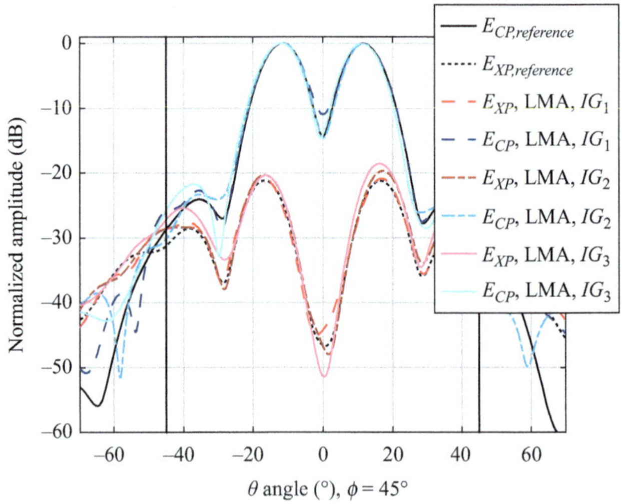

Figure 6.22 AUT FF pattern. From reconstructed aperture fields using NF amplitude-only information, nonlinear cost function minimization ($E_{CP,LMA}$: copolar, $E_{XP,LMA}$: crosspolar). $IG_1 : M_x = M_y = 1$, $IG_2 : M_x = M_y = f(sinc(x)sinc(y))$, $IG_3 : \{M_x = M_y = 0\} \cup \{M_x(x = y = 0) = M_y(x = y = 0) = 1\}$

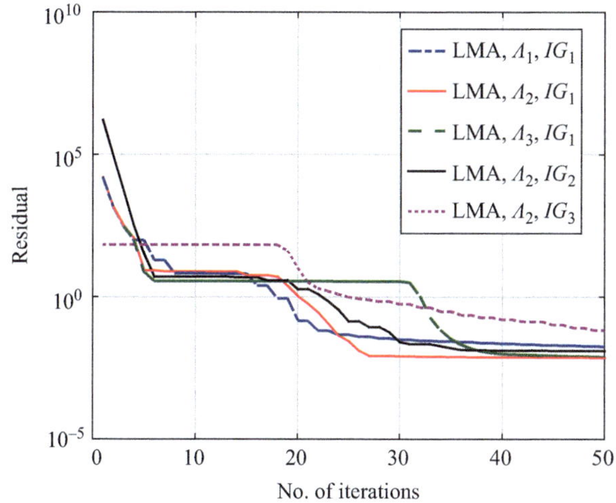

Figure 6.23 Levenberg–Marquardt algorithm (LMA) convergence for different values of initial guess (IG) and damping parameter Λ. *Convergence is shown for 50 iterations. Residual value is* $\delta = |RMS_{est} - RMS_{acq}|$. $IG_1 : M_x = M_y = 1$, $IG_2 : M_x = M_y = f(sinc(x)sinc(y))$, $IG_3 : \{M_x = M_y = 0\} \cup \{M_x(x = y = 0) = M_y(x = y = 0) = 1\}$. Λ_1: $Sf = 5$. Λ_2: $Sf = 2$. Λ_3: $Sf = 1.2$

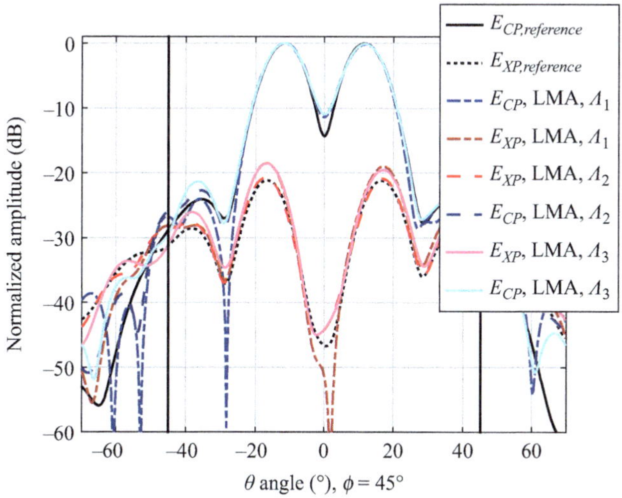

Figure 6.24 AUT FF pattern. From reconstructed aperture fields using NF amplitude-only information, nonlinear cost function minimization ($E_{CP,LMA}$*: copolar,* $E_{XP,LMA}$*: crosspolar).* Λ_1: $Sf = 5$. Λ_2: $Sf = 2$. Λ_3: $Sf = 1.2$

the reference one correspond to Λ_1, in agreement with the convergence results shown in Figure 6.24.

6.4 Conclusions

Phase recovery techniques are able to estimate the phase of a radiated field from amplitude-only acquisitions. Although this estimation is never completely error free, phaseless techniques are the only available approach at high frequencies at which only power acquisition is available. In addition, they are also suitable for the characterization of antennas in cost-effective acquisition setups, where only scalar equipment is available.

In this chapter, two of the most popular phaseless techniques, both derived from optics, have been presented in this chapter. Different enhancements for both techniques have also been described.

The first technique is known as indirect off-axis holography, and it is based on creating an interference between a reference signal and the field of the AUT. The key advantages of this approach are as follows:

- *Iteration free*: A direct calculation enables to retrieve the phase.
- *Antenna independent*: Valid for high directivity as well as low-directivity antennas.
- *Single acquisition surface*: It only requires acquiring the field at a single surface.

However, there are also a number of disadvantages, which are enumerated as follows:

- *Reference field characterization*: The accuracy of the approach is strongly dependent on the correct characterization of the reference antenna along the acquisition surface.
- *Modified setup*: The setup requires some additional components such as a power divider and waveguides to convey the signal to the reference antenna. In addition, some balance between the power of the reference and the AUT branches must be achieved to avoid one signal masking the other one at the receiver stage.
- *Sampling requirements*: The different spectrum terms increase the spatial bandwidth; and therefore, the sampling steps are more demanding than other techniques.
- *Surface acquisition*: The acquisition surface must be correctly illuminated by the reference antenna.
- *Single frequency*: The original approach entails an interferometry along the spatial domain at a given frequency. Consequently, the acquired data processing must be repeated at each frequency. Moreover, the reference field must be known at all the considered frequencies.

In this chapter, several enhancements have been described, mitigating some of the aforementioned problems. For example, the mechanical phase shifting enables one to simplify the setup bypassing the need of some of the components, and it also enables a simple characterization of the reference field as it is given by the spatial displacements. In addition, a modification to efficiently measure wideband antennas has been illustrated by translating the interferometry to the TD. Furthermore, this latter approach, thanks to the TD interferometry in contrast to the spatial interferometry, involves a spatial bandwidth which is the same as for conventional antenna metrology.

The second phaseless technique described in this paper is based on the application of an iterative scheme to the field amplitude measured at two different surfaces. In this case, an algorithm looks for a phase distribution so that the field from one surface can be propagated to the other acquisition surface (and vice versa) matching the amplitude distribution. This conventional scheme can be integrated with the SRM method so that the field distribution at each iteration is referred to the field radiated by the equivalent currents. The main advantages of this scheme are summarized as follows:

- *Simple setup*: This approach does not modify the setup of the measurement range. In fact, the setup is similar to conventional setups for antenna measurements but a power meter is used at the receiving antenna. Consequently, this approach is suitable for compact, low-cost sensors for *in situ* antenna measurement systems.
- *Arbitrary surface acquisition*: The method can be formulated with complete independence of the acquisition surfaces.
- *Enhanced convergence*: *A priori* information about the AUT (geometry, theoretical distribution of the aperture fields) can be considered to reduce the degrees of freedom of the radiated field. Consequently, the hybridization with SRM results in a convergence faster than the one achieved with conventional iteration between the two surfaces without further constraints.
- *Easy integration with other techniques*: In the case of including the SRM, an equivalent currents model of the AUT is recovered. Consequently, diagnostics and NF–FF transformations can be directly computed.

However, the weakest points of this approach are inherent to its iterative nature, which are stated as follows:

- *Dependency on the IG*: The IG phase has a strong impact on the risk of stagnation at a local minimum and in the number of required iterations to reach a given threshold. The use of nonlinear CFs avoids this dependency at the expenses of increasing the complexity of the algorithm.
- *Multiple acquisition surfaces*: The algorithm requires to measure the amplitude of the field at two (or more) acquisition surfaces. Furthermore, the acquisition surfaces must be chosen so that the amplitude distribution is as different as possible since phase information is (indirectly) estimated from this amplitude distribution differences. This fact can be challenging for high directivity antennas (i.e., reflectors) as amplitude distribution at close distances to the AUT usually exhibits moderate changes.
- *Computational burden*: The computational time of these iterative schemes is higher than in off-axis holography, in which is almost negligible. In addition, the overall

computational cost in terms of time and, also, memory is further increased in the case of using the SRM. However, the use of graphic processing units (GPUs), which has widely spread in recent years, enables to achieve very competitive times for small and moderate size antennas. Speed-up algorithms for SRM (see Chapters 2 and 3) can alleviate the computational cost in the case of electrically large antennas.

In summary, phase recovery is a powerful approach, which can provide a complete characterization of an antenna in those cases in which only amplitude can be measured. There is a large variety of approaches, some of them requiring a modification of the antenna measurement setups, whereas other ones requiring a more complex and intensive data processing. As a general rule of thumb, it is expected that iterative schemes can provide a good performance for low and medium directivity antennas, whereas larger antennas can require the use of off-axis holography due to the risk of stagnation and computational burden of the iterative schemes. Nonetheless, each approach has a number of advantages and drawbacks, which must be carefully pondered according to the AUT and the measurement facility.

References

[1] Lehto A, Tuovinen J, Boric O, *et al.* Antenna phase pattern measurements using three power meters. In: Microwave Conference, 1992. 22nd European. vol. 1; 1992. p. 515–520.

[2] Lehto A, Tuovinen J, Boric O, *et al.* Accurate millimeter wave antenna phase pattern measurements using the differential phase method with three power meters. IEEE Transactions on Antennas and Propagation. 1992;40(7): 851–853.

[3] Bowen JW, Hadjiloucas S, Walker GC, *et al.* Interferometric technique for measuring terahertz antenna phase patterns. IEEE Sensors Journal. 2013;13(1): 100–110.

[4] Costanzo S, Massa GD, and Migliore MD. Integrated microstrip probe for phaseless near-field measurements on plane-polar geometry. Electronics Letters. 2001;37(16):1018–1020.

[5] Bucci OM, Gennarelli C, and Savarese C. Representation of electromagnetic fields over arbitrary surfaces by a finite and nonredundant number of samples. IEEE Transactions on Antennas and Propag. 1998;46(3):351–359.

[6] Costanzo S, Di Massa G, and Migliore MD. A novel hybrid approach for far-field characterization from near-field amplitude-only measurements on arbitrary scanning surfaces. IEEE Transactions on Antennas and Propagation. 2005;53(6): 1866–1874.

[7] Costanzo S, and Di Massa G. Wideband phase retrieval technique from amplitude-only near-field data. Radioengineering. 2008;17(4):8–12.

[8] Gabor D. Microscopy by reconstructed wave-fronts. Proceedings of the Royal Society of London. Series A: Mathematical, Physical and Engineering Sciences. 1949;197(1051):454–487.

[9] Leith EN, and Upatnieks J. Reconstructed wavefronts and communication theory. Journal of the Optical Society of America. 1962;52(10):1123–1128.

[10] Aoki Y. Microwave holography by a two-beam interference method. Proceedings of the IEEE. 1968;56(8):1402–1403.
[11] Napier PJ, and Bates RHT. Antenna – aperture distributions from holographic type of radiation-pattern measurement. Proceedings of the Institution of Electrical Engineers. 1973;120(1):30–34.
[12] Bennett JC, Anderson AP, McInnes P, *et al.* Microwave holographic metrology of large reflector antennas. IEEE Transactions on Antennas and Propagation. 1976;24(3):295–303.
[13] Arboleya A, Laviada J, Ala-Laurinaho J, *et al.* Indirect Off-Axis Holography for Antenna Metrology. In: Naydenova PI, editor. Holographic Materials and Optical Systems. InTech; 2017.
[14] Junkin G, Huang T, and Bennett JC. Holographic testing of terahertz antennas. IEEE Transactions on Antennas and Propagation. 2000;48(3):409–417.
[15] Smith D, Leach M, and Sambell A. An indirect holographic method for determining antenna radiation patterns and imaging antenna fields. In: IEEE Antennas and Propag. Soc. Int. Symp.; 2002. p. 706–709.
[16] Schejbal V, Kovarik V, and Cermak D. Synthesized-reference-wave holography for determining antenna radiation characteristics. IEEE Antennas and Propagation Magazine. 2008;50(5):71–83.
[17] Laviada J, and Las-Heras F. Phaseless antenna measurement on non-redundant sample points via Leith–Upatnieks holography. IEEE Transactions on Antennas and Propagation. 2013;61(8):4036–4044.
[18] Laviada Martínez J, Arboleya-Arboleya A, Álvarez López Y, *et al.* Phaseless antenna diagnostics based on off-axis holography with synthetic reference wave. IEEE Antennas and Wireless Propagation Letters. 2014;13:43–46.
[19] Yaccarino RG, and Rahmat-Samii Y. Phaseless bi-polar planar near-field measurements and diagnostics of array antennas. IEEE Transactions on Antennas and Propagation. 1999;47(3):574–583.
[20] Las-Heras F, and Sarkar TK. A direct optimization approach for source reconstruction and NF-FF transformation using amplitude-only data. IEEE Transactions on Antennas and Propagation. 2002;50(4):500–510.
[21] Schmidt CH, Razavi SF, Eibert TF, *et al.* Phaseless spherical near-field antenna measurements for low and medium gain antennas. Advances in Radio Science. 2010;8:43–48.
[22] Razavi SF, and Rahmat-Samii Y. A new look at phaseless planar near-field measurements: Limitations, simulations, measurements, and a hybrid solution. IEEE Antennas and Propagation Magazine. 2007;49(2):170–178.
[23] Puskely J. Application of iterative Fourier method in cylindrical phaseless antenna measurement technique. Radioengineering. 2012;21:422–429.
[24] Bucci OM, D'Elia G, Leone G, *et al.* Far-field pattern determination from the near-field amplitude on two surfaces. IEEE Transactions on Antennas and Propagation. 1990;38(9):1772–1779.
[25] Bucci OM, D'Elia G, Leone G, *et al.* Far-field computation from amplitude near-field data on two surfaces: Cylindrical case. IEE Proceedings, Part H – Microwaves, Antennas and Propagation. 1992;139(2):143–148.

[26] Alvarez Y, Las-Heras F, and Pino MR. The sources reconstruction method for amplitude-only field measurements. IEEE Transactions on Antennas and Propagation. 2010;58(8):2776–2781.
[27] Schnattinger G, Lopez C, and Eibert TF. Fast near-field far-field transformation for phaseless and irregular antenna measurement data. Advances in Radio Science. 2014;12:171–177.
[28] Brown T, Jeffrey I, and Mojabi P. Multiplicatively regularized source reconstruction method for phaseless planar near-field antenna measurements. IEEE Transactions on Antennas and Propagation. 2017;65(4):2020–2031.
[29] Huang T, Bennett JC, and Junkin G. A holographic technique for millimetre-wave antenna near-field/far-field transformation. In: Antennas and Propag., 1995., 9th Int. Conf. on (Conf. Publ. No. 407). vol. 1; 1995. p. 41–44.
[30] Smith D, Skobelev SP, and Leach MP. A modified holographic technique for antenna measurements. In: 2008 8th Int. Symp. on Antennas, Propag. and EM Theory (ISAPE); 2008. p. 46–49.
[31] Räisänen AV, and Ala-Laurinaho J. Holographic principles in antenna metrology at millimeter and submillimeter wavelengths. In: 2015 9th European Conf. on Antennas and Propag. (EuCAP); 2015. p. 1–2.
[32] Arboleya A, Ala-Laurinaho J, Laviada J, *et al.* Millimeter-wave phaseless antenna measurement based on a modified off-axis holography setup. Journal of Infrared, Millimeter, and Terahertz Waves. 2016;37(2):160–174.
[33] Wang JJH. An examination of the theory and practices of planar near-field measurement. IEEE Transactions on Antennas and Propagation. 1988;36(6): 746–753.
[34] Yaghjian AD. An overview of near-field antenna measurements. IEEE Transactions on Antennas and Propagation. 1986;34(1):30–45.
[35] Arboleya A, Laviada J, Ala-Laurinaho J, *et al.* Phaseless characterization of broadband antennas. IEEE Transactions on Antennas and Propagation. 2016;64(2):484–495.
[36] Leach MP, Smith D, Skobelev SP, *et al.* An improved holographic technique for medium-gain antenna near field measurements. In: 2007 2nd European Conf. on Antennas and Propag. (EuCAP); 2007. p. 1–6.
[37] Schejbal V, Pidanic J, Kovarik V, *et al.* Accuracy analyses of synthesized-reference-wave holography for determining antenna radiation characteristics. IEEE Antennas and Propagation Magazine. 2008;50(6):89–98.
[38] Arboleya A, Álvarez Y, and Las-Heras F. Millimeter and submillimeter planar measurement setup. In: 2013 IEEE Antennas and Propag. Soc. Int. Symp. (APSURSI); 2013. p. 1–2.
[39] Krozer V, Loffler T, Dall J, *et al.* Terahertz imaging systems with aperture synthesis techniques. IEEE Transactions on Microwave Theory and Techniques. 2010;58(7):2027–2039.
[40] Bisognin A, Nachabe N, Luxey C, *et al.* Ball grid array module with integrated shaped lens for 5G backhaul/fronthaul communications in F-band. IEEE Transactions on Antennas and Propagation Special Issue on Antennas and Propagation Aspects of 5G Communications. 2017;65(12):6380–6394.

[41] Huang Y, Chan K, and Cheeseman B. Review of broadband antenna measurements. In: 2006 1st European Conf. on Antennas and Propag. (EuCAP); 2006. p. 1–4.
[42] Arboleya A, Laviada Jaime, Ala-Laurinaho J, Álvarez Y, Las-Heras F, and Räisä AV. Reduced set of points in phaseless broadband near-field antenna measurement: Effects of noise and mechanical errors. In: 2016 10th European Conf. on Antennas and Propag. (EuCAP); 2016. p. 1–5.
[43] Schejbal V, Kovarik V, and Cermak D. Synthesized-reference-wave holography for determining antenna radiation characteristics. IEEE Antennas and Propagation Magazine. 2008;50(5):71–83.
[44] Tennant A, Junkin G, and Anderson AP. Relaxation of mechanical tolerances in phase retrieval antenna metrology. In: 1993. 23rd European Microw. Conf. (EuMC); 1993. p. 877–879.
[45] Garcia-Fernandez M, Alvarez Y, Arboleya A, *et al.* Antenna diagnostics and characterization using unmanned aerial vehicles. IEEE Access. 2017;5: 23563–23575.
[46] Alvarez Y, Las-Heras F, and Pino MR. Reconstruction of equivalent currents distribution over arbitrary three-dimensional surfaces based on integral equation algorithms. IEEE Transactions on Antennas and Propagation. 2007;55(12): 3460–3468.
[47] Araque JL, and Vecchi G. Improved-accuracy source reconstruction on arbitrary 3-D surfaces. IEEE Antennas and Wireless Propagation Letters. 2009;8:1046–1049.
[48] Foged L, Scialacqua L, Saccardi F, *et al.* Application of the dual-equation equivalent-current reconstruction to electrically large structures by fast multipole method enhancement. IEEE Antennas and Propagation Magazine. 2014;62(4):2041–2051.
[49] Persson K, Gustafsson M, Kristensson G, *et al.* Radome diagnostics – Source reconstruction of phase objects with and equivalent currents approach. IEEE Transactions on Antennas and Propagation. 2014;56(5):264–273.
[50] Marquardt DW. An algorithm for least-squares estimation of nonlinear parameters. Journal of the Society for Industrial and Applied Mathematics. 1963;11(2):431–441.
[51] Nielson HB. Damping Parameter in Marquardt's Method. Department of Mathematical Modeling, Technical University of Denmark; 1999. Technical Report IMM-REP-1999-05.

Chapter 7
Link between simulation and measurements

Lucia Scialaqua[1], *Francesca Mioc*[1], *and Lars J. Foged*[1]

The analysis of deployed antenna performance in complex scenarios can turn out to be problematic due to the realistic details to be taken into account in the investigation. In these cases, both numerical modelling and measurements are fundamental tools to evaluate the antenna performance. The numerical modelling created with computational electromagnetic (CEM) tools requires a particular representation of the source antenna. This can be achieved by knowing the exact characteristics of the antenna. However, in many practical cases, a full-wave representation is unfeasible or unavailable. This is where recent antenna-measurement techniques are proven effective.

In this chapter, you will learn of tools and techniques that overcome the limitations due to unknown source-model characterization and the complexity of the environment to be measured. The approach, based on the equivalent current (EQC) representation of an antenna, links measurement equipment/techniques with commercially available numerical computational tools to provide measurements of source antennas for simulations in the most complex of test environments.

The chapter is divided into the following sections: Section 7.1 introduces the motivation of 'The Link' and its advantages, Section 7.2 explains the basic theory of 'The Link', Section 7.3 introduce the applications for antenna in free space as also covered in more detail in section 3. Section 7.4 cover the important extension of the method to flushed mounted antennas and Section 7.5 shows practical application examples. Section 7.6 concludes the contents of this chapter.

7.1 Introduction

For the placement of a radiating device upon a large structure, such as an antenna on a satellite, a radar on an aircraft or a sensor on a car, investigation and optimization of the scenario is necessary. While full measurements are still required for final validation of deployed antenna performance due to the conclusiveness and high reliability of measured data, numerical modelling is more and more used in the initial stages of antenna placement investigation and optimization. Due to the

[1]Microwave Vision Italy (MVI), Pomezia, Italy

increasing complexity of test scenarios, CEM simulation tools are implementing the domain decomposition technique (DDT) in which each domain can be solved separately with the most suited numerical solver. The DDT features can be exploited to use antenna measurements to characterize some of the features of the complex scenario. In some cases, when implementing an antenna supplied by a third party, the mechanical and electronic characteristics needed for a full-wave representation of the antenna may be unavailable, particularly in the format required by the CEM tool. To overcome this problem, the radiating antenna can be characterized by a true radiating measurement. From the measured radiation pattern, an EQC representation of the measured source antenna in near field (NF) can be determined and this can be imported into the CEM tool for simulations. The obtained EQC model is an electromagnetic complete representation of the radiation pattern of the antenna and can be used as an equivalent black box based on a Huygens' formulation in simulations on the basis of the DDT [1–5].

7.2 Theory

The technique based on EQC reconstruction [6–8] and implemented in the software* allows to create accurate representative electromagnetic models of measured antennas (EQC models). An EQC representative model can be imported by a number of commercial CEM tools for simulation and testing in full, complex scenarios. The EQC method provides the electromagnetic glue, often named 'The Link', bringing together numerical simulation and antenna measurement to understand deployed antenna performance. The main theory and concept of The Link are described in the following sections.

7.2.1 Domain decomposition technique

The DDT allows operators to divide a scenario under test into smaller more manageable chunks, optimizing and reducing the computational cost. Applying the DDT in a complex antenna-placement problem means, in a practical sense, that the characterization of the radiating antenna is to be obtained separately as a subproblem then integrated into the final full scenario. In the past, the use of DDT was strictly limited to simulated problems where different numerical methods had to be used to solve the various subproblems. This limitation has recently been overcome. The technique based on EQC reconstruction allows true measurements to characterize the subproblems using the DDT.

7.2.2 Black box representation of devices radiating into 3D space

While the development of black box electrical representation of microwave components has revolutionized the design of integrated circuits, the development of an

*INSIGHT webpage: http://www.mvg-world.com/en/products/field_product_family/antenna-measurement-2/insight

equivalent black box representation of devices radiating into 3D space, such as antennas, has been much less successful. The black box representation of a microwave device is based on the definition of physical ports on the device and the relationship between incoming and outgoing waves defined by *S*-parameters. The device details are often unknown to the user since the *S*-parameters representation, fully characterizing the device, can only be obtained by measurements, simulations or provided by the vendor. Similar to what happens with *S*-parameters of microwave components, a radiating device into 3D space can be fully characterized by an NF source data set. In this case, it is defined as an equivalent black box based on a Huygens' formulation which represents the radiation pattern of the device [9].

7.2.3 Definition of The Link

The Link is the intersection between antenna measurement equipment, CEM numerical modelling and an EQC model of an NF measurement of the source antenna suitable for the CEM solver. To illustrate the procedure, an example of a reflector system fed by a dual ridge horn has been considered. This system can be divided into two parts: the horn (source antenna) and the reflector, as shown in Figure 7.1.

The radiation pattern (NF and/or far field (FF)) of the horn by itself in isolation is initially measured. The measured data is then post-processed to obtain an equivalent black box based on a Huygens' formulation. The equivalent black box can then be installed on the reflector for the simulation with different CEM tools as shown in Figure 7.2.

The EQC representation of the source antenna also allows accurate results when the antennas are in close proximity or mounted directly on complex structures. The procedure is general and can be used with any antenna in complex environments of arbitrary shape and complexity [1–5].

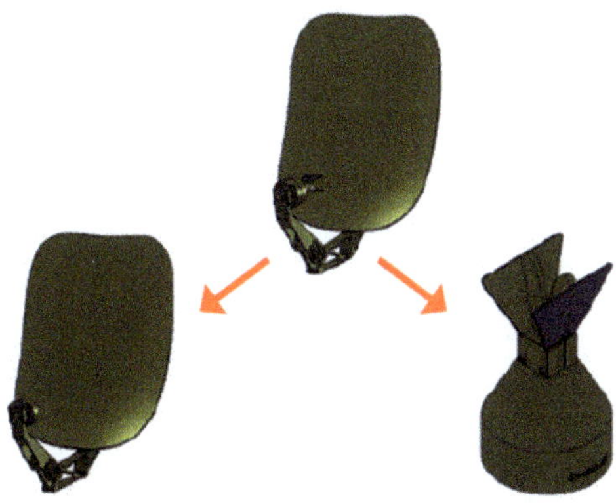

Figure 7.1 Reflector system: the MVG SR40-A reflector fed by the MVG SH4000 dual ridge horn

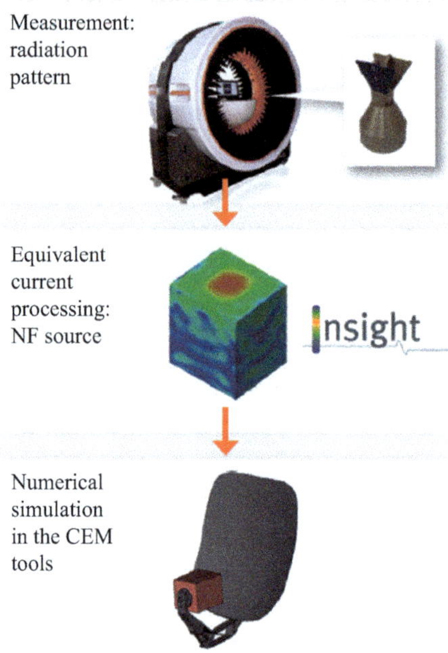

Figure 7.2 Illustration of 'The Link'. The source is measured in isolation, the measured NF or FF are processed creating a Huygens' box compatible with commercial CEM tools

7.2.3.1 Antenna-measurement equipment

An NF or an FF measurement of the antenna radiation pattern, over a spherical, cylindrical or planar scan surface, is equally applicable in the preparation of the EQC representation to be used in a CEM simulation. Measurements of the isolated antenna in many cases do not necessarily require large measurement equipment, and compact NF measurement ranges can be used to perform accurate and efficient measurements of the source antenna. Compact and portable measurement systems are in fact quite advantageous in this case as, like other instrumentation in the EM laboratory, they can be easily used in combination with and in proximity to the platforms where the simulations are carried out. In addition, if the system is a multi-probe system, measurement time can be proportionately reduced.

The measurements reported in this chapter and used for validation of 'The Link' have been performed in the compact and portable MVG SL measurement system [10] shown in Figure 7.3. The diameter of the SL probe array is 90 cm. In the spherical NF configuration, the device under test (DUT) is positioned on a foam mast in the centre of the system made from a very low dielectric, low-loss material. The full sphere measurement is performed by electronically scanning the probe array in elevation and rotating the DUT 180° in azimuth. Consequently, a full 3D measurement can be performed very rapidly [10].

Figure 7.3 MVG SL18 GHz spherical near-field multi-probe system covering the frequency range from 690 MHz to 18 GHz [10]

7.2.3.2 Equivalent current processing NF source modelling

An EQC representation of the isolated antenna is proposed as a measured NF model to represent the source antenna in a CEM tool. EQC representations can be computed from measured data by application of the inverse-source or EQC/source method [6–8]. The method has been completely described in Chapter 3 of this book. The key in using the inverse-source method is in its capability to reconstruct the EQC on arbitrary or generic 3D surfaces enclosing the antenna or DUT. For this reason, the method can be applied to all antennas without limitations of type or shape. An example of EQC reconstruction for a dual ridge horn is shown in Figure 7.4.

Equivalent current processing versus spherical waves

Since the EQC can produce a form-fitting representation of the antenna surface, the source antenna can be mounted more freely in any position on any large structure or final scenario. This method is recommended as opposed to spherical wave expansion. In spherical wave expansion, only the field radiated outside the minimal sphere of the source antenna can be computed [11]. This sphere of minimum radius fully enclosing the source antenna must not be intersected by the structure; therefore, the source antenna cannot be installed too close to the structure. As a result, this approach can be applied only to a limited number of practical test cases. When EQC processing is applied to measured data for numerical modelling, an equivalent black box based on a Huygens' formulation, enclosing the antenna, provides enough accuracy for the procedure. An example of the EQC representation of the dual ridge horn as an equivalent black box is shown in Figure 7.5.

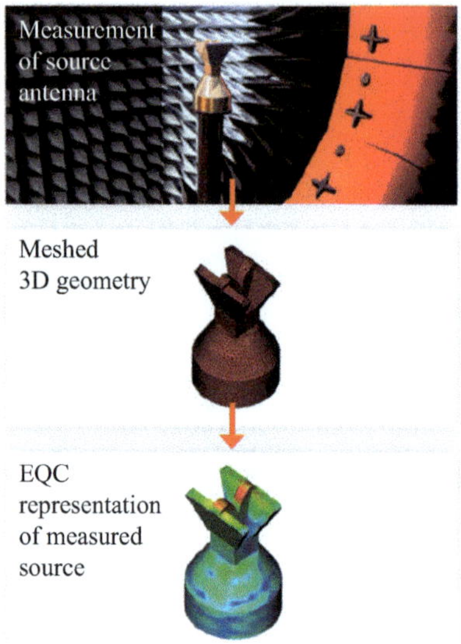

Figure 7.4 Equivalent current reconstruction from a measured radiation pattern of a dual ridge horn for diagnostic purposes

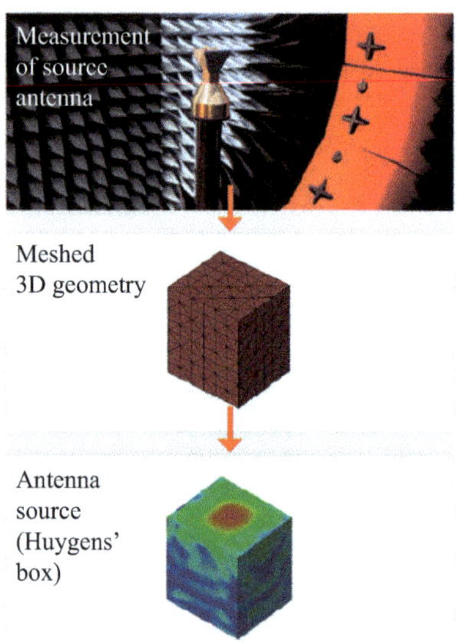

Figure 7.5 Equivalent currents reconstruction from a measured radiation pattern of a dual ridge horn using INSIGHT software

NF source modelling exchange data

NF source model, which is provided to the CEM tools, is a set of electric (E) and magnetic (H) values defined on the vertex of the mesh, discretizing the Huygens' box. The E and H values are calculated starting from the J and M currents determined during the EQC computation. The equivalent model is then converted to the different file formats managed by the CEM tools for importing such data.

Furthermore, the power of the equivalent model is an additional important parameter to be exchanged for antenna characterization, integrated in complex scenarios in numerical computations. Indeed, by this parameter, imported and managed by the CEM tool, it is allowed to calculate the effective antenna gain. Power is provided as a complex number:

$$Re_{POWER} + j \cdot Im_{POWER}$$

7.2.3.3 Numerical simulation in the CEM tools

Once the equivalent black box of the source antenna has been created by the EQC technique, it can be implemented into a number of CEM tools. The CEM solver will consider it as a full representation of the antenna in any scenario required for simulation. Applying the black box approach, the simulation is carried out with no more additional information. The advantages: no modifications to the source files are necessary, and the EQC representation can be used in any scenario to be tested, even the most complex. The EQC representation model can be imported by a number of commercial CEM solvers,[†] see Figure 7.6.

Figure 7.6 INSIGHT's EQC source model is exportable to most commercially available CEM tools

[†] www.cst.com, CST STUDIO SUITE^{TM}, CST AG; www.delcross.com/products-savant.php, Delcross; www.feko.info, Altair Engineering GmbH; www.ansys.com, HFSS, Ansy.; www.idscorporation.com/space, ADF, IDS.; www.wipl-d.com, WIPL-D, WIPL-D d.o.o.

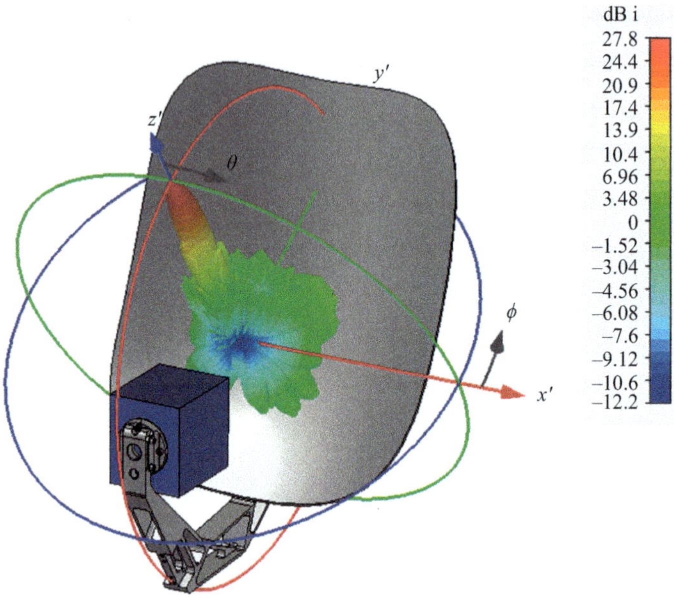

Figure 7.7 Simulated 3D far-field pattern obtained by The Link (measured source installed on the reflector at 8 GHz) (www.cst.com, CST STUDIO SUITE™, CST AG)

An example of a simulated FF radiation pattern (www.cst.com, CST STUDIO SUITE™, CST AG) obtained using The Link between measurements and CEM simulations is shown in Figure 7.7.

7.2.4 Validation of The Link

To validate The Link between antenna measurements and CEM simulations, different case scenarios have been tested. The objective of the validation was first to prove the accuracy of the technique and second to show its flexibility in application in several of the different commercially available computational electromagnetic simulation tools.

The validation was organized in two steps:

1. Free-space antenna validation.
2. Validation of antenna mounted on structures.

Three test cases were considered for validating antenna on structures (see Figure 7.8):

1. Reflector fed by a horn antenna.
2. Flush-mounted monocone antenna and open-ended waveguide on a flat structure.
3. Flush-mounted monopole antenna on a curved structure.

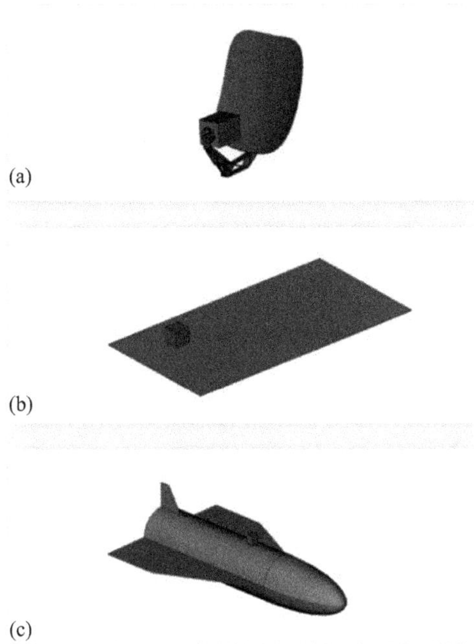

Figure 7.8 Test cases for the validation campaign of The Link between measurements and simulations. Reflector fed by a horn antenna (a); flush-mounted monocone antenna and open-ended waveguide on a flat structure (b); and flush-mounted monopole antenna on a curved structure (c). Equivalent current representations appear as red boxes

The EQC representations of the antennas, previously measured in isolation, were given to six CEM tool vendors. Comparisons were made between the results from the various vendors and against the reference measurements of the final scenario. The reference measurements consist of the antenna radiation patterns in the full final scenarios determined entirely in the measurement system.

The same information (an equivalent black box based on a Huygens' formulation modelling the source antenna) were provided to the vendors of six commercial CEM tools for comparison. In order to preserve the spirit and effectiveness of the validation campaign, no numerical results of the simulations were exchanged among software (SW) vendors during the entire campaign.

7.3 Antenna in free space

The first step of the validation is testing the import of the antenna in free space, comparing with the reference measurement. The antenna under test (AUT) used for validation of The Link (antenna in free space) is the dual ridge horn SH4000, operating in the band of 4–40 GHz, shown in Figure 7.9(a). This antenna has been selected since it is a well-known reference antenna, largely used for measurement

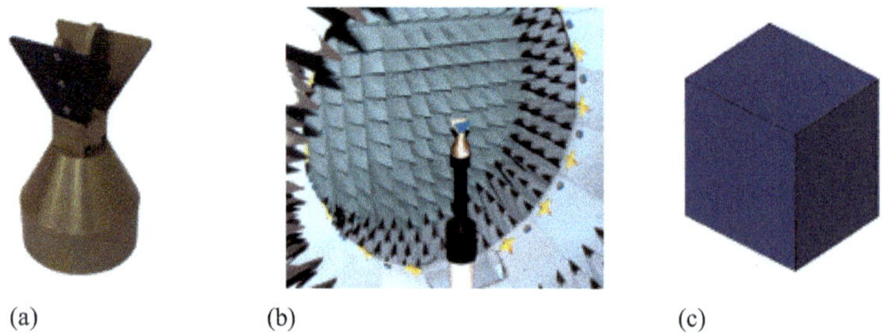

Figure 7.9 SH4000 dual ridge horn

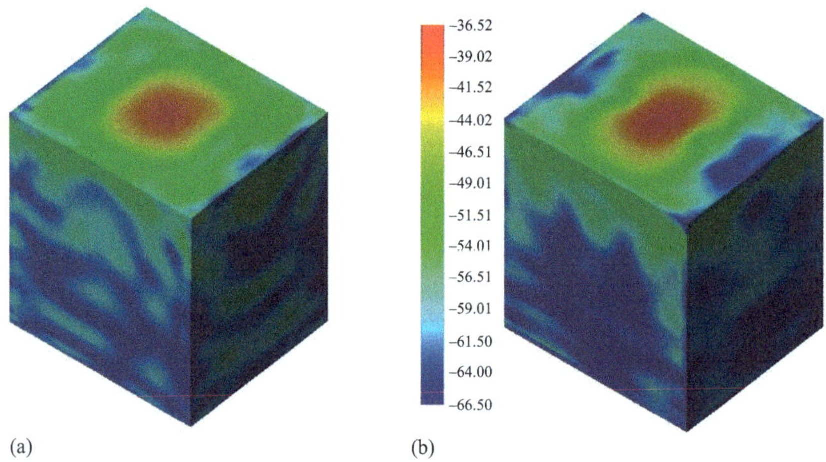

Figure 7.10 SH4000 dual ridge horn (isolated feeder) equivalent current representation from the measurement at 8 GHz: J electrical equivalent currents (a), M magnetic equivalent currents (b). Dynamic range of the plot is 30 dB

systems calibration, and it was the AUT during different measurement campaigns. The SH4000 has been initially measured in the SL18GHz spherical NF measurement system, as isolated source, as is shown in Figure 7.9(b). The reconstruction geometry is a box enclosing the antenna as shown in Figure 7.9(c).

Computation of the EQCs, representing the NF source from the SH4000 measured data, has been performed. The geometry of reconstruction is a box of dimensions [0.09 × 0.102 × 0.113] m and it is enclosing the SH4000, as is shown in Figure 7.9(c). Results from the validation of the procedure are illustrated in this section. Electric (*J*) and magnetic (*M*) EQCs computed from the measurement of the isolated feeder at 8 GHz are shown in Figure 7.10.

Preliminarily, a comparison has been performed for the SH4000 operating in isolation, between radiation pattern evaluated with measured EQC representation and full-wave simulation of the antenna (www.cst.com, CST STUDIO SUITE™, CST AG), as is shown in Figure 7.11.

Although Figure 7.11 shows the comparison results for (www.cst.com, CST STUDIO SUITE™, CST AG), this verification was performed with the same antenna for all the SW vendors involved in the validation campaigns. Good agreement between simulation and reference measurements was achieved verifying the accuracy of the correct definition of the E and H values for the Huygens' box (see 'NF source modelling exchange data' section) and the exchange data with all the SW vendors.

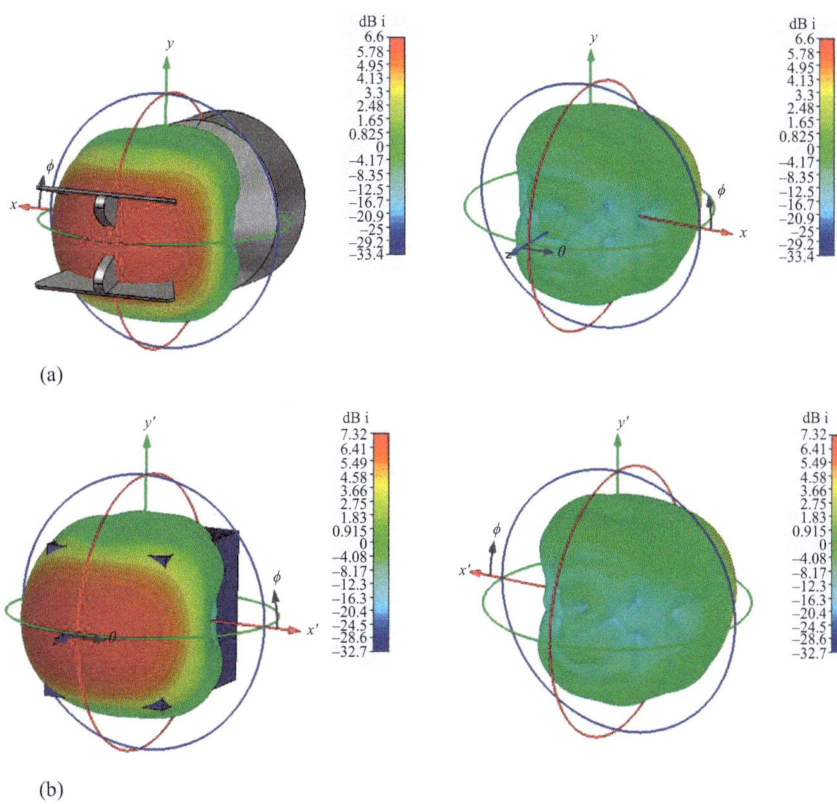

Figure 7.11 Comparison of the radiation pattern at 8 GHz between the simulated source (a) and the measured source (b)

7.4 Mounted antennas

Different applications of mounted antennas in complex environments have been examined to investigate the benefits and achievable accuracy of The Link between simulation and measurements:

1. SR40-A reflector antenna fed by an SH4000 horn,
2. SMC2200 monocone antenna on a rectangular flat plate,
3. OEW, open-ended waveguide, antenna on a rectangular flat plate and
4. SM6000 monopole on a mock-up of a space plane (curve surface).

The determination of the electromagnetic model of a freestanding source antenna is a straightforward calculation of a representative NF Huygens' box.

The electromagnetic model of the source antenna for flush-mounted applications is more difficult than the situation where the antenna is detached from the scattering structure. The proximity of scattering modifies the current distribution on the antenna itself.

An infinite ground plane boundary condition is a good approximation, but such a condition cannot be directly obtained on a realistic measurement scenario. However, this condition can be emulated from measurements of the source antenna on a finite ground plane, and the application of post-processing as discussed in [12]. The post-processing of the measured data eliminates the diffractive contributions from the edge of a finite ground plane creating the wanted infinite ground plane boundary condition. A circular ground plane with minimum 5λ diameter is considered adequate for most measurement source antennas.

The validation of The Link between measurements and numerical simulations is performed by comparison of radiation pattern of measured source simulation with system measurement.

7.4.1 Reflector fed by a horn antenna

The test object is an SR40-A reflector antenna fed by an SH4000 horn as shown in Figure 7.12. The freestanding SH4000 dual ridge horn and the complete reflector antenna have been measured in MVG, spherical NF, multi-probe systems, as shown in Figure 7.13.

EQCs constituting an accurate NF representation of the measured SH4000 have been computed processing on a $90 \times 102 \times 113$ mm box enclosing the SH4000.

The NF Huygens' box representation of the measured SH4000, obtained by the method of the EQCs, has been imported in different CEM tools (see footnote †). The complete reflector antenna has been simulated using a CAD model of the SR40 reflector and the Huygens' box fed. Each participant was responsible for generating a suitable mesh and the numerical stability of their solution as shown in Figure 7.14.

Simulation results are compared with measurement of the SR40 and SH4000 antenna at 8 GHz in the MVG SG-64 spherical NF facility in Paris as shown in Figure 7.13(a). The simulations, based on different numerical methods, are in very good agreement compared to each other. The agreement between simulation and

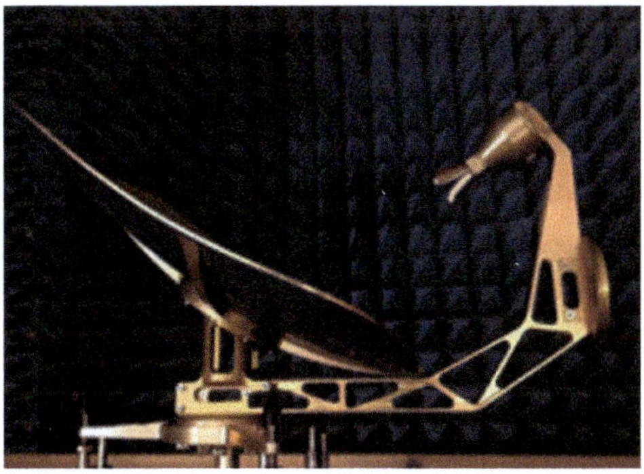

Figure 7.12 *The reflector antenna used in the validation experiment. SR40-A offset reflector and SH4000 dual ridge horn, operating from 4 to 40 GHz*

(a) (b)

Figure 7.13 *Reference measurements of the reflector antenna and feed in the MVG SG-64 spherical near-field facility in Paris (a). Measurement of the isolated SH4000 dual ridge fed in the MVG StarLab 18 GHz facility in Pomezia, Italy (b)*

measurements is good considering the approximation in the feed representation and uncertainties due to measurement, manufacturing and simulation.

The peak directivity values are reported in Table 7.1. MEAS is the measured reference from MVG. The simulated results are from the different SW vendors.[‡]

[‡]www.cst.com, CST STUDIO SUITE™, CST AG; www.delcross.com/products-savant.php, Delcross; www.feko.info, Altair Engineering GmbH; www.ansys.com, HFSS, Ansy.; www.idscorporation.com/space, ADF, IDS.

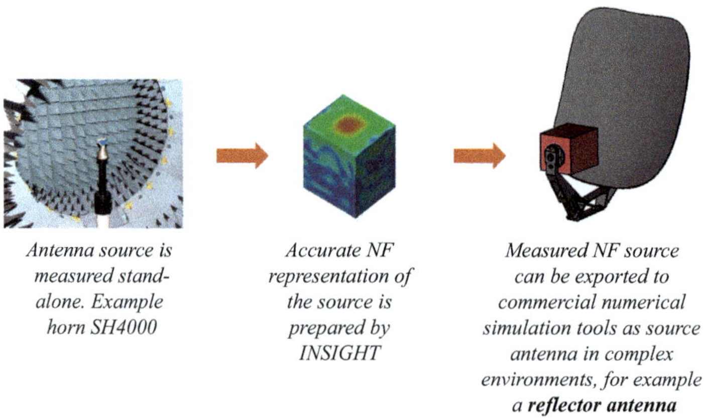

Figure 7.14 Use of a measurement of an SH4000 horn as feed in the simulation of a reflector antenna [4–40] GHz frequency band

Table 7.1 Peak directivity, SR40-A offset reflector fed by SH4000 dual ridge horn at 8 GHz

	MEAS	CST	Savant	FEKO	ADF	HFSS
Peak directivity (dB i)	27.42	27.69	27.66	27.83	27.60	27.51

The copolar pattern comparison, in the forward hemisphere, between measurement of the full antenna and simulations of the reflector with the measured source in different CEM tools (see footnote ‡) is shown in Figure 7.15. Very good agreement between measurements and simulation can be observed.

The agreement or correlation between simulations and measurements can be evaluated as the weighted difference between measured and simulated field using (7.1). The measured FF has been considered as reference field. The weighted difference for each simulation tool is overlaid with the measured E-plane and H-plane patterns in Figure 7.16.

The medium value of the weighted difference (7.1) expresses the correlation in a single value as shown in Table 7.2:

$$e_i(\theta,\phi) = \left|\frac{E(\theta,\phi) - \tilde{E}(\theta,\phi)}{E(\theta,\phi)}\right| \cdot \frac{|\tilde{E}(\theta,\phi)|}{|\tilde{E}(\theta,\phi)|_{\mathrm{MAX}}} \tag{7.1}$$

where $\tilde{E}(\theta,\phi)$ is the simulated pattern and $E(\theta,\phi)$ is the reference pattern.

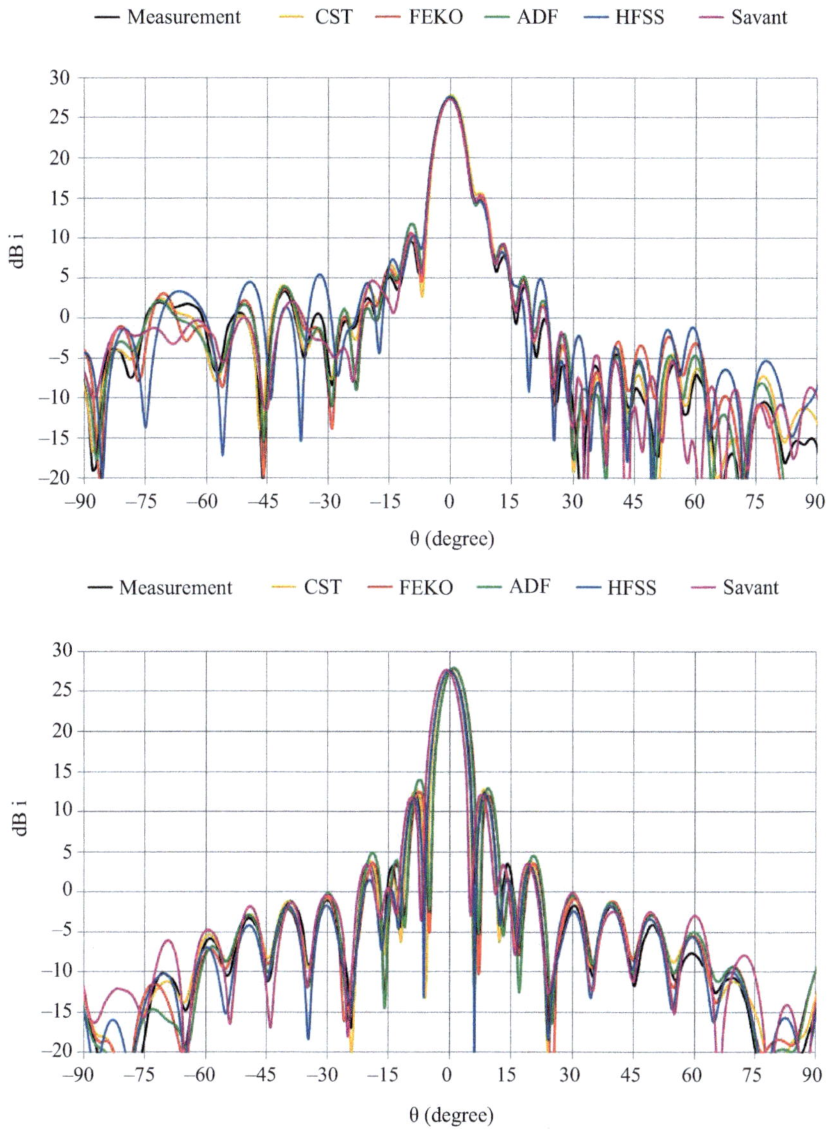

Figure 7.15 Directivity pattern for the simulation of the measured source with different numerical tools (see footnote ‡), with respect to the measurement of the reflector. E plane (top), H plane (bottom)

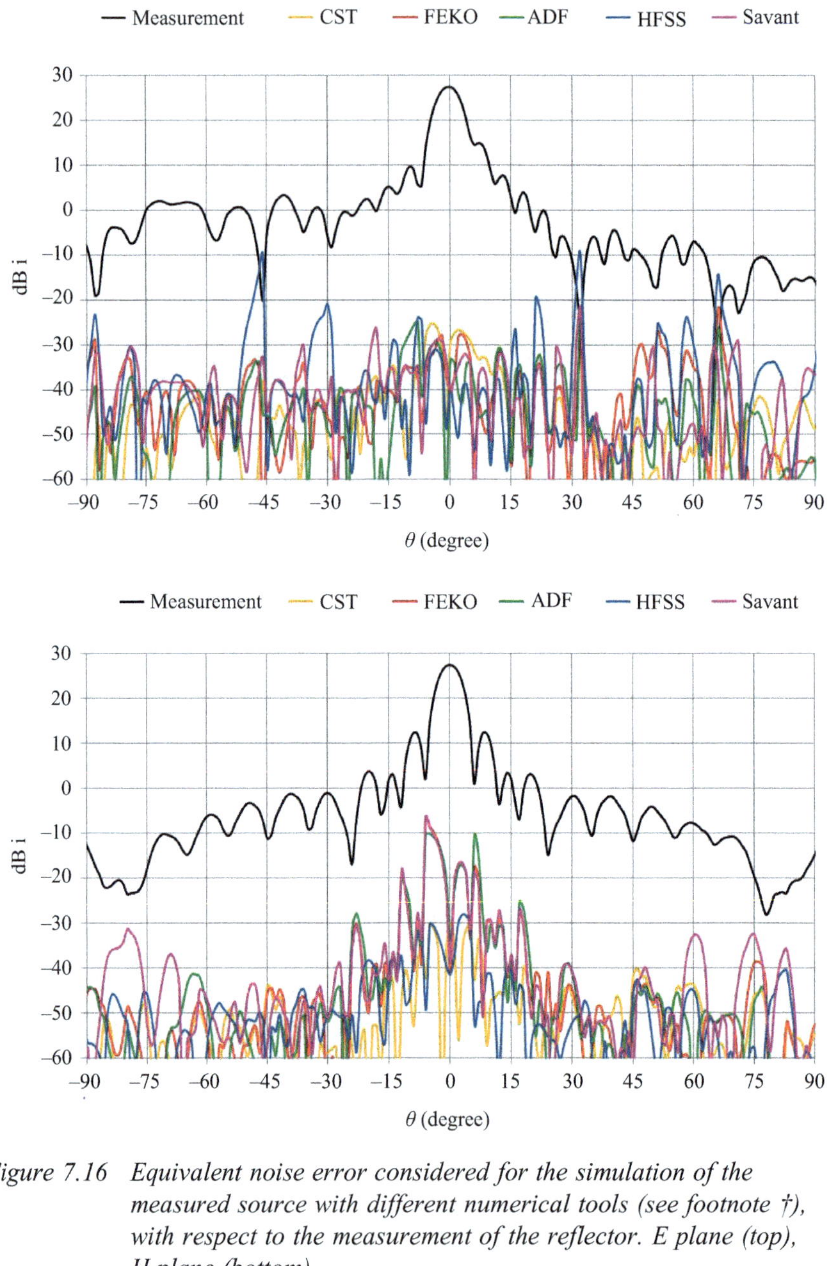

Figure 7.16 Equivalent noise error considered for the simulation of the measured source with different numerical tools (see footnote †), with respect to the measurement of the reflector. E plane (top), H plane (bottom)

Table 7.2 Equivalent noise error level, SR40-A offset reflector fed by SH4000 dual ridge horn at 8 GHz

	CST	Savant	FEKO	ADF	HFSS
$\phi = 0°$ plane (dB)	−42.3	−39.2	−39.7	−41.5	−36.1
$\phi = 90°$ plane (dB)	−48.0	−33.5	−34.4	−34.3	−46.7

Figure 7.17 Rectangular ground plane validation structure during measurement in the MVG spherical near-field antenna measurement system, SL18GHz, with the SMC2200 monocone

7.4.2 Flush-mounted antenna on a flat structure

Different applications of flush-mounted antennas in complex environments have been examined, a rectangular ground plane (flat structure) and a space-plane mock-up.

In this section, the validation of The Link using a flat structure is described. A simple ground plate of 30 × 60 cm has been chosen as initial validation scenario to minimize errors not directly related to the validation of the measurement/simulation link (see Figure 7.17). The tested antennas are the following radiators:

1. SMC2200 monocone antenna and
2. OEW antenna.

Initially, the SMC2200 antenna has been studied having polarization orthogonal to the ground plane. Later, the OEW has been investigated to verify the accuracy of the NF model of ground-plane aperture antenna.

The accuracy of the method is investigated by comparison with measurements and/or full-wave simulation of the full structure. The purpose of this investigation is to quantify the error deriving from the flat ground plane approximation employed in the measurement and processing of the measured source.

7.4.2.1 SMC2200 monocone antenna

The source antennas were mounted in a corner of the plate, positioned at a distance of 1.5λ and 2λ from the nearest edges (at the validation frequency 5.28 GHz). The validation structure during measurements in the StarLab SL18GHz spherical NF multi-probe system [10] is shown in Figure 7.17.

Preparation of near-field source

The evaluation of the electromagnetic model of the source antenna for flush-mounted applications is much more complex respect to the determination of the model of an antenna detached from the scattering structure. The proximity of scattering structure modifies the current distribution on the antenna itself.

An infinite ground plane boundary condition is a good approximation of the correct boundary conditions; however, it could be not directly obtained on a realistic measurement scenario. This condition can be emulated from measurements of the source antenna installed on a finite ground plane together with the application of measurement post-processing, as discussed in [12]. The post-processing of the measured data removes the diffractive contributions from the edges of the finite ground plane creating the wanted infinite ground plane boundary conditions. It is recommended to use a circular ground plane of at least 2λ diameter at the lowest limit of the frequency band.

In the validation examples, all antennas have been measured on a circular ground plane of diameter 7λ (at the test frequency of 5.28 GHz) in the StarLab SL18GHz spherical NF multi-probe system, as illustrated in Figure 7.18. After post-processing, to eliminate edge diffraction, the 3D electromagnetic model in the form of equivalent electric and magnetic currents associated to the source can be evaluated. It should be noted that, since an infinite ground plane condition is considered, the image of the source antenna is initially included in the EQC

Figure 7.18 Measurement of the SMC2200 monocone antenna on a limited ground plane in the MVG, SL18GHz spherical near-field multi-probe system [10]

Table 7.3 Peak directivity, SMC2200 on a flat plate at 5.28 GHz

	MEAS	CST	Savant	FEKO	ADF	HFSS	WIPL-D
Peak directivity (dB i)	6.2	5.7	5.9	5.8	6.0	5.5	5.8

Generic measured source antenna is measured on a minimum of representative ground plane.

Accurate NF representation of the source antenna is prepared by INSIGHT.

Rectangular plate: *link measurement and simulations*

Figure 7.19 Use of a measurement of an SMC2200 monocone as flushed mounted antenna on a rectangular plate [2.2–6] GHz frequency band

computation and then removed when determining the Huygens' box representation of the measured source as shown in Figure 7.19.

Results

The resulting pattern of the validation structures in Figure 7.17 has been computed using the measured SMC2200 monocone antenna as source model. The measured and simulated peak directivities at 5.28 GHz of the rectangular plate with SMC2200 monocone antenna are reported in Table 7.3. MEAS is the measured reference from MVG, and it is reported together with the simulated results (see footnote †) involved during the validation campaign. Very good agreement between measurements and simulation can be observed.

The measured reference is compared to the simulated copolar patterns from simulation using the measured source as shown in Figure 7.20. The agreement between simulation and measurements is very good considering the approximation due to the feed representation and uncertainties arise from measurement, manufacturing and simulation.

The agreement or correlation between simulations and measurements has been evaluated in terms of weighted difference (7.1) between measured and simulated field. The measured FF has been considered as reference field. An overlay of the weighted difference for each simulation tool with the measured patterns at $\phi = 0°$ (along the longest side of the plate) and at $\phi = 90°$ in the forward hemisphere is shown in Figure 7.21.

The medium values of the weighted difference have been calculated representing the correlation in a single value as shown in Table 7.4. The average correlation between simulations and measurement is ~30 dB, which is similar to what is obtainable with full-wave simulations of the antenna. This promising result

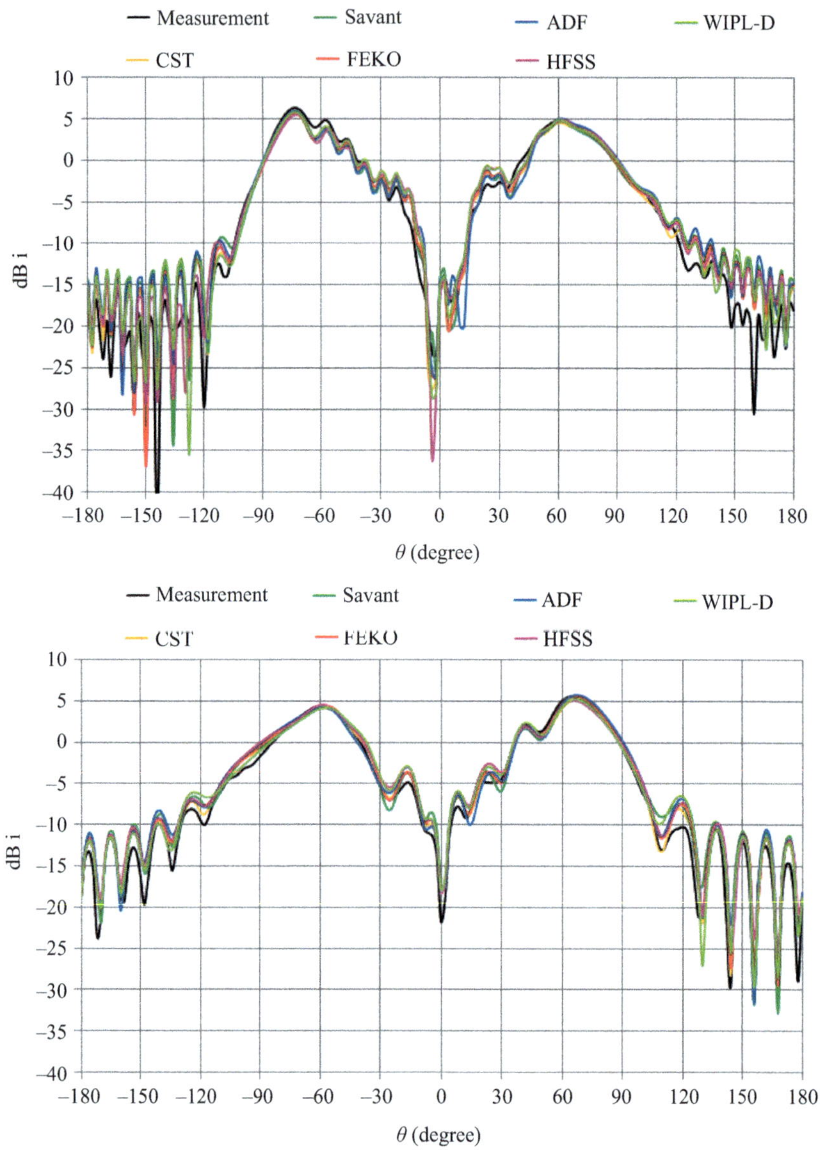

Figure 7.20 Directivity pattern for the simulation of the measured source with different numerical tools (see footnote †), with respect to the measurement of the full scenario, $\phi = 0°$ plane (top), $\phi = 90°$ plane (bottom)

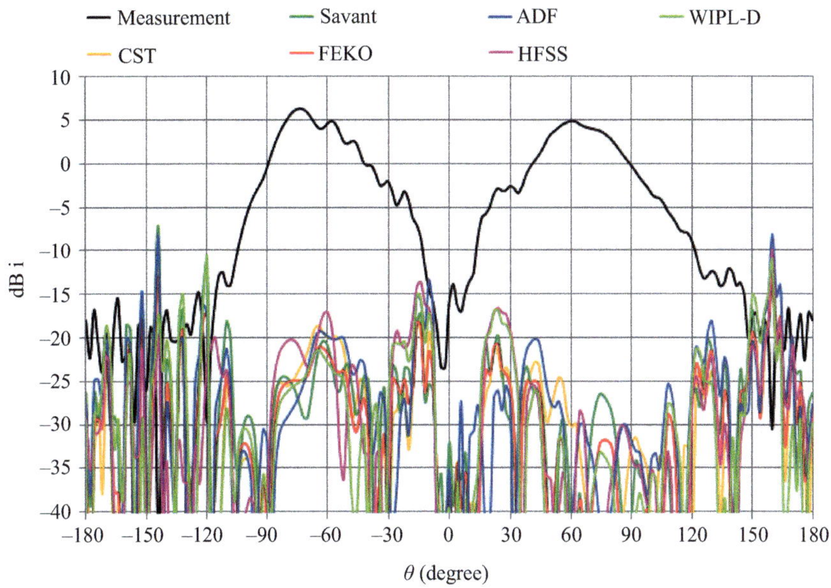

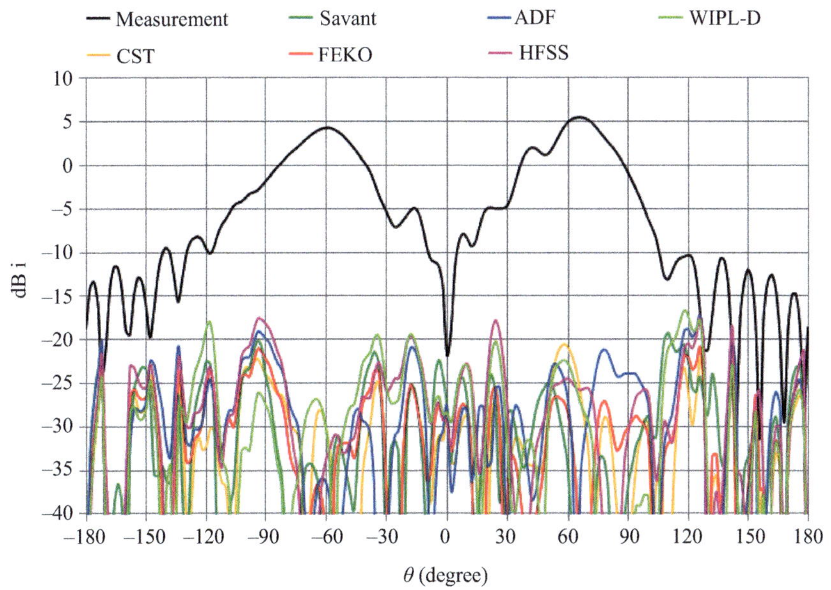

Figure 7.21 Equivalent noise error considered for the simulation of the measured source with different numerical tools (see footnote †), with respect to the measurement of the full scenario, $\phi = 0°$ plane (top), $\phi = 90°$ plane (bottom)

Table 7.4 Equivalent noise error level, SMC2200 on a flat plate at 5.28 GHz

	CST	Savant	FEKO	ADF	HFSS	WIPL-D
$\phi = 0°$ plane (dB)	−30.2	−30.1	−30.5	−28.7	−30.2	−31.0
$\phi = 90°$ plane (dB)	−33.2	−32.4	−33.4	−30.5	−30.8	−31.6

Figure 7.22 Rectangular ground plane validation structure during measurement in the MVG spherical near-field antenna measurement system, SL18GHz, with an open-ended waveguide (OEW)

confirms the accuracy of the measured source representation and the validity of the method implementing The Link between measurements and CEM tools.

7.4.2.2 Open-ended waveguide antenna

The second source antenna validated as mounted offset on the rectangular plate of 30 × 60 cm is an OEW antenna. The source antenna is located in a corner of the plate, in the same position of the monocone antenna described in Section 7.4.2.1. The validation structure during measurements in the MVG SL18GHz spherical NF multi-probe system [10] is shown in Figure 7.22. The longest side of the rectangular plate and the E-plane of the antenna are oriented along the RF x-axis (i.e. $\phi = 0°$ in spherical coordinates) in the measurement system.

Preparation of near-field source

For the OEW antenna the same working procedure (as for the monocone) is used to prepare the NF source. In this case a circular ground plane of 6λ in diameter at the test frequency of 4.38 GHz has been used. The measurement set-up and the radiation pattern are shown in Figure 7.23(a) and (b). An NF source model as Huygens' box has been created by the EQC technique. The working procedure of The Link for this test case is shown in Figure 7.19, same one as for the monocone antenna.

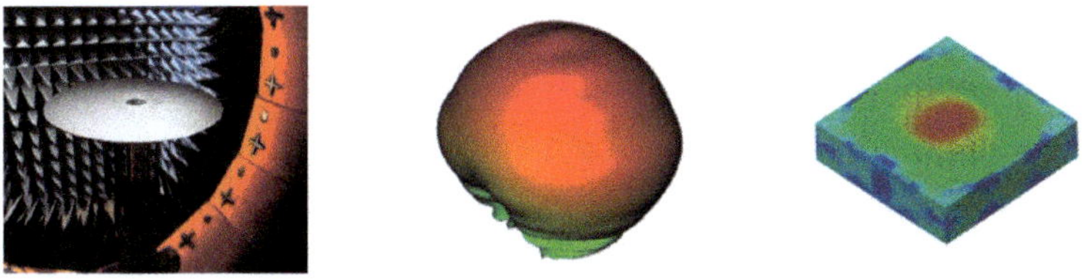

(a) (b) (c)

Figure 7.23 Measurement of the open-ended waveguide on a limited ground plane in the MVG, SL18GHz spherical near-field multi-probe system (a); measured radiation pattern at 4.38 GHz (b); EQC source/ Huygens' box (c)

Table 7.5 Peak directivity, OEW open-ended waveguide antenna at 4.38 GHz

	MEAS	CST	Savant	FEKO	HFSS	WIPL-D
Peak directivity (dB i)	8.09	8.15	8.22	8.09	8.34	8.26

Results

The resulting pattern of test case of the OEW antenna (source model) on the rectangular flat plate is reported in Figure 7.25. The measured and simulated peak directivities at 4.38 GHz of the rectangular plate with the source antenna are reported in Table 7.5. MEAS is the measured reference from MVG, while the other curves represent the simulations of the measured source on the rectangular plane by the different CEM tools [10,13–15]. Very good agreement between measurements and simulation can be observed.

The measured reference is compared to the simulated copolar patterns from simulation using the measured source. The agreement between simulation and measurements is very good as shown in Figure 7.24, considering the approximation due to the feed representation and uncertainties from measurement, manufacturing and simulation.

The agreement or correlation between simulations and measurements has been evaluated in terms of weighted difference (7.1) between measured and simulated field. The measured FF has been considered as reference field. An overlay of the weighted difference for each simulation tool with the measured patterns at $\phi = 0°$ and at $\phi = 90°$ in the forward hemisphere is shown in Figure 7.25.

The medium value of the weighted differences was calculated representing the correlation in a single value as shown in Table 7.6. The average correlation between simulations and measurement is ~30 dB, which is similar to what is obtainable with full-wave simulations of the antenna. This promising result confirms the accuracy of the measured source representation and the validity of the method implementing The Link between measurements and CEM tools.

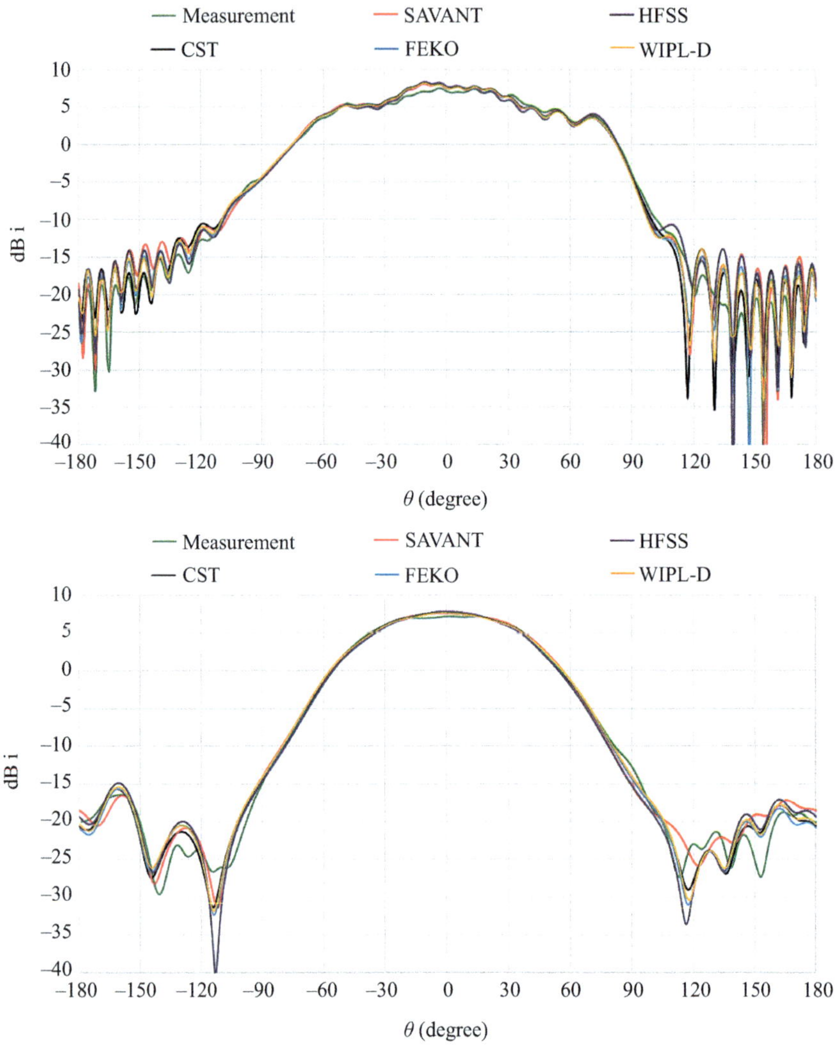

Figure 7.24 Directivity pattern for the simulation of the measured source with different numerical tools (see footnote †), with respect to the measurement of full scenario, $\phi = 0°$ plane (top), $\phi = 90°$ plane (bottom)

7.4.2.3 SM6000 monopole antenna

In order to further validate the proposed method with a more realistic and complex scenario, a flush-mounted antenna on a curved structure has investigated using a small monopole-like antenna mounted directly on the back of a space plane as shown in Figure 7.26. The purpose of this investigation is to quantify the error

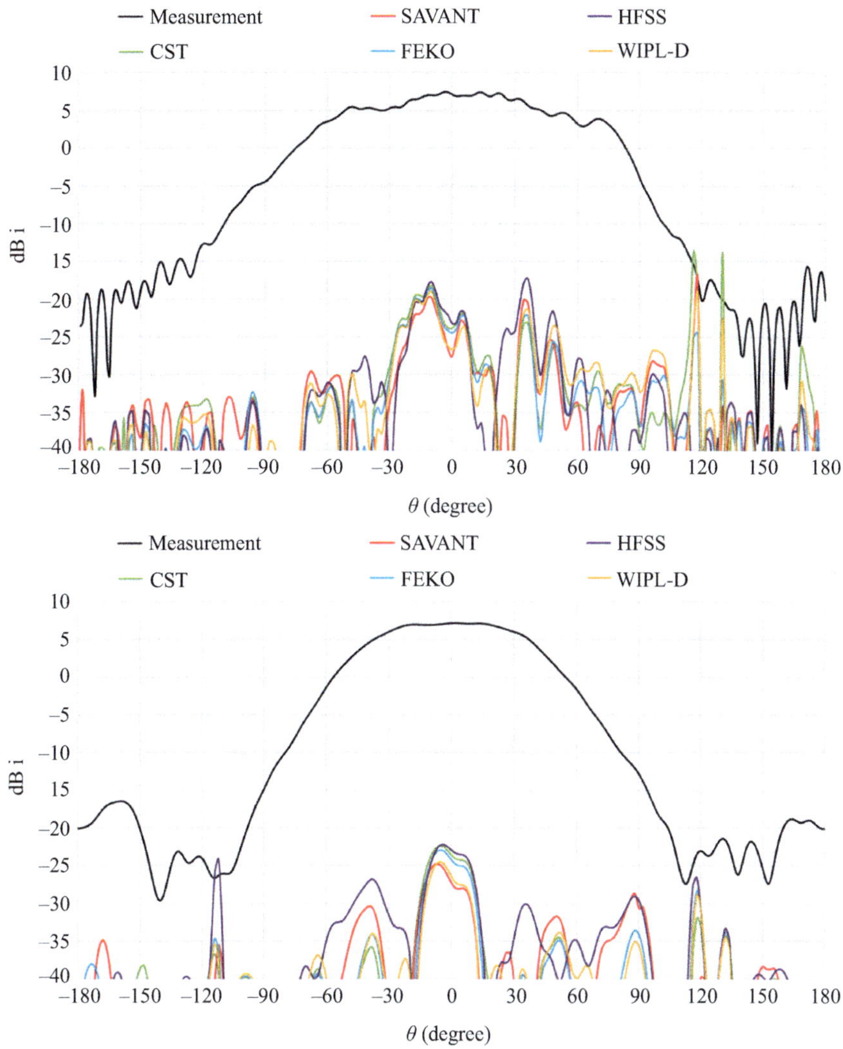

Figure 7.25 Equivalent noise error considered for the simulation of the measured source with different numerical tools (see footnote †), with respect to the full scenario, $\phi = 0°$ plane (top), $\phi = 90°$ plane (bottom)

Table 7.6 Equivalent noise error level, OEW open-ended waveguide antenna at 4.38 GHz

	CST	Savant	FEKO	HFSS	WIPL-D
$\phi = 0°$ plane (dB)	−32.3	−33.0	−32.9	−32.3	−31.8
$\phi = 90°$ plane (dB)	−39.4	−38.9	−38.5	−35.2	−38.9

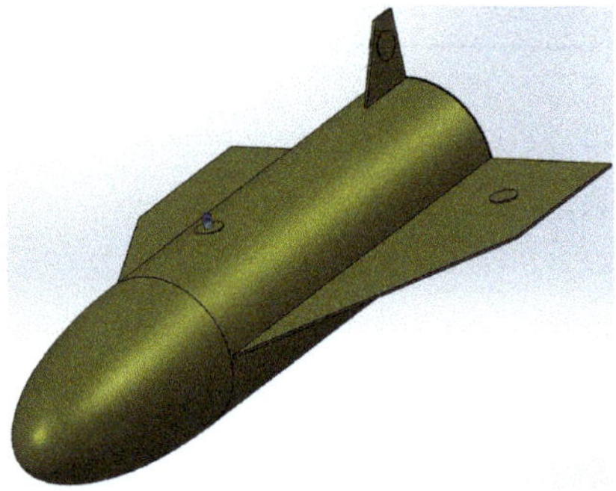

Figure 7.26 Space plane CAD model with the SM6000 monopole antenna

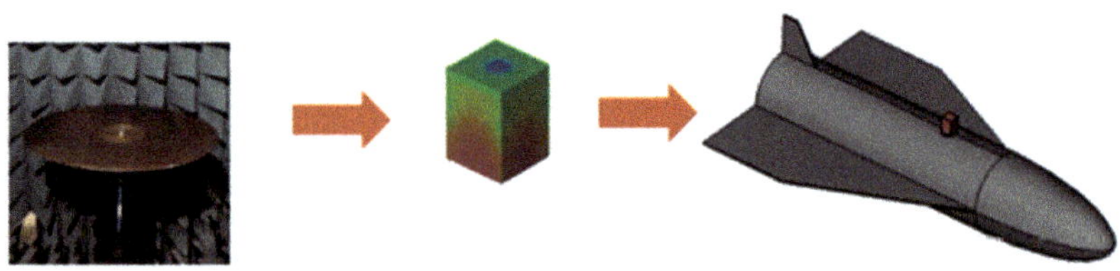

Figure 7.27 Use of a measurement of an SM6000 monopole as flushed mounted antenna on space plane mock-up at 6 GHz

derived from the flat ground plane approximation used in the measurement and processing of the measured source in a curved scenario. The source antenna is an SM6000 monopole working at 6 GHz.

Preparation of near-field source

To isolate this effect, the measured source has been preliminary derived using the procedure described in 'Preparation of near-field source' section. The measured source simulation has been processed using the EQC technique for importing in the numerical tool. The working procedure is shown in Figure 7.27.

Results

For this test case, a measurement of the full scenario was not available and comparisons were performed only between simulated results by the CEM tools (www.cst.com, CST STUDIO SUITE^{TM}, CST AG; www.feko.info, Altair Engineering GmbH; www.idscorporation.com/space, ADF, IDS). Very good agreement can be observed as shown in Figure 7.28.

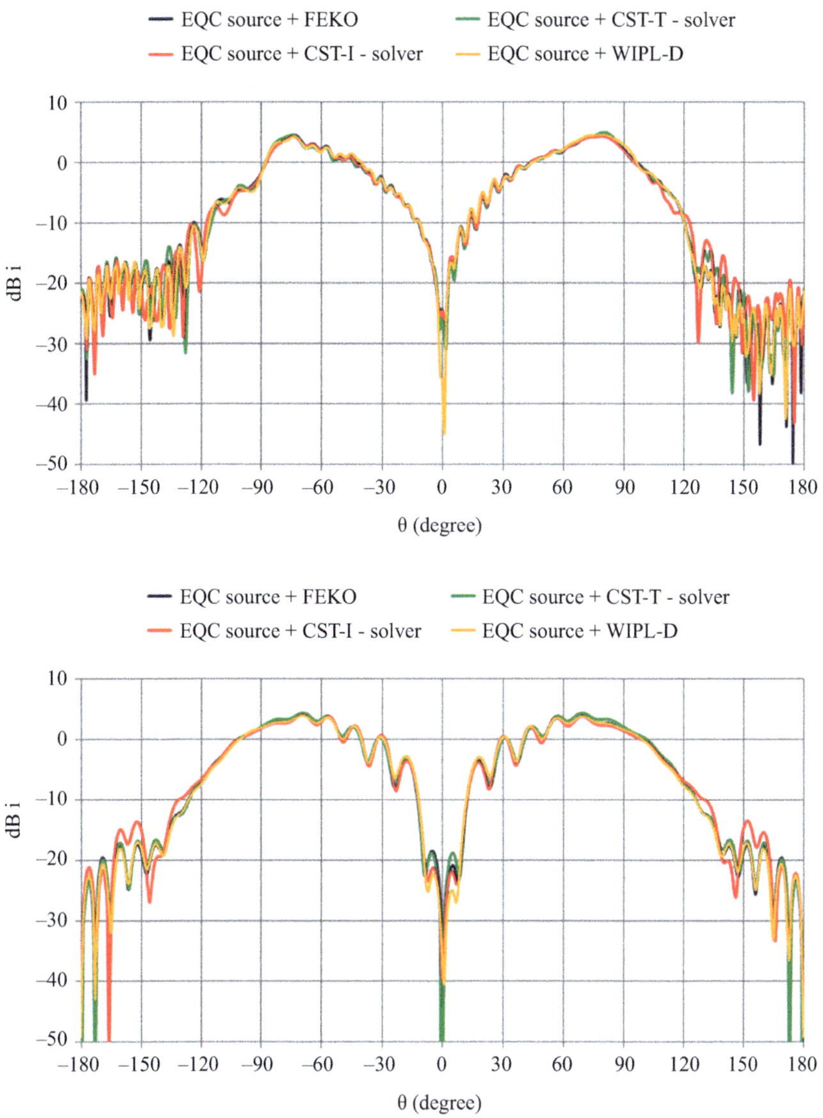

Figure 7.28 Directivity pattern of SM6000 monopole antenna on rectangular plate at 7.5 GHz, $\phi = 0°$ plane (top) and $\phi = 90°$ plane (bottom). Measurement and simulation using measured source: CST (www.cst.com, CST STUDIO SUITE^{TM}, CST AG), FEKO (www.feko.info, Altair Engineering GmbH), WIPL-D (www.wipl-d.com, WIPL-D d.o.o.). CST data have been simulated using both transient solver (T-solver) and the integral equation solver (I-solver)

7.5 Practical application of The Link

A practical application of The Link is described in this section. The scenario consists of an L-Band Global Navigation Satellite System (GNSS) antenna installed on a Sentinel satellite developed by RUAG.

7.5.1 Test object and complex scenario

The GNSS antenna, visible in Figure 7.29(a), is characterized by a pencil beam type pattern, with ±70° of coverage working at frequencies between 1,227 and 1,575 MHz. The antenna is a patch excited cup (PEC) with two stacked patches placed in a short cylindrical cup and four-points feed with capacitive coupling of the bottom patch and an isolated feed network (phase quadrature). The design work has been concentrated on the antenna radiation in the back direction and the possibility to, at the same time, keep a good coverage at low elevation angles.

The complex scenario is represented by the antenna installed on a Sentinel (https://sentinel.esa.int/web/sentinel/home) satellite mock-up, shown in Figure 7.29(b).

7.5.2 Validation workflow

The validation workflow, shown in Figure 7.30, consists of four steps: (1) FF radiating stand-alone measurements; (2) antenna characterization with EQC; (3) simulation in CEM tool; and (4) comparison with complex scenario measurements.

7.5.2.1 FF radiating stand-alone measurements

The GNSS antenna has been measured in stand-alone configuration, at 1,227 and 1,575 MHz, as visible in Figure 7.31.

7.5.2.2 Antenna characterization with EQC

EQC field representation, at 1,227 and 1,575 MHz, is calculated from the measured radiation pattern of the AUT by the inverse source method (INSIGHT webpage: http://www.mvg-world.com/en/products/field_product_family/antenna-measurement-2/insight) [1–9], implemented in INSIGHT, following the workflow depicted in Figures 7.32 and 7.33.

(a) (b)

Figure 7.29 GNSS antenna (a) and Sentinel satellite mock-up (b). RUAG courtesy

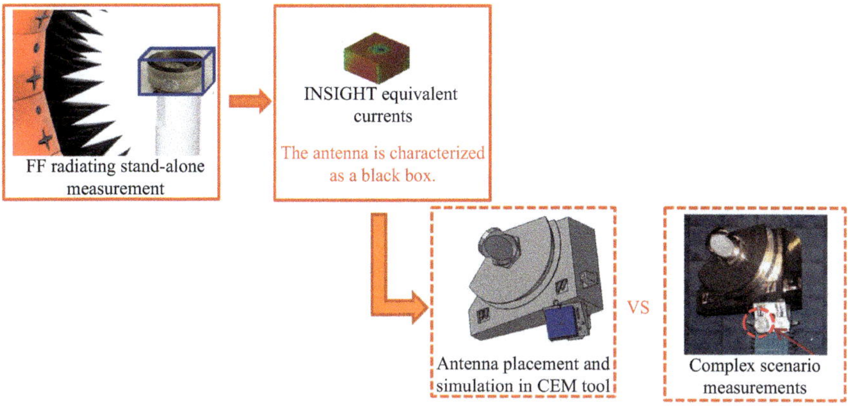

Figure 7.30 Validation workflow

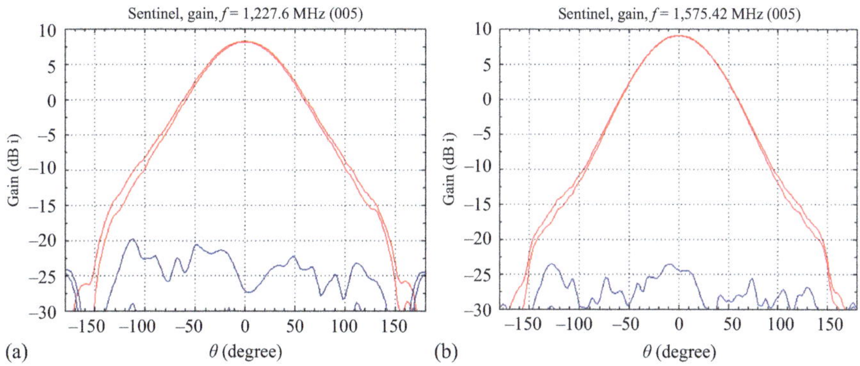

Figure 7.31 Gain pattern of the GNSS antenna at 1,227 MHz (a) and 1,575 MHz (b)

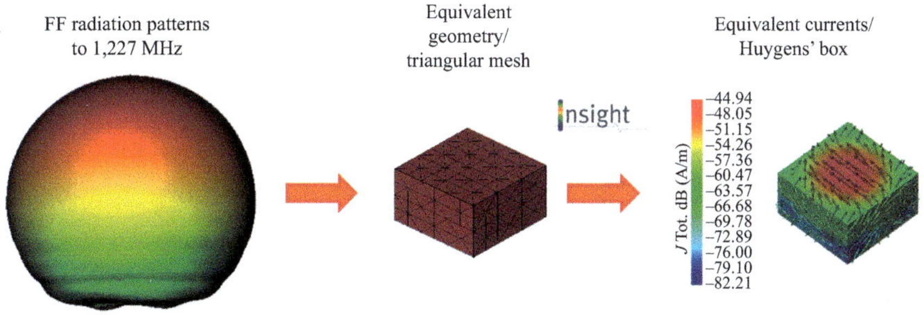

Figure 7.32 Antenna characterization in terms of EQC using INSIGHT at 1,227 MHz

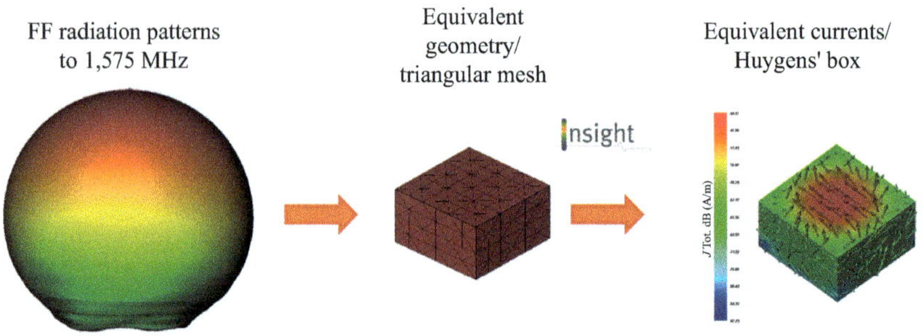

Figure 7.33 Antenna characterization in terms of EQC using INSIGHT at 1,575 MHz

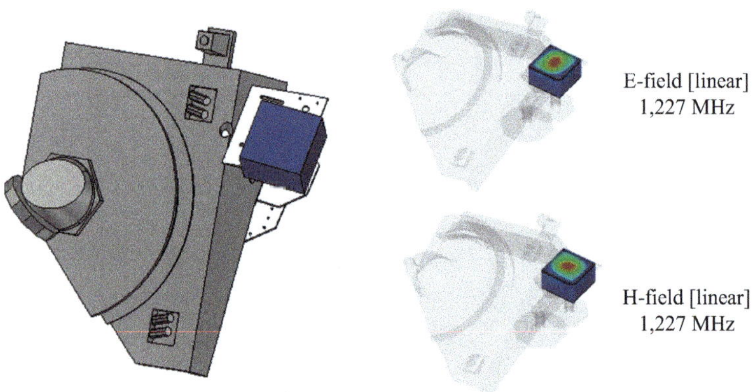

Figure 7.34 Placement and simulation in CEM tool at 1,227 MHz

7.5.2.3 Simulation in CEM tools

The EQC is computed in the form of Huygens' box enclosing the radiator to exchange the antenna source data in the numerical software (www.cst.com, CST STUDIO SUITE^{TM}, CST AG). The computed Huygens' box (measured data) representing the PEC antenna is positioned on the spacecraft model, as visible in Figures 7.34 and 7.35.

7.5.3 Results

The comparisons between complex scenario measurements and the field computed with CST, after importing the INSIGHT Huygens' box, are shown in the following plots.

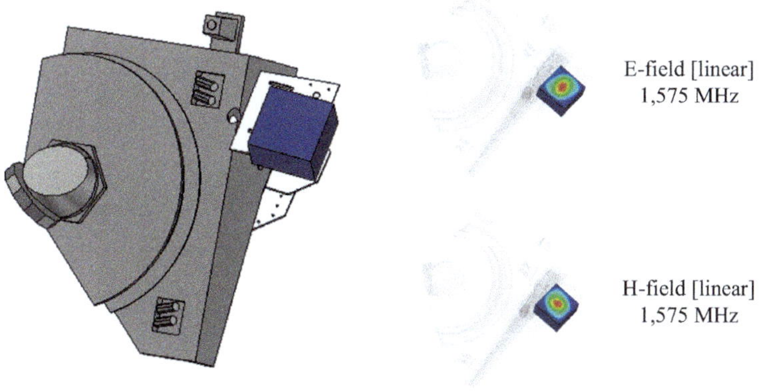

Figure 7.35 Placement and simulation in CEM tool at 1,575 MHz

The copolar (CO) and crosspolar (CX) components at 1,227 and 1,575 MHz, $\phi = 0°$ and $\phi = 90°$, of measurements and of CST simulation based on measured NF source (INSIGHT Huygens' box) are shown, together with the weighted difference on the copolar component (7.1) in Figures 7.36 and 7.37. The visual agreement is confirmed by the computation of the equivalent noise level reported in Table 7.7. The visual agreement between the patterns is confirmed by the equivalent noise level, which, in some cases, is around −39 dB, confirming the accuracy achieved in recent validation activities.

7.6 Conclusions

Post-processing of antenna measurements and numerical simulations has many common features. The outcome from most NF to FF transformations is a complete mathematical description of the measured reality. It is therefore an obvious extension to use this information to solve larger complex problems such as antenna placement by linking measurement and numerical simulation.

Typical antenna problems in industrial scenarios such as automotive, aerospace, telecommunication and defence are becoming increasingly complex systems. In these cases, a full numerical description, in terms of CAD and material description of the detailed scenario, may not be available as it may be composed of components from different vendors or otherwise different origins. The problem to be characterized may also be too large or simply not available for full measurement, hence the necessity for using numerical means for validation.

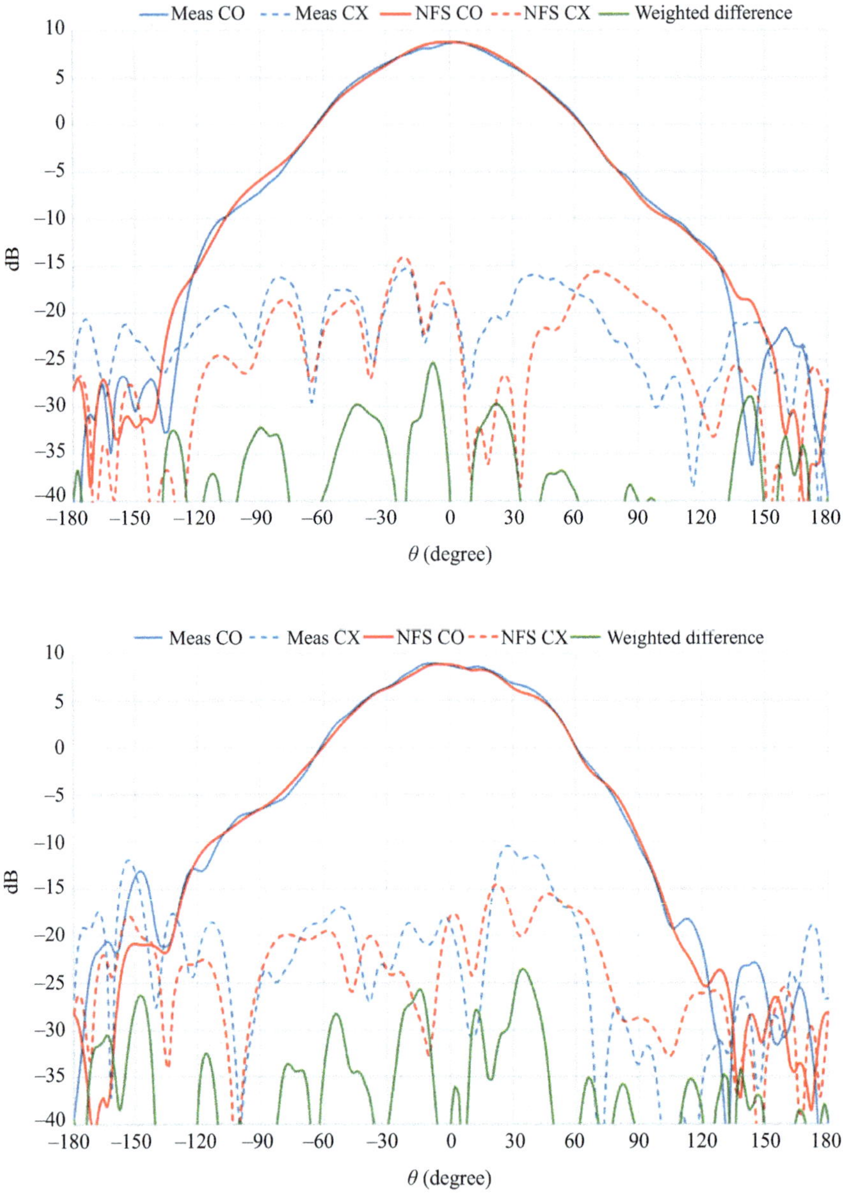

Figure 7.36 Copolar and crosspolar components, $\phi = 0°$ plane (top), $\phi = 90°$ plane (bottom), of measurements (MEAS) and of CST simulation based on INSIGHT NF measured Source (NFS) at 1,227 MHz

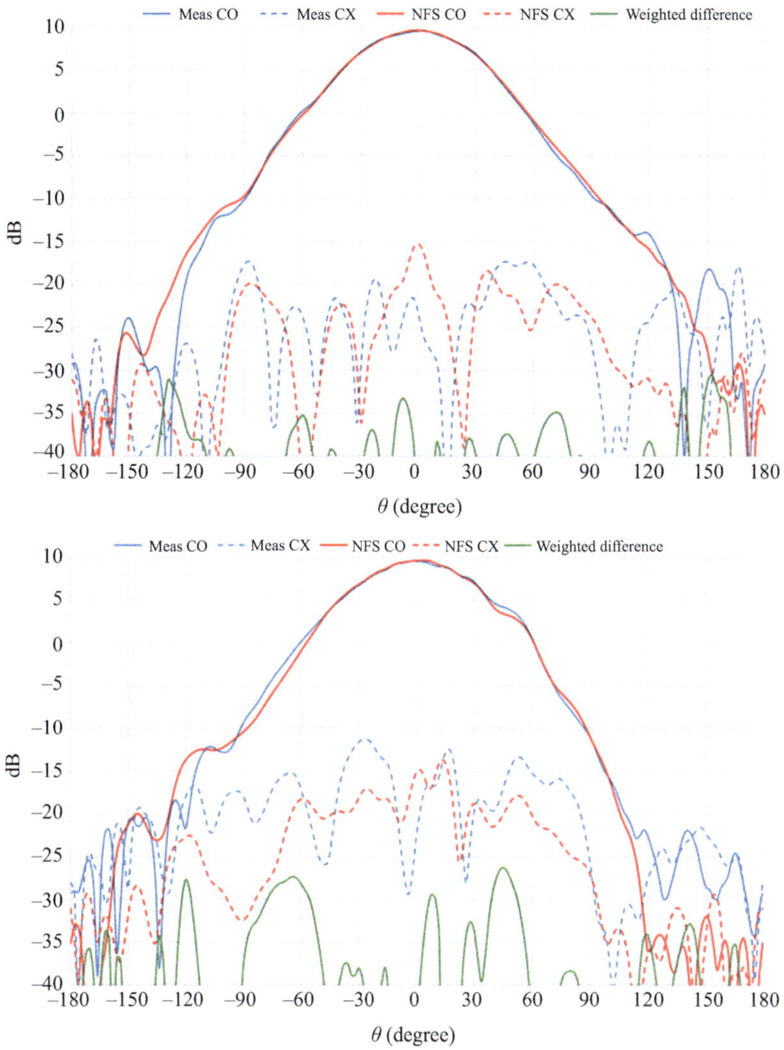

Figure 7.37 Copolar and crosspolar components, $\phi = 0°$ plane (top), $\phi = 90°$ plane (bottom), of measurements (MEAS) and of CST simulation based on INSIGHT NF measured Source (NFS) at 1,575 MHz

Table 7.7 Equivalent noise level

Freq. (MHz)	ϕ	Component	Equivalent noise level (dB)
1,227	0°	CO	−35.33
		CX	−34.96
	90°	CO	−33.18
		CX	−33.95
1,575	0°	CO	−39.10
		CX	−37.78
	90°	CO	−34.93
		CX	−33.96

'The Link' described here is ideal in such cases. It is an approximate technique, just like most numerical simulations, and offers a means to use measured data in a numerical simulation of a complex scenario with a good and, first of all, known accuracy. This enables the users of this technique to evaluate the solution within a known confidence level.

References

[1] L. J. Foged, L. Scialacqua, F. Saccardi, *et al.*, "Bringing Numerical Simulation and Antenna Measurements Together", IEEE Antennas and Propagation Society International Symposium, July 6–11, 2014.

[2] L. J. Foged, L. Scialacqua, F. Saccardi, and F. Mioc, "Measurements as Enhancement of Numerical Simulation For Challenging Antennas", EUCAP 2015, April 12–17, 2015.

[3] L. J. Foged, L. Scialacqua, F. Saccardi, and F. Mioc, G. Vecchi, and J. L. Araque Quijano, "Antenna Placement Based on Accurate Measured Source Representation and Numerical Tools", IEEE Antennas and Propagation Society International Symposium, July 19–24, 2015.

[4] L. J. Foged, L. Scialacqua, F. Saccardi, *et al.*, "Using Measured Fields as Field Sources in Computational EMC", 37rd Annual Symposium of the Antenna Measurement Techniques Association, AMTA, October 2015, Long Beach, CA, USA.

[5] L. J. Foged, L. Scialacqua, F. Saccardi, and F. Mioc, "Measured Antenna Representation of Flush Mounted Antennas for Computational Electromagnetic Solvers", EUCAP 2016, April 10–15, 2016, Davos, Switzerland.

[6] J. L. A. Quijano, and G. Vecchi, "Field and Source Equivalence in Source Reconstruction on 3D Surfaces", Progress In Electromagnetics Research (PIER), 2010, Vol. 103, pages 67–100, ISSN: 1559–8985.

[7] J. L. A. Quijano, and G. Vecchi, "Near- and Very Near-Field Accuracy in 3-D Source Reconstruction", IEEE Antennas and Wireless Propagation Letters, 2010, Vol. 9, pages 634–637, ISSN: 1536–1225.

[8] J. L. A. Quijano, and G. Vecchi, "Improved-Accuracy Source Reconstruction on Arbitrary 3-D Surfaces", IEEE Antennas and Wireless Propagation Letters, 2009, Vol. 8, pages 1046–1049, ISSN: 1536–1225.

[9] C. A. Balanis, Advanced Engineering Electromagnetics, New York, NY: John Wiley & Sons, Inc., 1989, chapter 7.

[10] L. J. Foged, and A. Scannavini, "Efficient Testing of Wireless Devices from 800 MHz to 18 GHz", Radio Engineering Magazine, 2009, Vol. 18, No. 4, pages 460–466.

[11] J. E. Hansen (ed.), Spherical Near-Field Antenna Measurements, London, United Kingdom: Peter Peregrinus Ltd., 1988, on behalf of IEE.

[12] L. J. Foged, F. Mioc, B. Bencivenga, M. Sabbadini, and E. Di Giampaolo, "Infinite Ground Plane Antenna Characterization from Limited Ground

Plane Measurements", Antennas and Propagation Society International Symposium APSURSI, July 11–17, 2010.

[13] J. L. Araque, L. Scialacqua, J. Zackrisson, L. J. Foged, M. Sabbadini, and G. Vecchi, "Suppression of Undesired Radiated Fields Based on Equivalent Currents Reconstruction from Measured Data", IEEE Antennas and Wireless Propagation Letters, 2011, Vol. 10, pages 314–317.

[14] J. L. A. Quijano, G. Vecchi, L. Li, *et al.*, "3D Spatial Filtering Applications in Spherical Near Field Antenna Measurements", AMTA 2010 Symposium, October, Atlanta, GA, USA.

[15] L. Scialacqua, F. Saccardi, L. J. Foged, J. L. Araque Quijano, G. Vecchi, and M. Sabbadini, "Practical Application of the Equivalent Source Method as an Antenna Diagnostics Tool", AMTA Symposium, October 2011, Englewood, CO, USA.

Chapter 8
Future trends in antenna measurements

Manuel Sierra Castañer[1] *and Lars J. Foged*[2]

This book summarizes new topics related to the antenna measurement improvement during the last few years. The authors have emphasized aspects considered more relevant: post-processing techniques using information about the antenna under test (AUT) in specific domains (electromagnetic sources, spherical or cylindrical modes, plane wave spectrum (PWS), etc.), combination of measurements and simulations and phase reconstruction from amplitude-only measurements. There is no doubt that other emerging techniques will become important in the future, in particular everything related to non-uniform gridding or techniques for time reduction during the acquisition process, but we have chosen to focalize on what we consider mature techniques at the moment of the writing of this book.

This last chapter of the book gives a summary of the different techniques from the different chapters of the book. Section 8.2 deals with the characteristics of the new antenna systems, basically digital systems applied for antenna reconfigurability, antenna beamforming, digital RADAR, massive MIMO, etc. Section 8.3 analyses the required electromagnetic parameters to be able to characterize these new antenna systems, and Section 8.4 summarizes new antenna measurement systems for those antennas and parameters. There is no doubt, that the algorithms and techniques exposed along this book will be necessary for the characterization of the future antenna systems; therefore, this last section tries to merge all these advances with the future requirements. Section 8.5 gives an outlook to the future in antenna measurements and Section 8.6 concludes the book.

8.1 Summary of the analysed post-processing techniques

Source reconstruction techniques are right now an important tool for understanding the performance of an electromagnetic device. The knowledge of these sources becomes very useful to discover the wrong performance of the device under test (DUT) but also some problems with the antenna measurement system. The first application was used since some years ago, while the second application has been

[1]Information Processing and Telecommunications Center, ETSI Telecomunicación, Universidad Politécnica de Madrid, Madrid, Spain
[2]Microwave Vision Italy (MVI), Pomezia, Italy

used in the last years. In particular, in Chapters 3 and 4 the knowledge of the sources allows the reduction of reflections or echoes in antenna measurement systems, the detection and cancellation of leakage, the reduction of noise or the extrapolation of the antenna radiation pattern, among others. Examples of these applications are shown extensively in those chapters, emphasizing not only the advantages, but also the limitations of those techniques. In particular two different techniques were employed for this application: Chapter 2 shows the theory for the calculation of equivalent currents on a surface enclosing the AUT. The method used for this rigorous calculation is the integral equation, based on an inverse method of moment. The main advantage of this method is the exactitude, and probably the main drawback is the required time and memory for the computation of the sources for electrically large surfaces. This method allows the reconstruction from all the practical antenna measurement systems (far-field (FF), spherical, cylindrical and planar near field (NF)). Of course, as much is the information, better is the accuracy of the sources. The examples shown in Chapter 3 are obtained using the commercial software Insight®, from Microwave Vision Group, and based on the theory presented in Chapter 2.

Chapter 4 shows another method based on holographic techniques for the calculation of the sources. This method is very simple, since it is based in Fourier transform for the calculation of the PWS, and its propagation through the z axis. The application of this method is direct for FF and PNF acquisitions, since there is a direct relation between the FF or PNF and the PWS. On the other hand, for spherical NF acquisitions, there is a relation between the spherical mode expansion and PWS, as it is explained in [1] by Capellin. Also, PWS can be calculated directly after the NF to FF (NFFF) transformation, in the three canonical acquisition grids: spherical, cylindrical and planar. The main advantages of this method are the mathematical simplicity, the calculation time and the required memory for the reconstruction of the electromagnetic sources. However, the main drawbacks come from the fact of the validity of the method for only half sphere (since the currents are reconstructed in a plane) and the appropriateness of the method for planar antennas, for the same reason. Another limitation is the reduction of the resolution of the method to half wavelength (in the best case), since the visible range of the PWS only gives this resolution. In order to improve this resolution, this method can be combined with iterative techniques, using some geometrical information of AUT, to obtain information in the invisible range. This combination increases the processing time and required memory but increases the accuracy of the method. In any case, Chapter 4 shows some examples for the reduction of echoes, leakage, noise or extrapolation of radiation pattern in some antennas using this method.

Chapter 5 studies the improvements in antenna measurements using the information of the AUT in spherical or cylindrical modes domain. Since the electromagnetic information of the radiated field by the AUT is confined in a finite number of modes, and this number depends on the dimensions of the AUT and geometry of the antenna measurement system, the transformation of the antenna radiated field to the spherical or cylindrical modes gives information of spurious signals that can be filtered out. Chapter 5 shows some examples of how this information can be employed in the improvement of the results of antenna measurements. In particular, MV-Echo®, from Microwave Vision Group, uses the decomposition of the acquired field in spherical modes to filter out spurious signals and other examples with cylindrical modes are also explained in that chapter. This technique is in fact very similar to the previous one: in source reconstruction, the domain where the filtered is applied is a spatial domain in the geometry of the AUT (and can be considered as a decomposition of the radiated filed in electromagnetic sources), while in these cases, the domains are related to the spherical or cylindrical modes decomposition of the radiated field. Since there is a relation between AUT geometry and modes, another filtering can be applied, reducing spurious signals.

Other techniques are based on a time-domain decomposition of the electromagnetic field. Frequency domain and time domain are two reciprocal domains of each other and they may be inter-transformed by using the fast Fourier transform or FFT. In antenna measurements, different techniques have been employed based on FFT or more complex as the Matrix Pencil. The filtering can be applied in time domain, getting again an improvement in the measurement results. We decided not to include such time gating methods in this book, since they have been widely studied elsewhere.

Chapter 6 explains the basis of phase recovery in amplitude-only antenna measurement. These techniques are becoming relevant for millimetre and sub-millimetre wave measurements, where the phase is difficult to obtain using standard methods. They are also necessary for special measurements where the reference signal is difficult to obtain, as in active antenna measurements. This chapter focused on two specific techniques, both derived from optics: in the first technique, an interferometry between the unknown radiated field and a known radiated field is used; and, under certain conditions, it is possible to retrieve directly the radiated field with a negligible error by means of simple operations in the spatial spectral domain (k-space). In the second technique, the phase recovery requires measuring the amplitude of the radiated field on at least two independent surfaces, and a post-processing consisting of an iterative algorithm is performed to find a phase distribution so that the propagation between planes matches the measured amplitudes. Also, an improvement of the convergence of the iterative schemes is made by means of the use of the sources reconstruction method and the use of optimization algorithms.

Chapter 7 establishes the basis of the link between simulation and measurements, including some application examples. The use of these techniques allows the calculation of the radiation parameters for very large antennas, or for antennas mounted on complex structures. Basically, the problem is divided into parts, some elements of the AUT are measured, and then, these components are merged with the more complex structure for simulating the whole system, as airplanes, satellites, etc. The improvement in the capability of the computers allows accurate calculations of the electromagnetic fields generated by these complex structures.

8.2 From passive to digital antenna technologies

Which are the antenna characteristics for 5G, active antennas, smart antennas, MIMO systems, RADAR systems, etc.? Until recently, the antenna and the radio-frequency (RF) system were often two separate systems. In most cases, you could find an RF connector where you could connect directly to your antenna, even, if these antennas could have some active elements such as amplifiers or phase shifters. With the appearance of terminal antennas for mobile systems, the antenna became difficult to separate from the rest of the RF system, and the over-the-air (OTA) antenna measurement techniques appeared.

Let us summarize the antenna evolution with Figure 8.1. Passive antennas are very easy to understand. A coaxial connector separates the antenna to the RF transceiver through a coaxial cable. An amplifier is located close to the antenna or in the transceiver room, and this transceiver ended with the analogue-to-digital converter. The different subsystems of the complete system are clearly defined. One of the main drawbacks of passive antennas is the feeding circuit. In the case of array antennas, the feeding system may be lossy, difficult to design and expensive. Of course, reflector antennas avoid the use of these feeding networks, but in this case, they are not planar structures. Another problem is that the amplification in transmission is placed before the antenna losses, requiring therefore high-power amplifiers. Also, on the reception side, low-noise amplifiers are placed after the antenna feeding network, introducing extra noise in the receiver.

The first step of the antenna evolution is the active antennas. In active antennas, each one of the antenna elements (or a sub-array of elements) can have an amplifier or a phase shifter. In this sense, the transmission power amplification is

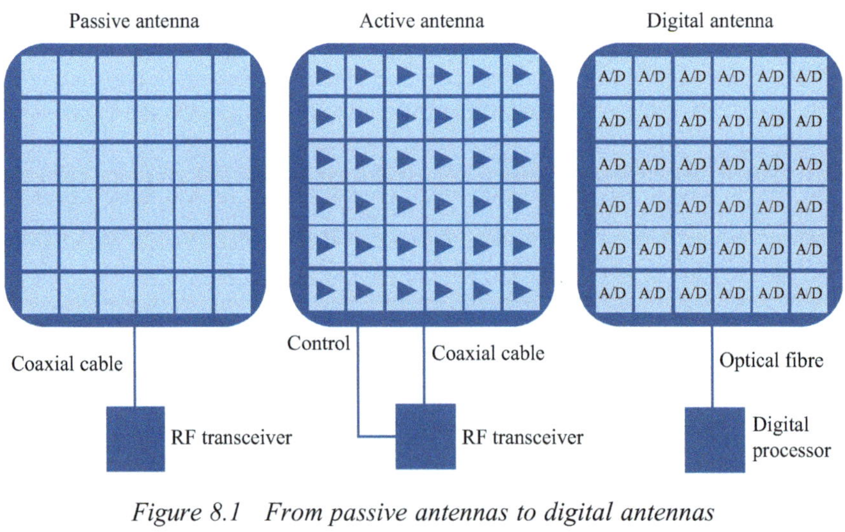

Figure 8.1 From passive antennas to digital antennas

distributed among different amplifiers, and the antenna noise is reduced on the reception side. However, there are two main problems: first the need of complex distribution network is not avoided. On the other hand, a control network has to be integrated in the antenna to set the amplifiers and phase-shifters. This kind of antenna design opens the possibility of radiation pattern reconfigurability, defining not only one radiation pattern but a set of different patterns to be configured depending on the antenna set-up. This is usually employed, for instance in RADAR systems, where the beam can explore in one or two directions.

However, the big advance comes in the last step, which we call digital antenna. Even in the literature, this kind of antennas are also called active antennas, we prefer to call them digital antennas, since they are conceptually different to the previous ones. The drastic cost reduction of analogue-to-digital converters and RF components allows the use of a dedicated RF transceiver per antenna array element. Of course, this is not yet rigorously true for millimetre or sub-millimetre wavelengths frequencies, where the costs are still high, but technology is changing rapidly. Now, instead of complex feeding networks, the end of the antenna element is a baseband cable that has to be connected to the digital processor through a bus. This digital processor is now in charge of all the array antenna performance: radiation pattern reconfigurability, beamforming or beam exploration, MIMO performance, massive MIMO (if there are enough number of antenna elements) or whatever the reader can imagine. Even, having several digital processors, different antenna pattern configurations can be simultaneously exploited.

In this sense, new antennas become completely different concept respect passive antennas. Not only in the sense of performance, where there are new functionalities now, but also in the requirement of integration of these digital antennas with the rest of the system components. The next section explores which electromagnetic parameters are now important and how to measure them.

8.3 Performance parameters for digital antennas

The performance parameters for antennas are well defined when the antenna is an isolated component. In IEEE standards, parameters such as antenna gain, antenna directivity, radiation pattern, axial ratio, polarization and VSWR are thoroughly described [2]. As most antennas are made from passive and linear materials, the determined performances are valid for the antenna both in transmission and reception, in accordance with the reciprocity theorem. Performances of isolated antennas can be measured using standard techniques in NF or FF measurement configurations [3–6].

Once antennas are integrated and inseparable from an RF system, the testing can no longer be performed at antenna level using traditional antenna performance parameters. Testing must be performed as a complete system, and the tested parameters are of system performance. Examples of such systems are modern wireless devices that perform data processing internally and do not allow users to access to the outputs/inputs of the antenna directly. In the early 2000s, the proliferation of wireless communication and the increased production of handsets and base-stations prompted the cellular communication industry to develop and standardize methods for OTA-radiated performance testing of integrated devices in a controlled laboratory environment, like a shielded anechoic chamber. Preceding that moment, the performances of wireless handsets and base-stations were mainly evaluated by 'in situ' methods. Performances were determined based on dropped calls and other quality factors on a standardized route on a wireless network, but also subject to changes in weather and traffic conditions, so results were rarely easily comparable.

In moving testing from the field to a controlled environment such as a laboratory, the testing was concentrated on the average endpoint performance as a component of a communication link budget. OTA testing of wireless device involves the determination of transmitter and receiver characteristics such as power, sensitivity and in some cases data throughput. The testing often comprises the effect of antenna, body of the wireless device and other objects often found in the close proximity of the device during use that can be represented by standardized phantoms: hand, head, torso, etc. The phantoms are representations of average size body parts, with average electrical parameters enabling reproducible and standardized testing of devices.

By measuring two spatial power quantities, in spherical coordinates, such as effective isotropic radiated power (EIRP) and effective isotropic sensitivity (EIS), both transmitter and receiver properties can be characterized, respectively. OTA testing of wireless devices is only slightly more complicated than traditional antenna testing. Such testing requires an external device, capable of setting up a communication and control link with the devices under test is needed. In the case of wireless device testing such as headsets, this is often a base-station emulator. In the case of base-station testing, it is often a control device on the fibre-optical link with the base station. The external device regulates the communication and maintains the test mode of the device. The definitions of EIRP and EIS are based on radiated power with respect to an isotropic antenna – just like isotropic gain. Similar to antenna gain, these parameters can be measured directly in a gain calibrated, OTA test set-up. The gain calibration can be performed following standard measurement techniques [3–6] where the gain by substitution method is the most common one to calibrate the OTA system. A reference antenna with a known gain is measured, providing a correction factor of the system. By the virtue of the gain calibration, the input/output isotropic power of the wireless device, thus EIRP and EIS performance parameters, can be measured directly at the input/output port of the test antenna or probe.

The integrated power pattern metrics of total radiated power (TRP) and total isotropic sensitivity (TIS) sometimes also are referred to as total radiated sensitivity or total receiver sensitivity (TRS), are chosen as the principal figures of merit in the 3GPP and CTIA standard [7–10] since they also represent the average directional line-of-sight performance of the transmitter and receiver, respectively. Specifically, TRP is a measure of the radiated power from a wireless device. It is defined as the integral of the power radiates by the wireless device in all directions over the entire sphere for both polarizations:

$$EIRP(\theta,\varphi) = TRP \times Gain_{DUT}(\theta,\varphi) \tag{8.1}$$

$$TRP = \frac{1}{4\pi}\iint EIRP(\theta,\varphi)\sin(\theta)d\theta d\varphi \tag{8.2}$$

If the EIRP is measured in discrete samples on a regular angular grid around the DUT with N samples in θ and M samples in φ for a total of $N \times M$ measurement points, the TRP can be approximated by a summation:

$$TRP \approx \frac{\pi}{2NM}\sum_{n=0}^{N-1}\sum_{m=0}^{M-1} EIRP(\theta_n,\varphi_m)\sin(\theta_n) \tag{8.3}$$

TIS or TRS refers to the average receiver sensitivity across the entire radiated sphere in order to achieve a specific error rate threshold for both polarizations, which represents the overall receiving capability of the DUT. If the EIS is measured in discrete samples on a regular angular grid around the DUT with N samples in θ and M samples in φ for a total of $N \times M$ measurement points, the TIS or TRS can be approximated by the following summation:

$$TRS \text{ or } TIS \approx \frac{2NM}{\pi}\frac{1}{\sum_{n=0}^{N-1}\sum_{m=0}^{M-1}(1/(EIS(\theta_n,\varphi_m)))\sin(\theta_n)} \tag{8.4}$$

The reported summation formulas for TRP, TIS and TRS are valid for regular angular grids. In case an irregular angular grid is used, a normalization factor on each angular term can be used.

For TRP and TIS measurements of general wireless handheld devices, the 3D sphere is sampled at 15° and 30°, respectively, as illustrated in Figure 8.2.

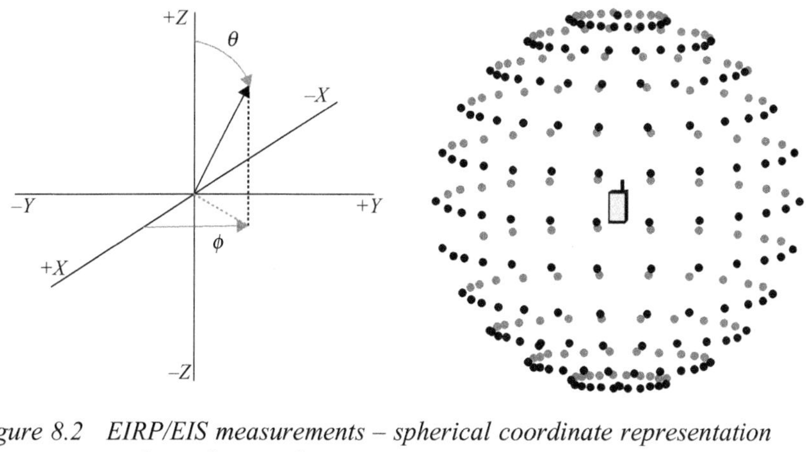

Figure 8.2 EIRP/EIS measurements – spherical coordinate representation and sampling grid

The integral values TRP for transmit power and TRS/TIS for sensitivity are system performances that do not state much about how the antenna is really working. These performance parameters do not separate the radiating elements from the rest of the subsystems, and the antenna system can have simultaneously several functions. Hence, it is difficult to define a radiation pattern of such a device. Even more, some of the new communications systems, like future 5G and IoT systems, are intended for a close range, NF communication and classical FF characterization may not be useful. Furthermore, in massive MIMO systems, each antenna element or sub-array receives and processes the signals. Therefore, the traditional antenna parameters such as gain, directivity and efficiency, as defined in IEEE standard or the integral values TRP and TRS/TIS, are less relevant for such systems. However, some testing is required to determine the correct performance of the communications system. Which are those parameters? Let us separate a typical digital antenna in transmission and reception to make the analysis simpler.

On reception side, the digital antenna elements receive the signal; the transceiver integrated in each antenna element has to receive it and convert to a digital signal. Then the digital processor collects all the signals and gets the best performance for the required functionality. In the simplest case, it gets a specific radiation pattern pointing to the transmitting user. In this simplest case, the radiation pattern does not depend on the antenna element side (or at least not only), but it mainly depends on the processing performed by the digital side. Therefore, probably it is not the radiation pattern of the parameter that better characterizes the antenna element's performance. However, what happens if the elements mismatched, or if one of the elements is disconnected, or just broken? How can we identify that? Since there is not an RF cable per element, the input impedance (or VSWR) is difficult to evaluate.

The appropriate electric currents on each element will give a good radiation performance on the electromagnetic side. Therefore, we can evaluate the antenna sources on each antenna element, measuring the radiation pattern in a fixed amplitude and phase scenario and post-processing the currents with a source reconstruction tool. Figure 8.3 shows the currents received by a square array antenna, where it is clearly possible to identify which elements are working improperly.

The knowledge of these currents gives information of the combined impact of these three factors:

- The currents received by each element from the impinging wave.
- The ohmic losses of each radiating element.
- The mismatch between each antenna element and the transceiver.

And, for a good antenna design, we need to reduce ohmic and mismatch losses for each one of the antenna elements. Also, ohmic losses become relevant for the calculation of antenna noise temperature, again important to improve the signal quality.

Of course, in the case of the currents, they can be separated in two orthogonal components, in the same way of the copolar and the cross-polar patterns in the classical radiation pattern representation.

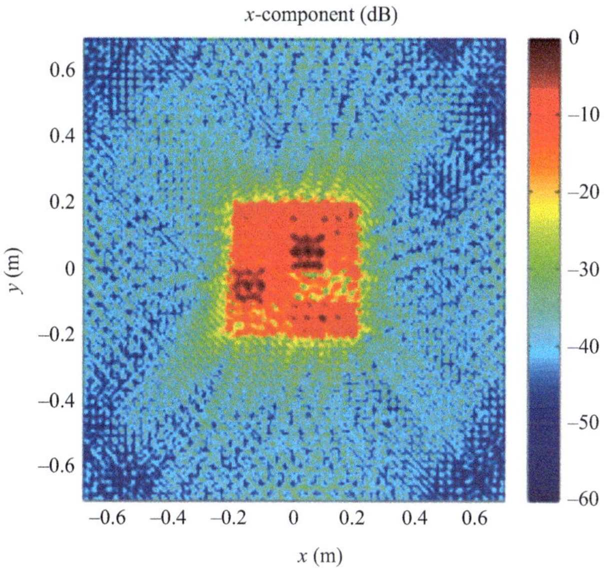

Figure 8.3 Electromagnetic currents of an antenna array

If we want to extract single numbers, as is done with the classical passive antennas, it would be necessary to integrate ohmic losses, mismatch losses and antenna noise temperature from each antenna element in one single value. Of course, this integration only gives a general case, since the actual performance depends on the digital processor. Again, let us go to a simple case: imagine the digital processor only takes the signal of one of the elements and multiplies by zero the others. The ohmic losses of these others can be neglected in the overall performance of the system. However, in the same way that a radiation pattern of a passive antenna is defined in an anechoic environment, it is interesting to have a reference case that can be the general case of all the elements averaged with equal amplitude and phase. Even more, if we think in a massive MIMO processing, the radiation pattern is not relevant, but it is the correct performance of the system and obviously would depend on the right performance of all the antenna elements.

In the case of transmission, again the radiation pattern should depend on the signals created by the digital processor. However, this is not the only subsystem to be considered: a mis-functionality in one of the elements makes that the generated radiation pattern was different from the desired one, or at least, the global performance of the system would be different. Therefore, it is again necessary to establish a reference configuration to establish the antenna parameters. Again, the simplest one would be the one where the weight of each element is uniform in amplitude and phase. This is a general case, but it can be used to understand the right performance of the electromagnetic side. Again, the currents can be determined by a source reconstruction tool applied to the radiation pattern measured in this reference scenario (theoretical uniform amplitude and phase). Again, the reconstructed currents give combined information of the following parameters:

- The currents transmitted by each element.
- The ohmic losses of each radiating element.
- The mismatch between each transmitter and antenna element.

In some communication systems, the same frequency is used for transmitting and receiving. Even in those cases, since the RF path is different for reception or transmission, both cases should be investigated.

What happens with the gain, directivity or radiation pattern? Obviously, with new antenna systems, it is impossible to define a single radiation pattern. Therefore, gain or directivity, dependent on the radiation pattern, is only valid for each particular state of the array. Even more, those parameters are thought for FF conditions and some of the new systems also work in NF. With this analysis, and in the reference case, the currents should be equal to amplitude and phase; therefore, a parameter as the standard deviation of the currents can be used to determine the quality of the antenna system.

In fact, the currents could be separated in two orthogonal components (one defining the copolar and the second one defining the cross-polar characteristic of the antenna system).

Are the losses or mismatch still important? Of course, antenna losses not only reflect the radiated or received power, but also the antenna noise temperature. And mismatch will have similar effect to the mismatch in passive antennas. The main difference is that an error in only one element can partially affect only the global performance of the antenna itself. Therefore, a global number can give a quality factor of the complete antenna system even if this number does not describe each individual case, in the same way the gain characterizes a passive antenna.

8.4 Antenna measurement systems for digital antennas

How to measure the digital antenna systems? What are the differences with respect to classical systems? Classical systems for the characterization of passive antennas, as was mentioned in Section 8.1, are FF antenna measurement systems such as compact antenna test ranges (CATRs), or NF systems in the three configurations: planar, spherical or cylindrical. The international measurement community is currently concentrating standardization activities on three types of testing methodology [7–9] that are largely derived from classical antenna testing metrology in NF and FF [3,6]. The direct FF techniques are, although the simplest measurement configurations, deemed lest practical due to size and dynamic range issues. The indirect FF systems include plane wave generators (PWGs) such as the reflector-based CATR and the array-based PWG. Finally, the classical NF measurement systems are considered with the spherical geometry as the most prominent, as it determines performance is the full 3D space. The three categories of measurement solutions are illustrated in Figure 8.4.

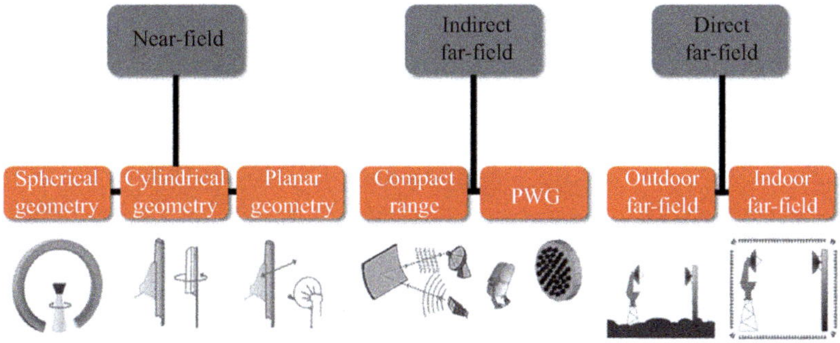

Figure 8.4 Different measurement option for future antenna systems under investigation by the measurement community

NF systems require the measurement of two tangential field components in amplitude and phase on a surface containing the AUT. The required number of field samples to perform the NFFF transformation is based on the size of the measured antenna and determined from standard sampling theory. Figure 8.5 shows the three canonical configurations: spherical, cylindrical and planar. Only, the spherical geometry completely encompasses the AUT and is therefore considered the most accurate. Yaghjian in [5] explained all the mathematical formulations of the NFFF transformation in the three canonical grids. Also, IEEE standard in [6] summarizes all the measurement procedure for this NF measurement systems. Again, the main objective of these methods is getting the FF parameters: gain, directivity, radiation pattern, axial ratio but also system parameters such as TRP and TRS/TIS can be determined in such systems.

One of the possible hardware solutions that accomplish with the previous requirements is based on multiprobe measurement systems. These systems were first based on modulated scattering technique [11] and later were improved by MVG in the new multiprobe systems for 5G applications such as the SG178 and StarLab 50 GHz. The radiated power and sensitivity profiles, EIRP and EIS in 3D space, can be determined accurately in such NF system by applying phase recovery techniques, as discussed in this book, followed by classical NFFF transformation [5]. An example of such system based on multi-probe technology is shown in Figure 8.6.

The measured radiated power profile EIRP, after NFFF, for three different scans are shown in Figure 8.7.

As discussed previously, the measurement of the radiated power profile EIRP, even at distances insufficient to be considered FF distance by classical standards, can be used to determine the TRP using (8.3) that is a simple integration of the EIRP values in 3D space. An example if such measurement, as part of a sampling study for TRP determination of 5G enabled devices at 24 GHz, is shown in Figure 8.8.

These reported NF measurements above are based on the application of the post-processing techniques outlined in the previous chapters. Those are phase recovery techniques for recovering the phase from amplitude measurements, source

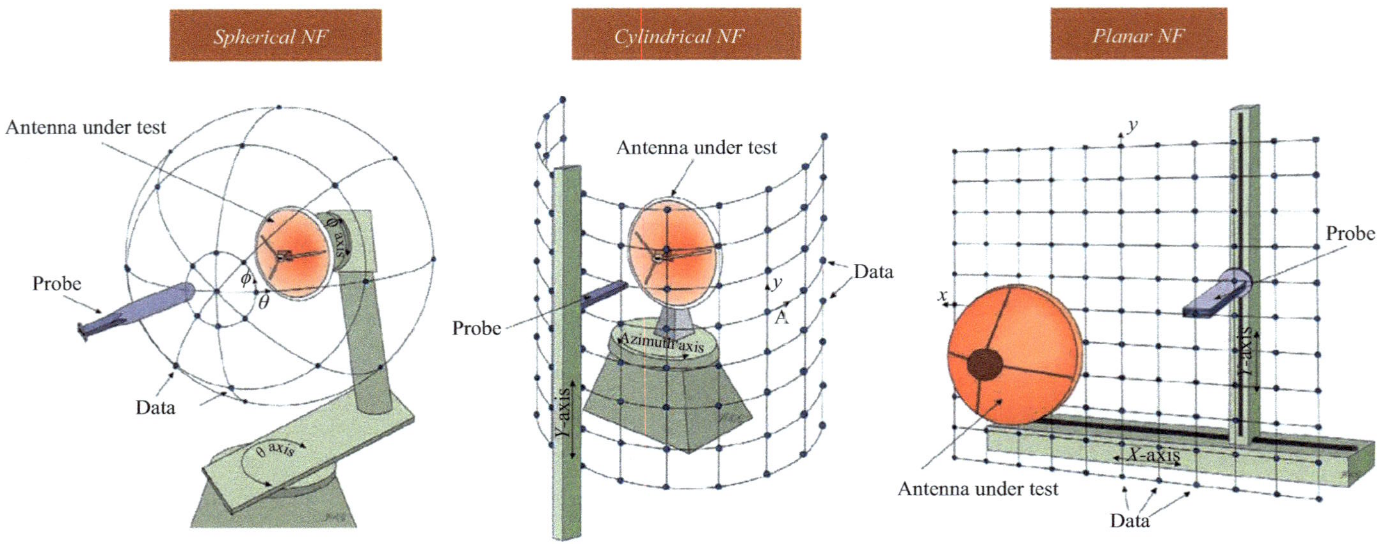

Figure 8.5 Near-field measurements configurations: spherical, cylindrical and planar

Figure 8.6 Multi-probe measurement system for full digital-base station antenna at sub-6 GHz frequencies

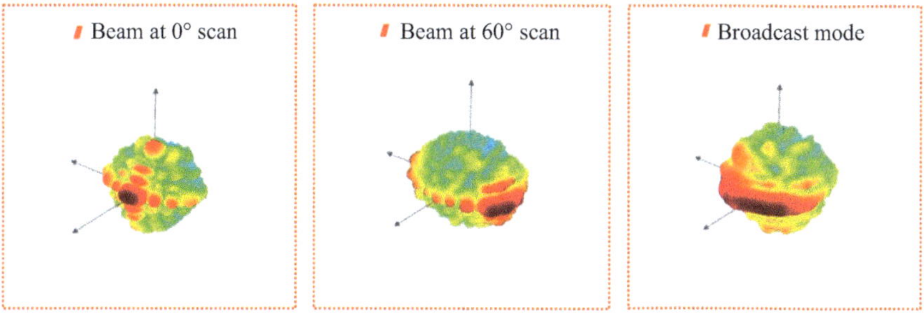

Figure 8.7 Examples of measured radiated power profile, EIRP, of the same massive MIMO device with three different beamforming configurations

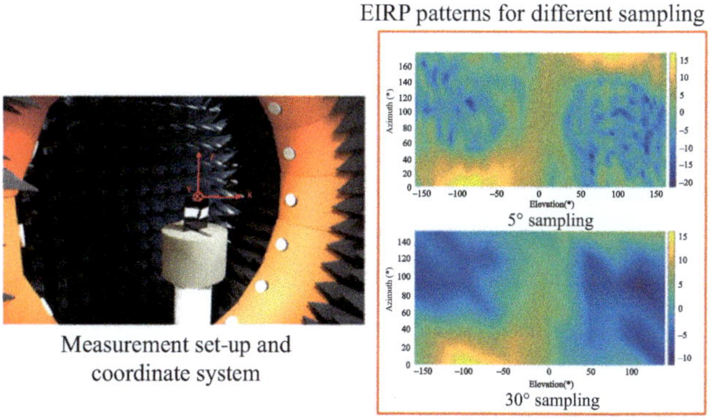

Upper Patch #4	
Resolution (degree)	TRP (dB m)
5	4.49
10	4.41
15	4.42
20	4.38
25	3.75
30	4.66

Figure 8.8 Examples of TRP measurement in a multi-probe system of a 5G-enabled mobile device @ 24 GHz

reconstruction techniques for the calculation of the electromagnetic sources, other post-processing techniques to improve antenna measurement results or combination of simulations and measurements for being able to characterize complex antenna systems.

The indirect FF systems include PWGs such as the reflector-based CATR and the array-based PWG. The CATR is an optical system based on a feed and one or more reflectors. It is the most common method to perform measurements of antennas based on plane wave generation. The properly shaped reflector(s) collimate the spherical wave radiated by the feed into an approximately plane wave over a finite volume in the test area called quiet zone (QZ). Common reflector shapes are parabolic or cylindrical. The size of the QZ area is typically half the size of the reflector, and the distance between the reflector and QZ is often three times the size of the reflector. The compact range enables the testing of medium to electrically large antennas at a significantly shorter distance than would be required in a traditional FF test range.

The generated plane wave from a compact range is approximately uniform over the QZ, due to the amplitude taper of the feed. In addition, sources of stray radiation will add in and out of phase to the QZ field. The main sources of stray radiation are direct radiation from the feed, edge effects of the reflector(s), chamber effects and mutual coupling between the reflector and AUT. The reflector edge effects are generally compensated by adding serrations to the reflector or edge shaping to direct diffractions away from the QZ. The bandwidth of a compact range system is generally only limited by the finite bandwidth of the feed and the efficiency of the edge treatment.

As an example, Figure 8.9 shows a passive reflector antenna (DTU-ESA VAST12 antenna) on the LEHA-UPM compact range (from Universidad Politécnica de Madrid). In this classical configuration, the antenna is measured in reception. The antenna RF connector is connected to the receiver and the signal is compared with a sample extracted from the RF synthesizer. This RF synthesizer generates a continuous wave with a constant power. The rotation of the AUT allows the measurement of the radiation pattern for each angular position (in FF). If an enough number of cuts of the antenna pattern are measured, the directivity could be calculated integrating the radiation pattern. Also, antenna gain can be calculated using several methods as the gain transfer method, three antenna methods or others. All these measurement methods are explained in [3] and the main set-ups can be found in [4].

If the CATR is gain calibrated, the radiated power from the device placed on the QZ can be measured directly at the feed level. This property makes the CATR the simplest method for characterizing digital antenna directly in the FF.

The PWG array is an array of elements with suitably optimized complex coefficients, generating a plane wave in the close proximity of the array [12]. The array can be real or synthetic. It can be implemented either as linear or planar arrays of elements in 1D or 2D systems for general antenna testing purposes. It typically consists of two main components: the radiating elements arranged in the structure

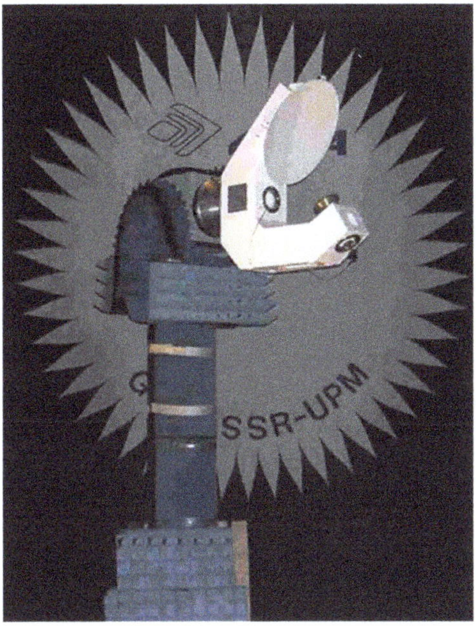

Figure 8.9 LEHA-UPM Gregorian compact antenna test range

of an array and the feeding system, whose architecture depends on the adopted technology that could be feeding by a standard beam forming network or feeding by an active system. The PWS can achieve FF testing conditions in a QZ located in a region close to the radiator, as in the case of CATRs.

The PWS enables direct measurements of FF AUT performance in a controlled indoor environment as an alternative to CATR. Main differences with CATR are due to non-existing diffraction and blocking effects and to the amount of mutual coupling as a function of the AUT distance. Furthermore, the PWS can have the advantage of reducing the physical size of the CATR, in particular at lower frequencies. Indeed, the physical size affects the dimension and associated costs of the anechoic chamber. The PWS is more flexible and, as a measurement system, has largely the same error contributions to measurement uncertainty as the CATR.

One of the other possible measurement strategies for active antennas we find is the reverberation chamber that became very popular [13] with the roll out of the 4G network. However, reverberation chamber is not directly considered in the development of the 5G tests standards. In a reverberating system, instead of free space environment, as used in conventional systems, isotropic environment was used. In this environment, integral parameters such as TRP and TRS could be determined. Later on, diversity gain and throughput in isotropic environments, and other more complex parameters were introduced. Figure 8.10 shows the reverberation chamber commercialized by Bluetest, one of the reference companies in this field.

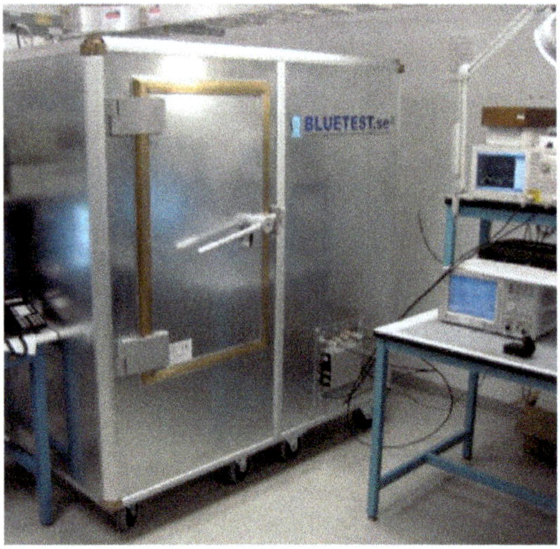

Figure 8.10 Reverberation chamber (Bluetest©)

As a common nominator, the new OTA antenna measurement systems for digital antennas should have the following characteristics:

- Use of the transmitter of the communications systems itself.
- No access to the reference signal, and extraction of the reference in a different way.
- Receiver not locked to the transmitter. This can be solved by amplitude only measurements or other ways of phase reconstruction.
- The receivers can be fabricated with modern (and low cost) software defined radio technology.
- Fast measurement systems are needed to be able to measure many different configurations and frequencies in a limited time.
- Post-processing techniques are needed to improve the measurement results and to extract the measurement parameters as defined in previous sections.

8.5 Outlook for the new developments in antenna measurement systems

The last section has summarized the contents of the previous chapters of the book, has introduced the new antenna systems, introducing new antenna parameters for testing these systems and finally has analysed the new challenges in the antenna measurement system to be able to characterize the electromagnetic performance of these new digital antennas.

The intention of this book was not the revision of all the latest advances in antenna measurement during the last years. In this sense, in the new frequency bands of sub-millimetre and terahertz, innovations appeared in the last few years.

For instance, very high accuracy positioning system for spherical grid acquisition for the new millimetre band applications, based on the modern robotic technology, was recently proposed [14]. The work done at Aalto University [15] designing amplitude and phase holograms as collimating elements to measure large satellite antennas at frequencies up to 650 GHz is also remarkable.

Also, new systems have been designed for the measurement of antennas installed on very large structures. For instance, the multiprobe StarBot system designed by MVG allows the measurement of antennas on airplanes (Figure 8.11).

Another example is the portable antenna measurement system, designed by Astrium GmbH, with support of the ESA's Advanced Research in the Telecommunication Systems programme [16]. This system allows the measurement of radiated signals in a non-uniform grid using a gondola. This solution is completed with the new NFFF algorithms for non-uniform grid, as the Fast Irregular Antenna Field Transformation developed by Technical University of Munich [17,18]. This solution, combined with post-processing tools, can afford the measurement of large satellite antennas.

A similar concept is the measurement of large structures using drones. In this sense, the works from Politecnico de Torino [19] and University of Oviedo [20], which both designed an antenna test system baed on a hexacopter with RF transmitter and dipole antennas, are remarkable.

Also, there has been a large improvement in the reduction of acquisition time, an important factor in antenna measurement field since the complexity of the antennas has been increasing during the last few years. In this sense, the work developed during the last years by University of Salerno and 'Federico II' of Naples [21], where different non-redundant acquisition schemes have been defined for planar, cylindrical or spherical scans depending on the shape and dimensions of the AUT, is remarkable.

Pursuing this time reduction, Microwave Vision Group, Technical University of Madrid and Aachen University published a comparison of different methods to

Figure 8.11 Microwave Vision Group StarBot system

perform single cut NFFF methods [22], that was introduced in Chapter 5 of this book.

8.6 Conclusions

Even if the recent advances in all the fields in antenna measurements have been impressive, the main conclusion the authors want to remark is the requirement of combination of hardware with software to be able to measure the more and more complex antenna systems. In this sense, the integration of radiating systems with digital processors implies that the knowledge of electromagnetic parameters is necessary to determine the correct performance of the radio systems. Those parameters can be obtained combining new hardware with post-processing tools.

In this sense, the conclusions we emphasize are as follows:

- New antennas are appearing. These antennas include the RF system, an analogue-to-digital converter in each element and a digital processor in charge of the array processing. In fact, antennas for cellular systems, RADAR, satellite tracking and other applications already include this technology.
- Electromagnetic performance has to be optimized in these new antenna systems to be able to exploit all the functionalities of the system.
- Classical antenna parameters are not the optimum ones for being able to test this electromagnetic performance. The knowledge of the electromagnetic sources becomes more important than the knowledge of the radiation pattern, since there are multiple radiation patterns for each antenna system.
- Classical antenna measurement systems are not the most appropriate for testing parameters of the digital antennas.
- New hardware, or evolution of existing one, as 5G multiprobe measurement system is required.
- This hardware has to be combined with the post-processing tools analysed along this book: phase recovery, source reconstruction, spatial or modal filtering, combination of measurements and simulations are each time more and more necessary to be able to afford the measurement of more complex antennas.

And, finally, technology is changing very quickly. We are still beginning to see what the future is, but there is no doubt that the evolution of new antennas will be very fast. And measurement systems have to change as fast as the antenna technology is changing. Post-processing techniques can be used to adapt the measurement hardware very well to the rapid evolution of the antenna technology.

References

[1] C. Capellin, "Antenna diagnostics for spherical near-field antenna measurements". Danmarks Tekniske Universitet, Copenhagen, Ph.D. Dissertation. 2007.

[2] "IEEE Standard Definitions of Terms for Antennas". 1993.
[3] IEEE Std. 149-1979, "IEEE Standards Test Procedures for Antennas". John Wiley. 1979.
[4] G.E. Evan, "Antenna Measurement Techniques". Norwood, MA: Artech House Inc. 1990.
[5] A.D. Yaghjian, "An overview of near-field antenna measurements", *IEEE Transactions on Antennas and Propagation*, vol. 34, no. 1, pp. 30–45, 1986.
[6] IEEE Std 1720-2012, "IEEE Recommended Practice for Near-Field Antenna Measurements". 2012.
[7] CTIA Certification Program. Test Plan for Wireless Device Over-the-Air Performance. Method of Measurement for Radiated RF Power and Receiver Performance. Version 3.7.1. February 2018. Washington DC.
[8] 3GPP TR 25.914, "Measurement of Radio Performances for UMTS Terminals in Speech Mode". 2018.
[9] 3GPP TS 34.114, "User Equipment (UE)/Mobile Station (MS) over the Air (OTA) Antenna Performance – Conformance Testing". 2016.
[10] 3GPP TS 37.114, "User-Equipment (UE) and Mobile Station (MS) GSM, UTRA, and E-UTRA over the Air Performance Requirements". 2018.
[11] J.C. Bolomey and F.E. Gardiol, "Engineering Applications of the Modulated Scatterer Technique". Boston, MA: Artech House. 2001.
[12] F. Scattone, D. Sekuljica, A. Giacomini, *et al.*, "Plane Wave Generator for Direct Far-field Over-The-Air Testing of Devices". Proceedings of the Antenna Measurement Technique Association (AMTA2018), Williamsburg, VA, United States, October 2018.
[13] P.-S. Kildal and K. Rosengren, "Correlation and capacity of MIMO systems and mutual coupling, radiation efficiency and diversity gain of their antennas: Simulations and measurements in reverberation chamber", *IEEE Communications Magazine*, vol. 42, no. 12, pp. 102–112, 2004.
[14] D.J van Rensburg "Factors Limiting the Upper Frequency of mm-Wave Spherical Near-field Test Systems". Proceedings of European Conference on Antennas and Propagation (EuCAP 2015), Lisbon, April 2015.
[15] T. Hirvonen, J. Ala-Laurinaho, J. Tuovinen, and A.V. Räisänen, "A compact antenna test range based on a hologram", *IEEE Transactions on Antennas and Propagation*, vol. 45, no. 8, pp. 1270–1276, 1997.
[16] H.J. Steiner, T. Fritzel, A. Geise, C. Schmidt, and M. Paquay. "First Results of Innovative Mobile Near-Field Antenna Measurement System for Extreme Large DUTs". Proceedings of EuCAP, 2015.
[17] T. Eibert and C. Schmidt, "Multilevel plane wave based near-field far-field transformation for electrically large antennas in free-space or above material halfspace", *IEEE Transactions on Antennas and Propagation*, vol. 57, no. 5, pp. 1382–1390, 2009.
[18] R.A.M. Mauermayer and T.F. Eibert, "Combining the Fast Irregular Antenna Field Transformation Algorithm with Asymptotic High Frequency Methods". Proceedings of EuCAP, 2015.

[19] G. Virone, A.M. Lingua, M. Piras, *et al.*, "Antenna pattern verification system based on a micro unmanned aerial vehicle (UAV)", *IEEE Antennas and Wireless Propagation Letters*, vol. 13, pp. 169–172, 2014.
[20] M. García Fernández, Y. Álvarez López, and F. Las Heras, "UAV-based Antenna Measurement and Diagnostics for Circularly Polarized Antenna Arrays". Proceedings of 2018 IEEE International Symposium on Antennas and Propagation & USNC/URSI National Radio Science Meeting, pp. 525–526. Boston, July 2018.
[21] F. D'Agostino, F. Ferrara, C. Gennarelli, R. Guerriero, and M. Migliozzi, "Experimental testing of nonredundant near-field to far-field transformations with spherical scanning using flexible modellings for nonvolumetric antennas", *International Journal of Antennas and Propagation*, vol. 2013, Article ID 517934, pp. 1–10, 2013.
[22] R. Cornelius, T. Salmerón-Ruiz, F. Saccardi, L. Foged, D. Heberling, and M. Sierra-Castañer. "A comparison of different methods for fast single-cut near-to-far-field transformation", *IEEE Antennas and Propagation Magazine*, vol. 56, no. 2, pp. 252–261, 2014.

Index

acquisition surfaces characteristics 201
Antenna Centre of Excellence 4
antenna-measurement equipment 242
antenna under test (AUT) 2, 91, 139–40, 201, 247, 275, 285
  AUT minimum sphere, optimization of 158
  FF pattern 230–1
  minimum sphere 167, 175
    and minimum cylinder 143–4
array antenna synthesis 30–5
autocorrelation term 204, 206, 210
automotive antenna ranges 179
azimuthal scanning 179

back-propagation 52, 92, 99–100
band-limited operator 122
base transceiver station (BTS) antenna 144
beam-forming network (BFN) 23, 50
broadband indirect off-axis holography 212
  application example 215–18
  method 212–15
BTS1940 antenna 70–1

commercial antenna diagnostic tools 4–5
Compact Antenna Test Ranges (CATRs) 1, 284, 289
Compact Payload Test Range (CPTR) 140
complex reference field 204–5
computational electromagnetic (CEM) tools 82, 239, 245, 264
  numerical simulation in 245–6
  simulation in 268
co-polar far-field pattern 110–11
cross-correlation 204
cross-polar far-field pattern 111
cylindrical modal coefficients (CMCs) 188–90
cylindrical near-field (CNF) measurements 3, 94, 96, 156
  noise reduction method for 108–9
cylindrical wave expansion (CWE) 139, 141
cylindrical waves, theory of 186–7

devices radiating into 3D space
  black box representation of 240–1
device under test (DUT) 242, 275
diagnostic technique 93
Diatool (TICRA) 5
differential evolution (DE) 202
differential phase measurement method 199
digital antennas
  antenna measurement systems for 284–90
  performance parameters for 279–84
domain decomposition technique (DDT) 240
Dual-Equation Formulation (DEqF) 15
dual reflectors 1
dual ridge horn 241, 244

echo reduction 73–7, 92
in single-cut measurements 190–3
with spherical wave expansion (SWE) 163–75
using holographic techniques 113–21
effective isotropic radiated power (EIRP) 280
effective isotropic sensitivity (EIS) 280
18-term error budget 2–3
electromagnetic compatibility (EMC) testing
application of EQC technique in 84–7
electromagnetics, theory of holographic techniques in 99–101
Equivalence Principle 5
equivalence theorem
application of 10
J_{eq}-only/M_{eq}-only formulation 12–13
zero interior fields 12
Love's form of 12–13
equivalent current (EQC) technique 45
advantages and limitations of 87–8
antenna characterization with 266–7
in electromagnetic compatibility (EMC) testing 84–7
EQC processing NF source modelling 243–5
in the improvement of antenna measurements accuracy 69
diagnostic and spatial filtering of supporting structure 69–73
echo reduction 73–7
truncation errors, reduction of 78–81
in the improvement of antenna system design 46
spacecraft antennas, analysis of coupling phenomena between 60–4
spherical higher order modes in the design of NF probes 64–9
verification of complex excitation coefficient of a space array antenna 56–60
wideband dual-ridge horn, advanced diagnostic of 47–9
X-band 32 × 8 array, design improvement of 50–5
power density determination of measured devices 81–3
equivalent magnetic current approach 96
equivalent noise level (ENL) 161–2, 182–3
equivalent principle 96
equivalent sampling 209–10
equivalent sampling cylinder (ESC) 145
equivalent sampling sphere (ESS) 145
error analysis 218–19

far-field (FF) distance 1
far-field (FF) pattern 106
far-field (FF) radiating stand-alone measurements 266
Fast Fourier Transform (FFT) 5, 94–5, 101, 133, 140
field filtering 91
field reconstruction 159
flush-mounted antenna on flat structure 255–65
Fourier Transforms (FTs) 91, 202
free space, antenna in 247–9
full-wave MoM simulation 15
future trends in antenna measurements 275
analysed post-processing techniques 275–7
digital antennas
antenna measurement systems for 284–90
performance parameters for 279–84

new developments in antenna measurement systems 290–2
from passive to digital antenna technologies 278–9

Gabor holography 199–200
GALILEO FOC Satellite antenna farm testing 60
Gauss–Newton algorithm (GNA) 224
Georgia Tech 1
Gerchberg–Papoulis algorithm 97, 99, 121
GIOVE-B array antenna 24
Global Navigation Satellite System (GNSS) antenna 266–7
Green's function 12–13

Hankel's function 21
high-quality absorbing materials 139
hologram 203
holographic techniques 91
detection and suppression of leakage 126–32
echo reduction using 113–21
extrapolation of radiation pattern and reduction of truncation errors 121–6
noise reduction using s 101–12
for reducing errors in antenna measurements 93
leakage 98
measurement scan truncation 96–8
noise 94
potential of source reconstruction techniques 93–4
reflections 94–6
theory of holographic techniques in electromagnetics 99–101
holography spatial filtering technique 140
horn antenna, reflector fed by 250–4
Huygens' formulation 241
improvements of antenna measurement accuracy
exploiting cylindrical mode expansion for 183
applications to echo reduction in single-cut measurements 190–3
single-cut measurements 188–90
theory of cylindrical waves 186–7
using spherical mode expansion 141
spherical wave expansion (SWE) theory 142–55
SWE as spatial filter 155–83
indirect off-axis holography 200, 202, 232
broadband indirect off-axis holography 212
application example 215–18
method 212–15
error analysis 218–19
improved schemes 205–6
method 203–5
modified hologram 206–7
synthesized reference-wave indirect holography 207
INSIGHT software 5, 244
EQC source model 245
interferometric techniques 199–201
IsoFilter (MI-Tech) 5, 96
iterative Fourier technique (IFT) 201–2
iterative methods 91, 124
iterative modal filtering (IMF) technique 79, 178–9, 183
iterative phase retrieval techniques 219
broadband indirect off-axis holography 212
application example 215–18
method 212–15
error analysis 218–19
fundamentals 219–21
phaseless sources reconstruction method 221

application example 225–32
description 221–5
synthesized reference-wave indirect holography 207
iterative techniques 199, 201–2

J-current SEqF (**J**-SEqF) 15

k-space 197, 203–4, 207, 210, 277
Ku-band reflector antenna 109–11

L-band radar antenna 119–21
leakage 98
detection and suppression of 126–32
Leith–Upatnieks Holography 200
Levenberg–Marquardt algorithm (LMA) 202, 224, 228, 231
The Link 239, 272
definition of 241
antenna-measurement equipment 242
equivalent current processing NF source modelling 243–5
numerical simulation in the CEM tools 245–6
practical application of 266
results 268–9
test object and complex scenario 266
validation workflow 266–8
validation of 246–7
literature backward-transformation technique 99
Love's equivalence 12–13
low-directivity antennas 204–5

Mars (NSI) 5
mathematical absorber reflection suppression 95–6
matrix pencil method 95
matrix–vector products (MVPs) 19
Maxwell's equations 12, 93
M-current SEqF (**M**-SEqF) 15
MEAS 251–2
measurement scan truncation 96–8
measurement surface 9
Method of Moments (MoM) codes in source reconstruction 19
metrology theory 2
Microwave Vision Group StarBot system 291
MIMO processing 283
modal expansion methods 93
modal/spatial filtering 156, 158–9
modified hologram technique 206–7
modified probes 199
monocones antennas 180–1
Moore–Penrose pseudo-inverse 14
mounted antennas 250
flush-mounted antenna on flat structure 255–65
open-ended waveguide (OEW) antenna 260–1
SM6000 monopole antenna 262–5
SMC2200 monocone antenna 256–60
reflector fed by a horn antenna 250–4
MVEcho (MVG) 5
MVG SGH820 165
MVG SH600 dual-ridge 147
MVG SH2000 horn antenna 191
MVG SH4000 dual-ridge horn 47
MVG SL18 GHz spherical near-field multi-probe system 243
MVG SR40-A reflector fed by the MVG SH4000 241

near field (NF) 121, 240, 276, 284–5
antenna measurements 129
source modelling exchange data 245
near-field-acquired noise 101
near-field measurements 93
near field to far-field (NFFF) transformation 30, 276, 285
near-field to near-field 29–30
near-to far-field algorithms 1–2, 4

NF–FF transformation 158, 175, 202, 219
NF–NF transformation 219
noise 92, 94
noise reduction 157
  using holographic techniques 101–12
nonredundant sampling theory 199–200
numerical modelling 239
numerical simulation in the CEM tools 245–6
Nyquist criteria 205

open-ended waveguide (OEW) antenna 205, 215, 260–1
opposite-phase technique 206–7
oversampling 141, 157

P_m power spectrum 150, 153–4
P_n power spectrum 150–1, 153, 156, 167–8, 173
particle swarm optimization (PSO) 202
pencil-of-function method, generalized 95
perfect electric conductor (PEC) 12, 179
perfect magnetic conductor (PMC) 12
phaseless sources reconstruction method (pSRM) 221
  application example 225–32
  description 221–5
phase recovery in antenna measurements 197, 232–4
  indirect off-axis holography 202
    broadband indirect off-axis holography 212
    error analysis 218–19
    improved schemes 205–6
    method 203–5
    modified hologram 206–7
    synthesized reference-wave indirect holography 207
  iterative phase retrieval techniques 219
    fundamentals 219–21
    phaseless sources reconstruction method (pSRM) 221–32
  main phase recovery techniques 199
    interferometric techniques 199–201
    iterative techniques 201–2
phase retrieval techniques 219–20
planar near-field (PNF) 2, 130, 145, 156
  measurements 96, 158, 165–8
plane wave expansion 4, 20–1, 99, 133
plane wave generators (PWGs) 284, 288
plane wave spectrum (PWS) 3, 5, 91–2, 94, 99–100
polynomial methods 95
power density determination of measured devices 81–3
power of the equivalent model 245
processing gain 159
Prony's algorithm 95
pyramidal-horn antenna 116, 118

quiet zone (QZ) 288

radio frequency (RF) energy 126
real coded genetic algorithm (RCGA) 202
reciprocity and the reception formula 20–2
reconstruction problem, uniqueness of the solution to 13–17
reconstruction surface 10
reflections 94–6
reflectivity level 94, 113
reverberation chamber 290
right-handed circularly polarized (RHCP) pattern 56

scalar equipment 198
scalar Helmholtz equation 186
scaled-model technique 180–1

SGH820 74–6, 165
SH4000 dual ridge horn 248, 250–2
signal-to-interference ratio (SIR) 164
signal-to-noise ratio (SNR) 35, 94
simulated 3D far-field pattern 246
simulation and measurements, The Link between 239
  antenna in free space 247–9
  mounted antennas 250
    flush-mounted antenna on flat structure 255–65
    reflector fed by a horn antenna 250–4
  practical application of The Link 266
    results 268–9
    test object and complex scenario 266
    validation workflow 266–8
  theory 240
    black box representation of devices radiating into 3D space 240–1
    definition of The Link 241–6
    domain decomposition technique (DDT) 240
    validation of The Link 246–7
single-cut measurements 188–90
  echo reduction in 190–3
Single-Equation Formulations (SEqFs) 15
single probe NF systems 141
SM6000 monopole antenna 262–5
SMC700 181
SMC2200 antenna 181, 255–60
source reconstruction problem 9
  array antenna synthesis 30–5
  field transformations, application in 27
    near-field to far-field 30
    near-field to near-field 29–30
  formulation of 9
    equivalence theorem, application of 10–13
    reconstruction problem, uniqueness of the solution to 13–17
  investigation and diagnostic tool, application as 23–7
  numerical solution of 18
    discretization 18
    linear system, setup of 18–19
    Method of Moments (MoM) codes 19
  numerical tests on the accuracy of source reconstruction 35–9
  probe correction 19
    formulation, with probe correction 22–3
    reciprocity and the reception formula 20–2
source reconstruction techniques 9, 19, 93–4
sources reconstruction method (SRM) 197, 221, 233–4
space array antenna, verification of complex excitation coefficient of 56–60
spacecraft antennas, analysis of coupling phenomena between 60–4
S-parameters representation 241
spatial filtering 140, 156
  based on SWE 157–9
  oversampling in 164
spectral-truncation operator 122
spherical mode expansion
  improvements of antenna measurement accuracy using: *see* improvements of antenna measurement accuracy
spherical near-field (SNF) 3–4, 94
  reduction of truncation errors in SNF measurements 175–83
  SNF-to-far-field transformation 96
  with stationary probe and rotating AUT 168–75
spherical near-to-far-field transformation algorithms 1

spherical reference coordinate system 142
spherical wave coefficients (SWC) 100, 143, 147–55, 178
  computation of 146–7
spherical wave expansion (SWE) 24
  as spatial filter 141, 155
    echo reduction with SWE 163–75
    noise reduction in SNF measurements using the SWE 159–63
    reduction of truncation errors in SNF measurements 175–83
    spatial filtering based on SWE 157–9
  theory 142–55
    AUT minimum sphere and minimum cylinder 143–4
    computation of the SWC 146–7
    sampling criteria 145–6
    understanding the SWC 147–55
spherical waves, equivalent current processing versus 243–4
spherical wave vector functions (SWF) 143–4
SR40-A reflector antenna 250–1
StarGate128 measurement system, base station antenna in 71
synthesized reference field
  application example 210–11
  by means of electronic shifts 207–8
  by means of mechanical shifts 208–10
  indirect off-axis holography with 206
synthesized reference-wave indirect holography 207

3D grid 208–10
time-domain filtering postprocessing techniques 140
time-gating techniques 95
TM-modes 150
total isotropic sensitivity (TIS) 280
total radiated power (TRP) 280
total radiated sensitivity/total receiver sensitivity (TRS) 280
translation operator 21
transmission formula 146
truncation errors, reduction of 78–81

vector Helmholtz equation 186
vector instrumentation 198

wideband dual-ridge horn, advanced diagnostic of 47–9

X-band 32 × 8 array, design improvement of 50–5

zero interior fields 12
zero-padding (ZP) 158, 175